Jinhu Lü · Xin Chen · Guanrong Chen

Chaotic Systems with Multiscroll Attractors

Theories, Methods and Applications

Springer

Jinhu Lü
School of Automation Science
and Electrical Engineering
Beihang University
Beijing, China

Xin Chen
Institute of Artificial Intelligence
Beihang University
Beijing, China

Guanrong Chen
Department of Electrical Engineering
City University of Hong Kong
Kowloon, Hong Kong, China

ISBN 978-3-032-09853-5 ISBN 978-3-032-09854-2 (eBook)
https://doi.org/10.1007/978-3-032-09854-2

This Springer imprint is published by the registered company Springer Nature Switzerland AG
The registered company address is: Gewerbestrasse 11, 6330 Cham, Switzerland

We dedicate this book to our families for their patience, love, and support over the years.

Foreword

The chaotic Chua's circuit [1] constructed in 1983 has a three-piecewise linear resistor function [2], which can generate a double-scroll chaotic attractor.

In 1993, Johan Suykens and Joos Vandewalle [3] extended the three-piecewise linear resistor function to multi-piecewise linear functions, thereby generating multi-scroll chaotic attractors.

In 1996, Arena et al. [4] implemented a generalized n-double-scroll Chua's circuit using a 3-state-controlled cellular neural network, significantly expanding the application range of the circuit.

In 1997, Suykens et al. [5] further developed a unified method for generating a more comprehensive family of n-scroll attractors.

In 2000, Yalcin et al. [6] experimentally verified the existence of 3-scroll and 5-scroll chaotic attractors in generalized Chua's circuits and successfully constructed a circuit with a 6-scroll attractor.

In 2001, a significant progress was made by Wallace Tang et al. [7], who introduced sine functions as replacements for the piecewise-linear functions in Chua's circuit to generate multiscroll attractors.

In 2002, Yalcin et al. [8] introduced step functions to constructing circuits to generate a family of grid-scroll attractors.

In 2002, Zhong et al. [9] successfully constructed an analog circuit and observed 10-scroll attractors on the oscilloscope.

In 2003, Fujiwara et al. [10] developed a fast, compact and CMOS-compatible 3-scroll attractor prototype chip, paving the way for new applications of chaotic attractors in integrated circuits.

In 2005, a breakthrough was made by Lü et al. [11], in designing and experimentally realizing circuit systems capable of generating up to 1000-scroll attractors. That overcame the technical hurdle due to the constraints of voltage requirements and circuit complexity, allowing the largest number of scrolls that could be experimentally generated to be limited to only 10 before 2005.

In 2006, Jinhu Lü and Guanrong Chen [12] presented a comprehensive summery of the theory, design and circuit implementation frameworks of various generators of multiscroll chaotic attractors.

To that end, in the past 20 years or so, research and applications of multi-scroll chaotic attractors have been developed swiftly, extending from multiscroll and multi-directional multiscroll to multiwing and multi-directional multiwing, as well as multi-fold, mutitorus, grid-scroll and multi-directional grid-scroll attractors, and so on, with more and more potential applications in signal processing and information security.

This book summarizes the above-reviewed research progress and results, presenting the theoretical foundations, generation methods, circuit implementations, and practical applications of chaotic circuits and systems with various multiscroll attractors. It is a festive exploration in chaos theory and technology, with valuable knowledge and information useful for future development.

March 2025

Leon O. Chua
University of California, Berkeley
Berkeley, CA, USA

References

1. Chua, L.O.: The genesis of Chua's circuit. Archiv für Elektronik und Übertragungstechnik **46**, 250–257 (1992)
2. Chua, L.O., Matsumoto, T., Komuro, M.: The double scroll. IEEE Transactions on Circuits and Systems **32**, 798–818 (1985)
3. Suykens, J.A.K., Vandewalle, J.: Generation of n-double scrolls ($n = 1, 2, 3, 4, ...$). IEEE Transactions on Circuits and Systems, Part-I **40**, 861–867 (1993)
4. Arena, P., Baglio, S., Fortuna, L., Manganaro, G.: How state controlled CNN cells generate the dynamics of the Colpitts-like oscillator. IEEE Transactions on Circuits and Systems, Part-I **43**, 602–605 (1996)
5. Suykens, J., Huang, A., Chua, L.O.: A family of n-scroll attractors from a generalized Chua's circuit. Archiv für Elektronik und Übertragungstechnik **51**, 131–138 (1997)
6. Yalcin, M., Suykens, J., Vandewalle, J.: Experimental confirmation of 3- and 5-Scroll attractors from a generalized Chua's circuit. IEEE Transactions on Circuits and Systems, Part-I **47**, 425–429 (2000)
7. Tang, W.K.S., Zhong, G.Q., Chen, G., Man, K.F.: Generation of N-scroll attractors via sine function. IEEE Transactions on Circuits and Systems, Part-I **48**, 1369–1372 (2001)
8. Yalcin, M., Suykens, J., Vandewalle, J., Ozoguz, S.: Families of scroll grid attractors. International Journal of Bifurcation and Chaos **12**, 23–41 (2002)
9. Zhong, G., Man, K.F., Chen, G.: A systematic approach to generating n-scroll attractors. International Journal of Bifurcation and Chaos. **12**, 2907–2915 (2002)
10. Fujiwara, T., Horio, Y., Aihara, K.: An integrated multi-scroll circuit with floating-gate MOSFETs. Proceedings of the 2003 International Symposium on Circuits and Systems. Bangkok, Thailand (2003)
11. Lü, J., Yu, S., Leung, H., Chen, G.: Experimental verification of multidirectional multiscroll chaotic attractors. IEEE Transactions on Circuits and Systems, Part-I **53**, 149–165 (2006)
12. Lü, J., Chen, G.: Generating multiscroll chaotic attractors: Theories, methods and applications. International Journal of Bifurcation and Chaos **16**, 775–858 (2006)

Acknowledgements

We thank our mentor Prof. Leon O. Chua for his inspiration and long-term support of our research on multiscroll chaotic attractors, with unwavering encouragement for our work.

We acknowledge many colleagues and friends who have contributed in various ways to the completion of this book. These include, in particular, Profs. Simin Yu from Guangdong University of Technology, China, and Wallace K. S. Tang from City University of Hong Kong, China, who contributed many technical designs and implementations of various multiscroll chaotic attractors introduced in this book.

We also acknowledge many colleagues and collaborators who contributed to the development of the multiscroll chaotic attractors techniques and methodologies in the past, with significant discoveries and advancements that are presented in our book. These include, in particular, Profs. Kazuyuki Aihara, M. A. Aziz-Alaoui, Ahmed S. Elwakil, Luigi Fortuna, Giuseppe Grassi, Fengling Han, Michael Peter Kennedy, Serdar Ozoguz, Johan Suykens, Joos Vandewalle, Müstak E. Yalcin, and Xinghuo Yu, among others.

Contents

Chapter 1
Introduction

Abstract Over the past three decades, theoretical design and circuit implementation of various chaos generators have been a focal subject of increasing interest due to their promising applications in various real-world chaos-based technologies and information systems. This chapter introduces the background and history of multiscroll chaotic attractors, and prepares some mathematical theories which are useful in the design and analysis of multiscroll chaotic attractors to be developed in the book.

1.1 A Brief History of Chaos

Since the mid-20th century, research has focused on studying how complex system states transit from order to chaos, and vice versa, as well as understanding the nature and characteristics of chaos.

Poincaré [1] is recognized as a pioneer in the study of chaos. While investigating whether the stability of the solar system could be mathematically proven, he discovered that even models with just three celestial bodies could produce significantly random results. This phenomenon was recognized later as a form of chaos in conservative systems.

In 1963, Lorenz [2] formulated the now-famous Lorenz equation when he studied a natural convection system in the atmosphere with temperature gradients. That was the first example of a deterministic equation in a dissipative system that could generate chaotic solutions. Since then, a lot of investigations on chaos were developed based on the Lorenz system model. In 2000, a paper titled "The Lorenz attractor exists" published in *Nature* [3] rigorously demonstrated the mathematical existence of the Lorenz attractor. In general, a stable equilibrium point, a stable limit cycle, or a set of stable points in the phase space are referred to as an "attractor" that attacks nearby system trajectories. If such an attractor of positive measure satisfies some chaotic properties, to be further discussed below, it is called a "chaotic attractor".

In 1971, Ruelle and Takens [4] applied chaos theory to explain the mechanism of turbulence formation. They independently discovered a set of new attractors in complex dynamical systems, demonstrating that the motions associated with these

J. Lü et al., *Chaotic Systems with Multiscroll Attractors*,
https://doi.org/10.1007/978-3-032-09854-2_1

attractors are chaotic. They referred to these new attractors as strange attractors, which could be chaotic or non-chaotic.

Chaos was then considered representing the third major revolution in the 20th-century physics, following quantum mechanics and relativity [5]. The term “chaos” first appeared in scientific literature as a mathematical concept introduced by Li and Yorke in 1975 [6]. Over the past two decades thereafter, it has rapidly evolved to become a modern science with rich nonlinear physical means and profound mathematical implications. Today, chaos is regarded as a frontier science that integrates physical principles with mathematical theory; it examines concepts such as the sensitive dependence on initial conditions, topological transitivity and mixing, positive Lyapunov exponents, fractional dimensions, and singular attractors [7].

Since the 1980s, research has focused on studying how systems transit from order to chaos, and vice versa, as well as understanding the properties and characteristics of chaos. Additionally, the development of multi-scale classification theory and symbolic dynamics facilitated research and theoretical analysis on various chaotic systems and structures. Grassberger and Procaccia [8, 9] applied the phase-space reconstruction method to analyze the statistical characteristics of strange attractors in experimental systems. By examining time-series data, they calculated important chaotic feature quantities, such as fractional dimensions and Lyapunov exponents [10]. Their work has significantly advanced the chaos theory, driving it to move into the stage of practical applications.

Lately, chaos was found to have broad applications across various fields, including acoustics, optics, turbulence, chemical reactions, biological and ecological evolutions, astronomy, weather forecasting, information technologies, and even trading in the stock market, among many others.

Chaos theory continues to develop. While many unresolved issues remain, from time to time new discoveries are reported, new breakthroughs take place, and new applications emerge. It is an ever-young scientific subject with countless significant theoretical and practical implications.

1.2 Main Features of Chaos

The time evolution of any variable that influences the long-term development of a dynamical system provides insights into the long-term behavior of all other variables within the system through internal coupling interactions. Consequently, one can analyze the chaotic behavior of a system by studying the time series of any of its key variables that affects its long-term evolution.

A characteristic quantity of an attractor describes some aspects of the attractor, which can be categorized into “microscopic” and “macroscopic” levels. The microscopic level pertains to the number and types of the unstable periods that form the framework of a strange attractor. In contrast, the macroscopic level refers to the average feature quantity of the entire attractor, such as the correlation dimension and the Lyapunov exponent. In particular, some invariants of an attractor, such as correlation dimension and Lyapunov exponent are essential in characterizing the system's chaotic properties [10]. Whereas the Lyapunov exponent, which is the system's characteristic exponent, measures the average exponential divergence of nearby trajectories and quantifies the "degree of chaos" within a conservative dynamical system.

1.2.1 *Basic Concepts and Calculation Methods*

1.2.1.1 Lyapunov Exponent

Chaotic motion is characterized by its extreme sensitivity to initial conditions. When two trajectories of a chaotic system start from very close initial values, they diverge exponentially over time. A positive Lyapunov exponent quantitatively describes this divergence phenomenon [7].

First, consider the discrete-time case. In a one-dimensional (1-D) dynamical system, $x_{n+1} = F(x_n)$, the behavior of two initial points–whether they move apart from or come closer to each other–depends on the value of the derivative $\left|\frac{dF}{dx}\right|$. If the derivative value is $\left|\frac{dF}{dx}\right| > 1$, the two points will separate through iterations; if the value is $\left|\frac{dF}{dx}\right| < 1$, the two points will converge to each other. However, during the continuous iteration process, the value of $\left|\frac{dF}{dx}\right|$ also changes, leading to situations where the points may either separate or come together from time to time. To effectively represent the entire process in which two adjacent points eventually separate or converge exponentially, one can calculate their average over time, which is iterations in the discrete-time setting.

For this purpose, assume that separating or approaching rate of the two arbitrary initial points is changing at an exponential rate, denoted λ. Then, the distance between the two points, which were originally separated by a tiny $\varepsilon > 0$, in terns of iteration step number n, is

$$\varepsilon\, e^{n\lambda(x_0)} = \left|F^n(x_0 + \varepsilon) - F^n(x_0)\right|. \tag{1.1}$$

Dividing both sides by ε and then taking limits $\varepsilon \to 0$ and $n \to \infty$, one obtains from Eq. (1.1) that

$$\begin{aligned} \lambda(x_0) &= \lim_{\varepsilon\to 0, n\to\infty} \frac{1}{n} \ln \left| \frac{F^n(x_0+\varepsilon) - F^n(x_0)}{\varepsilon} \right| \\ &= \lim_{n\to\infty} \frac{1}{n} \ln \left| \frac{dF^n(x)}{dx} \right|_{x=x_0}. \end{aligned} \tag{1.2}$$

By putting all points along the trajectory together, one obtains

$$\lambda = \lim_{n\to\infty} \frac{1}{n} \sum_{i=0}^{n-1} \ln \left| \frac{dF^n(x)}{dx} \right|_{x=x_i}, \tag{1.3}$$

where λ is independent of the selection of the initial value.

This λ is called the Lyapunov exponent of the underlying dynamical system. It represents the average exponent in the exponential separation rate over the entire process.

As discussed earlier, if condition $\lambda < 0$ is satisfied, it indicates that adjacent points will eventually converge and approach a single point. This convergence is essential for maintaining stable equilibrium points and periodic motion such as limit cycles. However, if condition $\lambda > 0$ is satisfied, it means that adjacent points will ultimately separate from each other, reflecting the local instability of the trajectory. If the trajectory is also globally bounded or dissipative, the trajectory cannot escape to infinity but will repeatedly stretch and fold thereby creating a chaotic attractor as a result. Therefore, condition $\lambda > 0$ can be used as a criterion, a necessary condition, for the forming and existence of a chaotic attractor of the system.

For a general n-D dynamical system, the Lyapunov exponents are defined as follows.

Let F be an n-D map defined on $R^n \rightarrow R^n$, which determines an n-D discrete-time system:

$$x_{k+1} = F(x_k), \quad k = 0, 1, 2, \ldots . \tag{1.4}$$

Set the initial condition of the system as an infinitely small n-D sphere S_n, which undergoes natural deformation during the evolution process. This tiny sphere will transform to be an ellipsoid through the iteration process, with all the principal axes arranged in the order of their lengths. Thus, the i-th Lyapunov exponent is defined, based on the rate of increase of the length $P_i(k)$ of the i-th principal axis of S_n at the k-th iteration step, $k = 0, 1, 2, \ldots$, by

$$\lambda_i = \lim_{k \to \infty} \frac{1}{k} \ln \left[\frac{P_i(k)}{P_i(0)} \right], \quad i = 1, 2, \ldots, n. \tag{1.5}$$

These Lyapunov exponents are related to how the trajectories behave in the phase space, whether contract or expand. When a Lyapunov exponent is negative, the trajectory is stable and contracts, meaning that it is less sensitive to initial conditions. Conversely, when the Lyapunov exponent is positive, the trajectory diverges at an exponential rate, indicating its high sensitivity to initial conditions.

Arrange all Lyapunov exponents in their magnitudes, as follows:

$$\lambda_1 \geq \lambda_2 \geq \lambda_3 \geq \cdots \geq \lambda_n. \tag{1.6}$$

In the spectrum of Lyapunov exponents of a chaotic system, the minimum Lyapunov exponent is negative, indicating the rate of trajectory contraction, while the maximum Lyapunov exponent is positive, reflecting the rate of trajectory divergence. Thus, to verify whether or not a system is chaotic, it suffices to verify its maximum Lyapunov exponent $\lambda_{\max}$: if $\lambda_{\max} > 0$ then the system is chaotic, provided that the system is also globally bounded.

The sum of all Lyapunov exponents, $\sigma := \sum_{i=1}^{n} \lambda_i$, can be seen as an indication of the overall average convergence or divergence rate of the system trajectories. For a typical 3-D chaotic system, its three Lyapunov exponents have the signs $(-, 0, +)$ with the sum $\sigma < 0$, implying that the system is locally diverging from the neighborhood of an equilibrium point while globally bounded.

1.2.1.2 Poincaré Surface of Section

When a section is properly chosen in the phase space, it allows for the observation of the system's motion characteristics and changes, as long as the section does not tangent to the trajectory or contain the trajectory. This section is referred to as a Poincaré section, which is a plane in the 3-D phase space or a surface in the n-D phase space.

A continuous trajectory in the phase space, and the points on the corresponding Poincaré section where they intersect, are called the intersection points [11]. Let the intersection points obtained from the recorded trajectory be $B_0, B_1, \ldots, B_i, \ldots$. Then, the system trajectory can be understood and analyzed from such sequence of points moving continuously on the Poincaré section. This reduces the mapping trajectory to a lower-dimensional point sequence:

$$B_{i+1} = T B_i, \quad i = 0, 1, 2, \ldots,$$

where T is called the Poincaré map.

By analyzing the intersection points on the Poincaré section, it is possible to determine whether chaotic behavior is present. If there is only one point or a finite number of points on the Poincaré section, the trajectory motion is periodic with period equal to the number of points. If the Poincaré section has a closed curve, the motion is quasi-periodic. If the Poincaré section has a dense points, typically with a fractal structure, the motion is considered chaotic.

1.2.1.3 Phase-Space Reconstruction

The purpose of phase space reconstruction is to restore chaotic attractors in higher-dimensional phase spaces. As one of the characteristics of chaotic systems, chaotic attractors reflect the regularity of chaotic systems, meaning that a chaotic system state will eventually fall into a specific trajectory surrounding an irregular limiting point set of positive measure, where the set is invisible while the trajectory is, which is called a chaotic attractor.

Takens [12] found that, from the trajectory data of a perhaps unknown dynamical system, the dimension of the system phase space can be well estimated, referred to as the embedding dimension. Specifically, based on a measured data time sequence, $\{x(i)\}_{i=1}^{\infty}$, define a delayed vector $y(i) = (x(i), x(i-d), \ldots, x(i-(m-1)d))$, where d is a delay integer for example $d = 1$, so as to establish a new phase space in $\{y(i)\}_{i=1}^{\infty}$. Here, m is referred to as the embedding dimension. Takens [12] proved that, if $m \geq 2d + 1$ then all regular trajectories including attractors of the underlying dynamical system can be recovered from this embedding dimension space reconstructed using the measured data obtained from the underlying dynamical system.

Takens [12] showed that the embedding space trajectory maintains differential homeomorphism with the underlaying dynamical system in the reconstructed R^m-space.

Definition 1.1 Let (N_1, ρ_1) and (N_2, ρ_2) be two metric spaces, and suppose that there is a map, $\varphi : N_1 \to N_2$, satisfying

1. φ is surjective;
2. $\rho_1(x, y) = \rho_2(\varphi(x), \varphi(y))$ holds for all $x, y \in N_1$.

Then, (N_1, ρ_1) and (N_2, ρ_2) are said to be isomorphically isometric.

Definition 1.2 If (N_1, ρ_1) is isometrically isomorphic to a subspace (N_0, ρ_0) of another metric space (N_2, ρ_2), then (N_1, ρ_1) is said to be embeddable into (N_0, ρ_0).

Takens Theorem [8] Let M be a d-dimensional manifold and $\varphi : M \to M$ be a diffeomorphism, assume that $y : M \to R$ has a continuous second derivative, and define

$$\phi(\varphi, y) : M \to R^{2d+1},$$

where $\phi(\varphi, y) = \left(y(x), y(\varphi(x)), y\left(\varphi^2(x)\right), \ldots, y\left(\varphi^{2d}(x)\right)\right)$. Then, $\phi(\varphi, y)$ is an embedding of M into R^{2d+1}.

For a given time series, $\{x_1, x_2, \ldots, x_{n-1}, x_n, \ldots\}$, if the embedding dimension m and time delay d are appropriately chosen, then according to Takens Theorem, the dynamical properties of the attractor of the underlying dynamical system can be recovered in the sense of topological equivalence from the reconstructed phase space

$$Y(t_i) = (x(t_i), x(t_i + d), x(t_i + 2d), \ldots, x(t_i + (m-1)d)), \tag{1.7}$$

where $i = 1, 2, \ldots$.

A commonly used method for embedding phase space reconstruction is proposed by Grassberger and Procaccia [8], simply called the G-P algorithm, as follows:

1. Using a time series $\{x_1, x_2, \ldots, x_{n-1}, x_n, \ldots\}$, start with a small value m_0, reconstruct the embedding phase space (1.7).
2. Calculate the correlation function

$$C(r) = \lim_{N\to\infty} \frac{1}{N^2} \sum_{i,j=1}^{N} \theta\left(r - \left|Y(t_i) - Y\left(t_j\right)\right|\right), \tag{1.8}$$

where $\left|Y(t_i) - Y(t_j)\right|$ represents the distance between the phase points $Y(t_i)$ and $Y(t_j)$, $\theta(z)$ is the Heaviside function, and $C(r)$ is a cumulative distribution function, representing the probability that the distance between two points on the attractor in the phase space is less than r.

3. For an appropriate range of r, the dimension D of the attractor should satisfy a logarithmic linear relationship with the cumulative distribution function $C(r)$, namely,

$$D(m) = \frac{\ln C(r)}{\ln r}. \tag{1.9}$$

Thus, the correlation dimension estimate $D(m_0)$ corresponding to a certain initial value m_0 can be obtained by data fitting.

4. Increase the embedding dimension $m_1 > m_0$, and repeat steps 2 and 3 until the corresponding dimension estimate $D(m_0)$ no longer changes significantly within a predefined error range as m increases. The resulting D at this point is the desired estimation of the correlation dimension of the attractor.

Clearly, in the reconstructed phase space, the selection of the time delay d and embedding dimension m are of great importance, which however are quite difficult to determine in general.

1.2.2 *Fractal Dimensions of Chaos*

A strange attractor in the phase space is a set of points with a geometric shape that emerges from the repeated convergence and separation of state trajectories of the dynamical system in consideration. This convergent and separating process creates a pattern of the point set that stretches and folds back on itself, exhibiting a high complexity.

In a dissipative system, dynamical motion of the system states in the phase space contracts, resulting in a strange attractor with a certain geometric dimension smaller than that of the phase space and the geometric properties of such a strange attractor can be analyzed through its geometric dimension. Typically, smaller parts of the strange attractor in smaller scales resemble the whole in a specific manner, strange attractors are classified as some kind of fractals. A general common feature of such fractals is their fractal dimension, a concept developed by Benoit Mandelbrot in the 1970s [13].

1.2.2.1 Hausdorff Dimension

Consider a non-empty set U in the n-D Euclidean space, where the diameter of U defines its geometric dimension, usually different from n. If there exists a countable collection of sets U_i with diameters not exceeding δ in the n-D space that covers F, namely,

$$F \subset \bigcup_{i=1}^{\infty} U_i,$$

then this collection is called a δ-coverage of F.

Let F be a subset of $\mathbf{R}^n$ and let s be a nonnegative number. For any given $\delta > 0$, define

$$A_\delta^s(F) = \inf\left\{\sum_{i=1}^{\infty} |U_i|^s : \{U_i\} \text{ is the } \delta\text{−coverage of } F\right\}. \tag{1.10}$$

Now, consider the coverage of F using sets with diameters not exceeding δ and minimize the sum of these diameters raised to the power of s. As δ decreases, the lower bound of Eq. (1.10) approaches a limit, denoted as

$$A^s(F) = \lim_{\delta\to 0} A_\delta^s(F), \tag{1.11}$$

and this limit, if exists, is defined to be the s-D Hausdorff measure of F, which is typically 0 or ∞ for general geometric objects in the phase space.

It follows from Eq. (1.10) that, for any given set F and any fixed $\delta < 1$, the function $A_\delta^s(F)$ is non-increasing with respect to s. Therefore, in Eq. (1.11), $A^s(F)$ does not increase as s increases. Consequently, if $t > s$ and $\{U_i\}$ is a δ-coverage of F, then the following inequality holds:

$$\sum_t |U_i|^t \le \delta^{t-s} \sum_t |U_i|^s, \quad i = 1, 2, \ldots.$$

Taking the sums and bounds gives

$$A_\delta^t(F) \le \delta^{t-s} A_\delta^s(F).$$

Further, let $\delta \to 0$. It follows that, for any $t > s$, if $A^s(F) < \infty$ in Eq. (1.11), then $A^t(F) = 0$. So, there is a critical value of s in $A^s(F)$ that causes $A^s(F)$ to "jump" from ∞ to 0. This critical value of s is called the Hausdorff dimension of F, denoted as $D_H(F)$.

1.2.2.2 Correlation Dimension

Grassberger and Procaccia [8] proposed the G-P algorithm for calculating the correlation function (1.8) of a finite sequence of data $\{x_1, x_2, \ldots, x_{n-1}, x_n\}$ obtained from a dynamical system, which leads to the correlation dimension (1.9). This correlation dimension satisfies

$$\lim_{r\to 0} C(r) \propto r^D \tag{1.12}$$

for some constant D, called the correlation dimension.

The correlation dimension D of any point set A in the phase space, comparing to its Hausdorff dimension D_H, satisfies $D \le D_H(A)$.

In practical numerical calculation of the correlation dimension D from Eq. (1.12), small values of $r > 0$ will be used. It is noted, however, that if r is taken to be too small then the result obtained using Eq. (1.9) would not be accurate due to numerical instability. In practice, it is common to increase the length n of the time series from a small value while keeping D constant, which corresponds to a straight line segment in the double logarithmic relationship $\ln C(r) \sim \ln r$. The slope of the best fitting straight line between them, except for lines with a slope of 0 or ∞, provides an accurate estimation of the correlation dimension D.

1.2.2.3 Similarity Dimension

Start with a rectangle on a plane. Increasing its size by ℓ times in all directions yields a larger rectangle, which is equivalent to ℓ^2 original rectangles. Similarly, for a cube in the 3-D space, by the same expansion one obtains a larger cube that is equivalent to ℓ^3 original cubes. Generally, for a d-dimensional hypercube in the d-dimensional space, enlarging the dimensions in each direction by a factor of ℓ yields a larger hypercube with a d-dimensional volume that is ℓ^d times of the original one. Taking the logarithm of $N = \ell^d$ gives

$$D_S = \frac{\ln N}{\ln \ell}, \tag{1.13}$$

which defines the similarity dimension. This measure is suitable for fractal geometry with self-similarity properties, also known as self-similarity dimension.

1.2.2.4 Lyapunov Dimension

For an attractor in the phase-space portrait of a dynamical system, the direction corresponding to a positive Lyapunov exponent supports the expansion of the attractor, the direction with a negative Lyapunov exponent supports its contraction, and the direction with a zero Lyapunov exponent supports a periodic motion. The attractor's effective dimension is obtained by balancing the effects of expansion and contraction.

Let the Lyapunov exponents be arranged in descending order, as $\lambda_1 \ge \lambda_2 \ge \lambda_3 \ge \cdots$. One effective dimension of an attractor in the phase space is the Lyapunov dimension [14], also referred to as the Kaplan-Yorke dimension, which takes into account both expanding and contracting directions of the attractor. Specifically, for each $k = 1, 2, \ldots,$ compute

$$D_L(k) = k + \frac{S_k}{|\lambda_{k+1}|},$$

where $S_k = \sum_{i=1}^{k} \lambda_i$, for $k = 1, 2, \ldots, K$, such that K is the first value of k that ensures both $S_K > 0$ and $S_{K+1} < 0$. Then, using this integer K, the Lyapunov dimension of the attractor is defined as $D_L := D_L(K)$, namely,

$$D_L = K + \frac{S_K}{|\lambda_{K+1}|}. \tag{1.14}$$

It was conjectured [14] that the Lyapunov dimension and self-similarity dimension are always equal. As a matter of fact, for many systems, including chaotic systems, $D_L = D_S$.

1.2.3 Lyapunov Exponents of Chaotic Systems

Due to the defining nature of the Lyapunov exponents, they are useful for determining and describing the global averaged dynamical behaviors of a dynamical system. As mentioned above, the direction corresponding to a positive Lyapunov exponent supports the expansion of the attractor in the phase space, while the direction with a negative Lyapunov exponent supports its contraction and the direction with a zero Lyapunov exponent supports a periodic motion. Therefore, how to find, or calculate, the Lyapunov exponents of a given dynamical system is of extremely importance in applications.

There are many effective methods for calculating Lyapunov exponents, which generally fall into two categories: Wolf methods and Jacobian methods. The Wolf method is suitable for time series that are free of noise and exhibit highly nonlinear evolution of small vectors in tangent space [15]. In contrast, the Jacobian method is more appropriate for time series with significant noise and cases where the evolution of small vectors in tangent space is nearly linear [7, 16, 17]. Barana and Tsuda [18] proposed a p-norm algorithm that bridges the gap between the Wolf and the Jacobian methods. However, due to the complexity involved in selecting and calculating the p-norm, its practical implementation can be quite challenging.

1.2.3.1 Nicolis Method

In 1983, Grassberger et al. [8, 9] demonstrated that chaos can be confirmed if the maximum Lyapunov exponent is greater than zero. This means that at least the system dynamics are expanding along the direction of the maximum Lyapunov exponent, which is sufficient for chaos to exist if the system dynamics are also globally bounded.

For numerical computation, there is a Nicolis method [19] available for calculating the maximum Lyapunov exponent.

Using $\|\cdot\|$ to represent the distance in the phase space. Then, the maximum Lyapunov exponent λ_1 in the spectrum (1.6) satisfies

$$\begin{aligned}\|\delta x_0\| e^{n\lambda_1} &= \left\| F^n (x_0 + \delta x_0) - F^n (x_0) \right\| \\ &\approx \| J (x_{n-1}) J (x_{n-2}) \cdots J (x_0) \delta x_0 \| , \quad n = 1, 2, \ldots ,\end{aligned}$$

where $\delta > 0$ is a small variation and $J(x)$ is the Jacobian matrix of the system at x.

To calculate λ_1, denote δx_0 as $d_0 e_0$, where d_0 represents modulus and e_0 represents a unit vector. Similarly, denote $J(x_0)e_0$, $J(x_1)e_1, \ldots, J(x_{n-1})e_{n-1}$, by their magnitudes $d_1, d_2, \ldots, d_n$ with unit vectors $e_1, e_2, \ldots, e_n$, respectively. Then, the maximum Lyapunov exponent λ_1 can be calculated by

$$\lambda_1 = \lim_{n\to\infty} \frac{1}{n} \ln |d_n d_{n-1} \cdots d_2 d_1| . \tag{1.15}$$

1.2.3.2 Wolf Method

The method for extracting the Lyapunov exponent from a univariate time series is still based on the reconstructed phase space of the time series. Wolf et al. proposed an approach that estimates the Lyapunov exponent based on the evolution of phase trajectories, phase planes, and phase volumes [15]. These methods are collectively known as the Wolf method, which is widely applied in chaos research and Lyapunov exponent-based chaotic time series prediction.

Given a chaotic time series $x_1, x_2, \ldots, x_k, \ldots$, with embedding dimension m and time delay τ, the reconstructed phase space is Eq. (1.7).

Select an initial point $Y(t_0)$ and let its distance to the nearest neighbor $Y_0(t_0)$ be L_0. Track the time evolution of these two points until time t_1, at which their separation exceeds a predefined threshold $\varepsilon > 0$, i.e.,

$$L_0^t = |Y(t_1) - Y_0(t_1)| > \varepsilon.$$

Retain $Y(t_1)$ and find another nearby point $Y_1(t_1)$ such that

$$L_1^t = |Y(t_1) - Y_1(t_1)| < \varepsilon,$$

while ensuring the angle between them is as small as possible. Continue this process until $Y(t)$ reaches the end of the time series at N. The total number of iterations during this tracking process is M, and the largest Lyapunov exponent is given by

$$\sigma = \frac{1}{t_M - t_0} \sum_{i=0}^{M} \ln \frac{\Delta L_i}{L_i}. \tag{1.16}$$

Here, ΔL_i represents the change in the length of the line segment after the evolution of the system.

If the second-largest Lyapunov exponent is to be computed, a point and its two nearest neighbors forming a triangle must be tracked. If this triangle becomes too skewed or its area grows too large, a new triangle is selected, minimizing the angle between its two sides and those of the original triangle. The process continues until the endpoint is reached. The second-largest Lyapunov exponent is then given by

$$\sigma_1 + \sigma_2 = \frac{1}{t_M - t_0} \sum_{i=0}^{M} \ln \frac{\Delta A_i}{A_i}, \tag{1.17}$$

where ΔA_i represents the change in triangle area after the evolution of the system, and A_i represents the area of the triangle. Based on this, σ_2 can be obtained, and similarly, $\sigma_3, \sigma_4, \ldots$ can be computed.

In principle, this process can continue until the last Lyapunov exponent σ_n is determined. However, due to practical limitations such as the finite length N of the available data and the influence of noise, only the largest Lyapunov exponent can be reliably estimated.

1.2.3.3 Jacobian Method

Since stochastic processes (such as Lévi flows) can also exhibit fractional dimensions, fractional dimensionality is not an essential characteristic of chaos, but rather a necessary condition for its existence. To further clarify the chaotic nature of a system, it is essential to compute its Lyapunov exponent. The Jacobian method is a technique developed for practical applications to calculate the Lyapunov exponent [7, 16, 17].

Consider the following system of differential equations:

$$\dot{x} = F(x), \tag{1.18}$$

where $\dot{x} = \frac{dx}{dt}$ and $x \in \mathbf{R}^m$. The evolution of the tangent vector e at a point $x(t)$ in space can be described by

$$\dot{e} = T(x(t))e, \tag{1.19}$$

where $T = \frac{\partial F}{\partial x}$ and T is the Jacobian matrix of F. The solution to Eq. (1.19) can be expressed as

$$e(t) = U(t, e(0)), \tag{1.20}$$

where $U : e(0) \mapsto e(t)$ is a linear operator mapping the initial perturbation to its state at time t.

The asymptotic behavior of this mapping U can be characterized by an exponential rate:

$$\lambda\left(x(0), e(0)\right) = \lim_{t\to\infty} \frac{1}{t} \ln \frac{\|e(t)\|}{\|e(0)\|}. \tag{1.21}$$

Therefore, the Lyapunov exponent of system Eq. (1.21) can be expressed as the average of the repeated process described above

$$\lambda = \lim_{k\to\infty} \frac{1}{k\Delta t} \sum_{j=1}^{k} \ln \frac{\|e((j+1)\Delta t)\|}{\|e(j\Delta t)\|}. \tag{1.22}$$

Given a time series $\{s_1, s_2, \ldots, s_l, \ldots\}$, with embedding dimension m and time delay τ, the reconstructed phase space is defined as

$$x(t_k) = \left(s(t_k), s(t_k+\tau), \ldots, s(t_k+(m-1)\tau)\right), \quad k = 1, 2, \ldots, N \tag{1.23}$$

where the discrete time interval is Δt.

Let the unit tangent vector at point $x(t)$ in the phase space at time t (where $t = k\Delta t, k \in \{1, 2, \ldots, N-1\}$) be e. After a time interval Δt, this tangent vector evolves into e'.

Since the evolution of the tangent vector over a short time interval is approximately a linear process, we can approximate the evolution of e into e' by a linear combination of the evolution values of other tangent vectors at the trajectory point $x(t)$.

Consider the small vectors x_i formed by the differences between the point $x(t)$ in space and other points on the trajectory. These small vectors construct a matrix X, where the vectors x_i are sufficiently small, given by

$$X = (x_1, x_2, \ldots, x_n), \quad x_i = x(l_i\Delta t) - x(t), \tag{1.24}$$

where the small vectors x_i satisfy the condition

$$|x(l_i\Delta t) - x(t)| < \varepsilon, \quad l_i \in \{1, 2, \ldots, N\}.$$

Select a specific linear combination y that satisfies the following condition

$$Xy \approx e, \quad y \in \mathbf{R}^n.$$

Clearly, after a time interval Δt, this linear combination y ensures the following equation holds

$$X'y \approx e', \quad y \in \mathbf{R}^n.$$

Here, X' consists of the evolved vectors of the tangent vectors in X after a time interval Δt, given by

$$X' = \left(x_1', x_2', \ldots, x_n'\right),$$

where

$$x_i' = x\left((l_i + 1)\Delta t\right) - x\left(t + \Delta t\right).$$

If the above evolution process exhibits good linear characteristics, then the following holds

$$Xy \approx e, \quad X'y \approx e',$$

where $y = X^+ e, \quad X^+ = \left((X^T X)^{-1}\right) X^T$. The term $\|y\|_2$ is minimized (it is the smallest element).

Usually, the following method is used to compute

$$TX \approx X', \quad TE \approx E', \quad T = X'X^T.$$

where $E = \{e_1, e_2, \ldots, e_d\}$ is the standard orthogonal basis. The difference $(TX - X')$ in each row is minimized using the 2-norm, i.e.,

$$E' \approx TE = X'X^T E = X'Y \approx E.$$

Thus, we have the relations

$$XY = E, \quad X'Y = E', \quad Y = X^+ E.$$

Here, Y is the matrix form of y, and each column of y has the smallest 2-norm. Finally, the Lyapunov exponent can be obtained from Eq. (1.21) as follows:

$$\lambda = \frac{1}{N\Delta t} \sum_{j=0}^{N-1} \ln \frac{\|e\left(j+1\right)\Delta t\|}{\|e\left(j\Delta t\right)\|}. \tag{1.25}$$

1.2.4 Shilnikov Criterion

The most widely used criterion for verifying the existence of chaos is the Shilnikov Theorem [20, 21]. On one hand, this theorem is relatively easy to use for determining the existence of chaos in many autonomous systems; for instance, based on this theorem one can confirm the presence of Smale horseshoe chaos in the piecewise Chua's circuit system and the piecewise Lorenz system. On the other hand, it provides a theoretical basis with a effective method for constructing chaotic systems in some applications.

Consider a 3-D autonomous dynamical system,

$$\dot{x}(t) = \zeta(x), \quad t \in \mathbf{R},\ x \in \mathbf{R}^3, \tag{1.26}$$

where $\zeta : \mathbf{R}^3 \to \mathbf{R}^3$ belongs to C^r with $r \geq 2$.

Assume that this 3-D autonomous system has two distinct equilibrium points O_1 and O_2. If the following conditions are satisfied, then the system Eq. (1.26) as well as its slightly perturbed system are chaotic in the sense of having Smale horseshoe:

1. The two equilibrium points O_1 and O_2 are saddle-foci, and the eigenvalues of the Jacobian of the system at these two equilibrium points are $\gamma_i, \sigma_i \pm j\omega_i, i = 1, 2$, which satisfy the Shilnikov inequality

$$|\gamma_i| > |\sigma_i| > 0, \quad i = 1, 2.$$

   Additionally, one of the following conditions holds:

$$\sigma_1\sigma_2 > 0 \quad \text{or} \quad \omega_1\omega_2 > 0.$$

2. There exists a heteroclinic loop connecting the equilibrium points O_1 and O_2. This heteroclinic loop is composed of two heteroclinic orbits $\Gamma_i, i = 1, 2$.

Here, a heteroclinic orbit between two equilibrium points of an autonomous system is an orbit that converges to one equilibrium point as $t \to \infty$ and converges to the other equilibrium point as $t \to -\infty$, as illustrated by Fig. 1.1. The main difficulty in applying the Shilnikov criterion lies in the proof of the existence of a heteroclinic orbit in the system, which usually is assumed in analyzing a chaotic system in engineering applications.

Under the above conditions, the system has infinitely many Smale horseshoes, therefore is chaotic in this sense of Smale horseshoes.

Using a switching controller, the equilibrium points of a given system can be shifted and even exchanged. Take a 3-D linear switching system as an example, with the switching plane $S : x = 0$. The heteroclinic loop connecting two switching planes is shown in Fig. 1.2a. This can generate a double-wing or double-scroll chaotic attractor, as illustrated by Fig. 1.2b.

If, under switching control, an equilibrium point is shifted to $P_1(x_1, y_1, z_1) \in V_1$, then the corresponding 1-D stable manifold in the phase space is given by a line equation denoted $E^S(P_1)$, and the corresponding 2-D unstable manifold in the phase space is given by a plane equation denoted $E^U(P_1)$.

Similarly, if an equilibrium point is shifted to $P_2(x_2, y_2, z_2) \in V_2$, then the corresponding 1-D stable manifold in the phase space is given by a line equation denoted $E^S(P_2)$, and the corresponding 2-D unstable manifold in the phase space is given by a plane equation denoted $E^U(P_2)$.

Similarly, the heteroclinic loop connecting multiple switching planes, as shown in Fig. 1.3a, can generate a multi-wing or multi-scroll chaotic attractor, as illustrated by Fig. 1.3b.

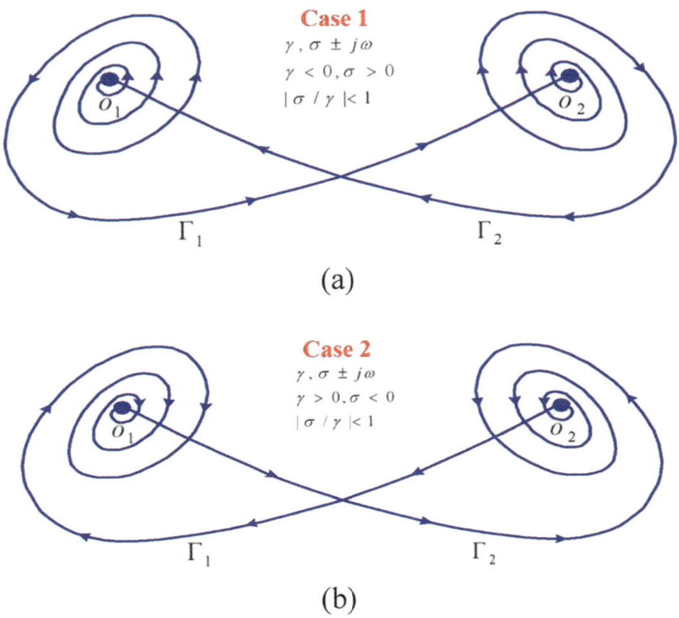

Fig. 1.1 Shilnikov criterion. **a** Heteroclinic orbit between two case-1 saddle-focus equilibria. **b** Heteroclinic orbit between two case-2 saddle-focus equilibria

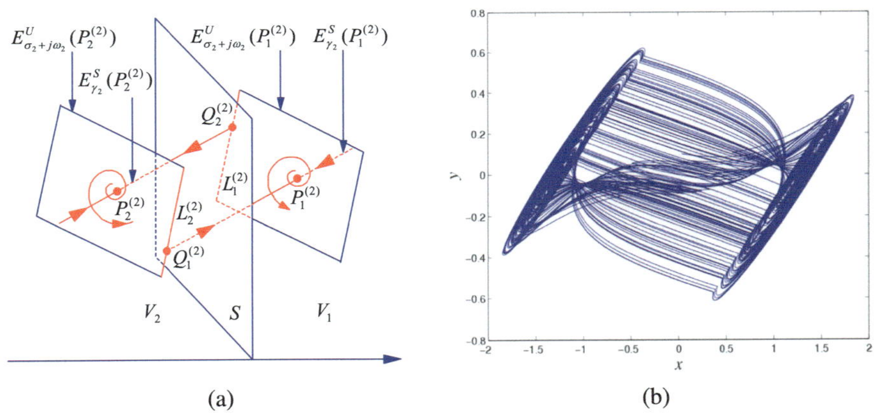

Fig. 1.2 Shilnikov criterion for a double-scroll attractor. **a** Single heteroclinic orbit. **b** Double-scroll attractor

1.2.5 Criteria of Chaos

Only dissipative chaotic systems but not Hamiltonian chaotic systems are studied in this book, where a system is dissipative if the system flow divergence is negative.

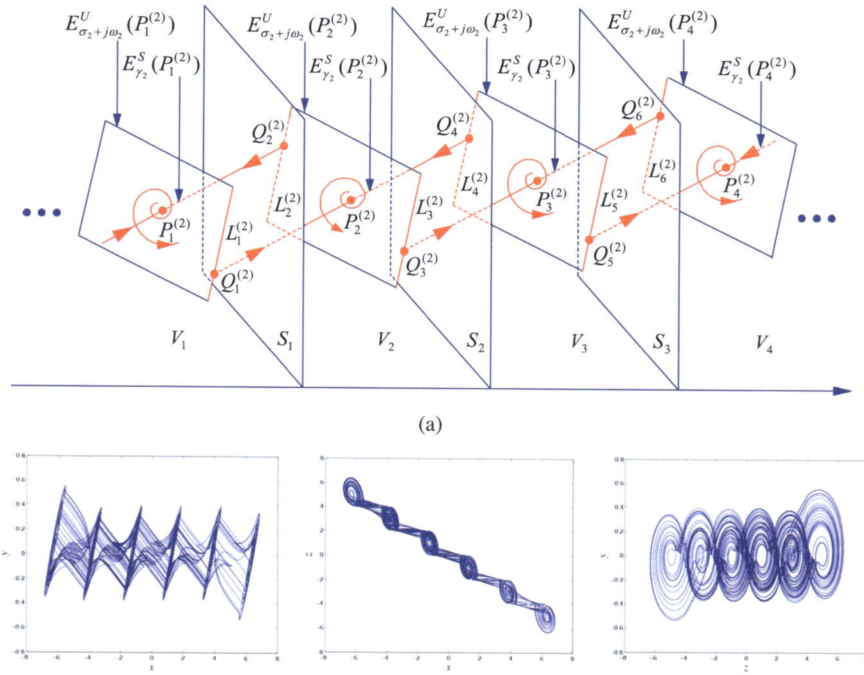

(b)

Fig. 1.3 Shilnikov criterion for a multi-scroll attractor. **a** Multiple heteroclinic orbit. **b** Multi-scroll attractor

Among the many characteristics of discrete-time chaotic systems, two are easy to use to verify the existence of chaos in engineering systems [7]:

1. The system has a positive Lyapunov exponent.
2. The system orbits are globally bounded.

Here, the first condition indicates that the system orbits are locally diverging and cannot be periodic, which is the repulsion effect, while the second condition guarantees that the system orbits cannot diverge to infinity, which is the attraction effect. Thus, the combination of these two effects generate the chaotic dynamical behavior.

It is noted that, the continuous-time autonomous chaotic systems, represented by differential equations rather than maps, are not the main subject for study in this book. Among the many characteristics of such systems, there are three easily-verified ones:

1. The system simultaneously possesses positive, negative, and zero Lyapunov exponents.
2. The orbits are globally bounded, which can be verified if the absolute value of the negative Lyapunov exponent is lager than the positive Lyapunov exponent.

3. The system is dissipative.

Typically, a continuous-time autonomous chaotic system has these three properties.

1.3 Several Typical Chaotic Systems

1.3.1 Chua's Circuit

The Chua's circuit [22–24], as illustrated by Fig. 1.4, is described by

$$\begin{cases} \dfrac{dV_{C1}}{dt} = \dfrac{1}{RC_1}\left(V_{C_2} - V_{C_1}\right) - \dfrac{1}{C_1} g\left(V_{C_1}\right), \\ \dfrac{dV_{C2}}{dt} = \dfrac{1}{RC_2}\left(V_{C1} - V_{C2}\right) + \dfrac{1}{C_2} i_L, \\ \dfrac{di_L}{dt} = -\dfrac{1}{L} V_{C_2}, \end{cases} \tag{1.27}$$

where V is voltage, i is current, R is resistance, C is capacitor, L is inductance, and

$$g\left(V_{C_1}\right) = m_1 V_{C_1} + \frac{1}{2}\left(m_0 - m_1\right) \times \left(\left|V_{C1} + b_1\right| - \left|V_{C_1} - b_1\right|\right) \tag{1.28}$$

is a piecewise linear (PWL) resistance function, as shown in Fig. 1.4b.

Chua's circuit system (1.27) has three equilibria: one at the origin, one approximately located at the center of the upper hole, denoted P_+, and one located at a point symmetric about the origin, denoted P_-. Each equilibrium point has one real eigenvalue and two complex conjugate eigenvalues. At P_+ (and P_-), associated with

$$\frac{1}{C_1} = 10,\ \frac{1}{C_2} = 0.5,\ \frac{1}{L} = 0.5,\ R = \frac{10}{7},\ m_0 = -0.1,\ m_1 = 4,\ b_1 = 1, \tag{1.29}$$

the real eigenvalue is $\lambda_p \approx -6.37$, while the other ones are $\omega_p \pm j\rho_p \approx 0.01 \pm j1.82$.

At the origin, the eigenvalues are $\lambda_o \approx -33.07$ and $\omega_o \pm j\rho_o \approx 0.21 \pm j1.86$.

It is clear that the real eigenvalue at the origin, denoted λ_o, totally dominates the others. This indicates that the rate of expansion at the origin is exceptionally strong.

Now, let $x = V_{C1}/b_1$, $y = V_{C2}/b_1$, $z = Ri_L/b_1$, $\tau = t/C_2R$, $\alpha = C_2/C_1$, $\beta = C_2R^2/L$, $\overline{m_0} = m_0R$, and $\overline{m_1} = m_1R$. Then, system (1.27) can be rewritten in a dimensionless form as

$$\begin{cases} \dot{x} = \alpha\left(-x + y - f(x)\right), \\ \dot{y} = x - y + z, \\ \dot{z} = -\beta y, \end{cases} \tag{1.30}$$

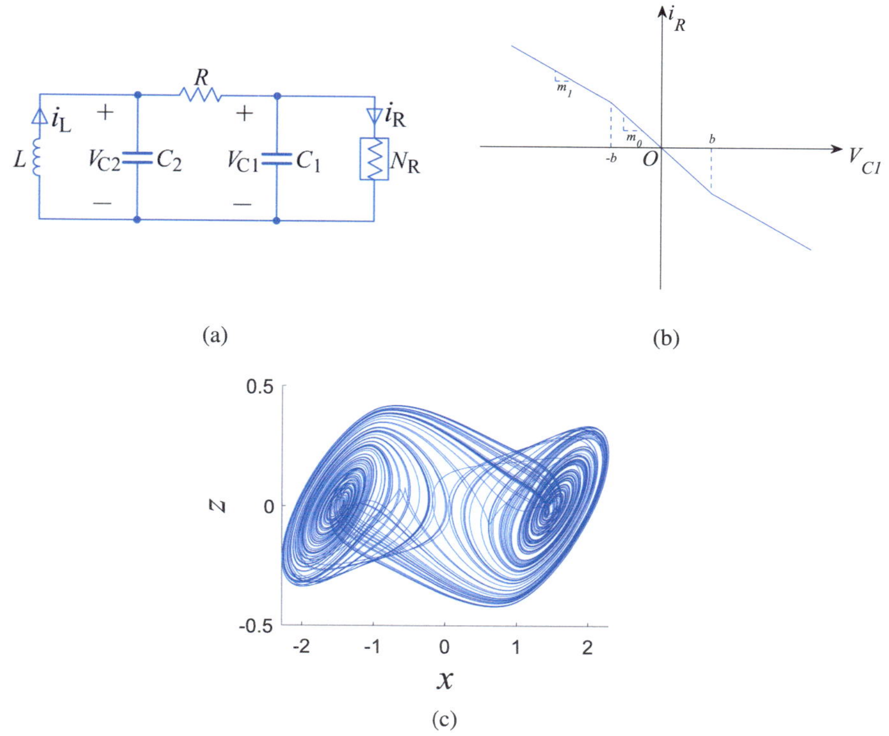

Fig. 1.4 Chua's circuit [22–24]. **a** Circuit diagram. **b** PWL resistance function. **c** Double-scroll chaotic attractor

where

$$f(x) = \overline{m_1}x + 1/2\,(\overline{m_0} - \overline{m_1})\,(|x+1| - |x-1|)\,.$$

With $\alpha = 10.0$, $\beta = 15$, $\overline{m_0} = -8/7$, $\overline{m_1} = -5/7$, the system produces a double-scroll chaotic attractor, as shown in Fig. 1.4c.

Figure 1.5 shows the maximum Lyapunov exponents of system (1.30).

1.3.2 Lorenz System

In 1963, Lorenz [2] discovered the extreme sensitivity to initial conditions in a deterministic system of equations describing weather forecasting, referred to as a chaotic system and named Lorenz system later on.

The Lorenz system is described by

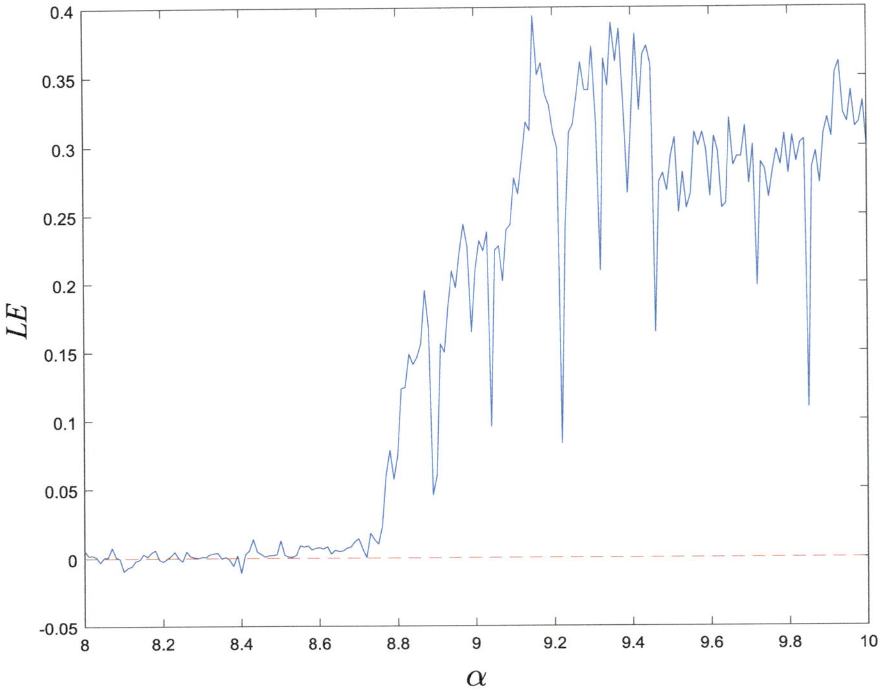

Fig. 1.5 Maximum Lyapunov exponents of system (1.30) with $\alpha \in [8, 10]$

$$\begin{cases} \dot{x} = a\,(y - x)\,, \\ \dot{y} = cx - xz - y, \\ \dot{z} = xy - bz, \end{cases} \tag{1.31}$$

where x, y, and z are constant parameters.

Fix $a = 10$ and $b = 8/3$, and let c vary. For $0 < c < 1$, the system has a unique equilibrium point $A(0, 0, 0)$, which is stable. All trajectories tend to zero as $t \to \infty$, meaning that physically no flow convection occurs. When $c = 1$, it is a critical case where convection is about to take place. For $c > 1$, the system has three equilibrium points and one remains to be $A(0, 0, 0)$, which becomes unstable; the other two are

$$P_{\pm}\left(\sqrt{b(c-1)}, \pm\sqrt{b(c-1)}, c-1\right).$$

These two equilibrium points are stable when $1 < c < \frac{a(a+b+3)}{a-b-1} \approx 24.74$.

From a physical perspective, the system steady-state corresponds to convection, which has already occurred. As $t \to \infty$, whether the trajectory tends to the point P_- or the point P_+ is inconsequential. If it tends to P_+, it means that the temperature is higher in the upper region, leading to convection from top to bottom. Conversely, if it tends to P_-, the opposite happens.

From a mathematical perspective, when $c \in (1, 24.74)$, there are two distinct possible cases. At the equilibrium point $A(0, 0, 0)$, the three eigenvalues consist of one positive value and two negative values. As a result, the stable manifold at point A forms a surface S, while the unstable manifold is a curve. Through calculations, it is found that, when $1 < c < 13.926$, the unstable manifold eventually spirals toward the equilibrium point P_- or P_+ on the same side.

As c increases from 1, the spiraling loops gradually expand. When $c = 13.926$, the unstable manifold asymptotically approaches the origin A indefinitely, forming a heteroclinic orbit.

When $c > 13.926$, the unstable manifold wraps around to the opposite side and ultimately tends to the equilibrium point P_- or P_+ on the opposite side.

When $24.06 < c < 24.74$, the equilibrium points P_- and P_+ remain stable. However, calculations indicate that the system's dynamical behavior in this range is significantly more complex than that with $c < 24.06$. This complexity is due to intrinsic bifurcations occurring at a value slightly less than 24.74.

When $c > 24.74$, the equilibrium points P_- and P_+ become unstable, meaning that a steady convection is no longer possible. Consequently, the system enters the chaotic region.

Within this chaotic region, for certain values of c, the system exhibits statistically steady chaotic behavior, exhibiting a strange attractor. In other cases, a stable periodic solution may emerge, forming a closed trajectory where the system cycles back onto itself.

The Lorenz system (1.31) contains two quadratic terms xz and xy, enabling interactions between different components of the system, which generates a strange attractor, named the Lorenz attractor, as shown in Fig. 1.6.

Figure 1.7 show the maximum Lyapunov exponents of the Lorenz system (1.31) with $a = 10,\ b = 8/3$, and $c \in [0, 350]$.

1.3.3 Chen System

Consider the controlled Lorenz equation

$$\begin{aligned} \dot{x} &= a\left(y - x\right), \\ \dot{y} &= cx - xz - y + u, \\ \dot{z} &= xy - bz, \end{aligned} \tag{1.32}$$

where a, b, and c are constant parameters such that the Lorenz system is not in the chaotic state without the control input u. Let u be a linear feedback controller:

$$u = k_1 x + k_2 y + k_3 z, \tag{1.33}$$

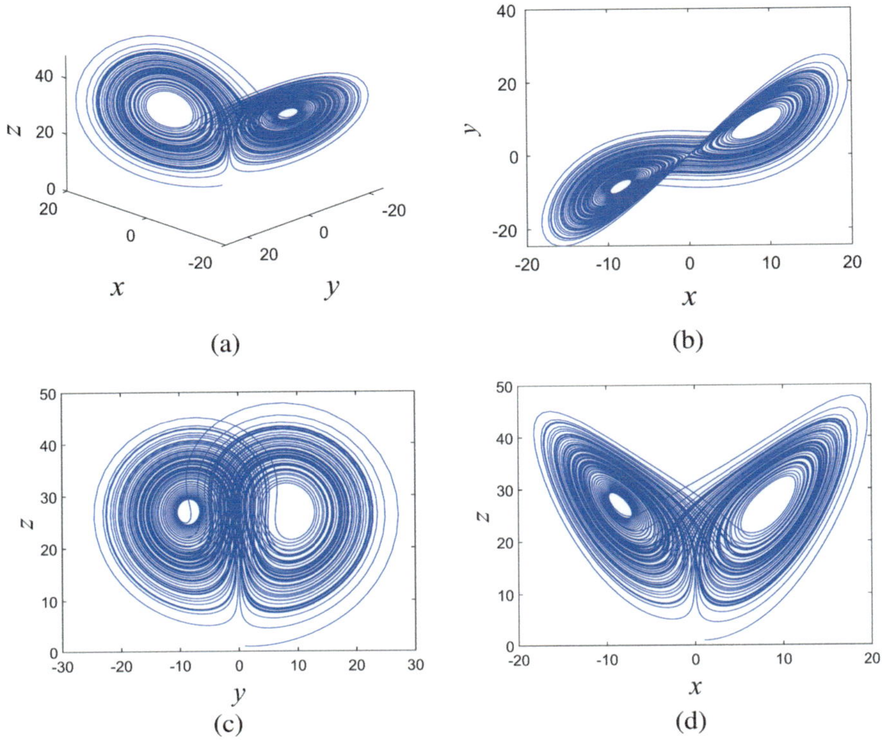

Fig. 1.6 Lorenz attractor. **a** 3-D plane. **b** x-y plane. **c** y-z plane. **d** x-z plane

wherek_1, k_2, k_3 are constant parameters, referred to as control gains, to be determined such that the controlled non-chaotic Lorenz system (1.32) with Eq. (1.33) becomes chaotic.

To find the equilibria of the controlled Lorenz system described by Eqs. (1.32) and (1.33), let

$$\begin{aligned} a(y-x) &= 0, \\ (c+k_1)x - xz + (k_2-1)y + k_3 z &= 0, \\ xy - bz &= 0. \end{aligned} \tag{1.34}$$

It is easy to verify that $(0, 0, 0)$ is a trivial equilibrium and the two nonzero equilibria are

$$x = y = \frac{1}{2}k_3 \pm \frac{1}{2}\sqrt{k_3^2 + 4b(c + k_1 + k_2 - 1)}. \tag{1.35}$$

To proceed, consider the controlled system Jacobian evaluated at an equilibrium point $\left(x', y', z'\right)$:

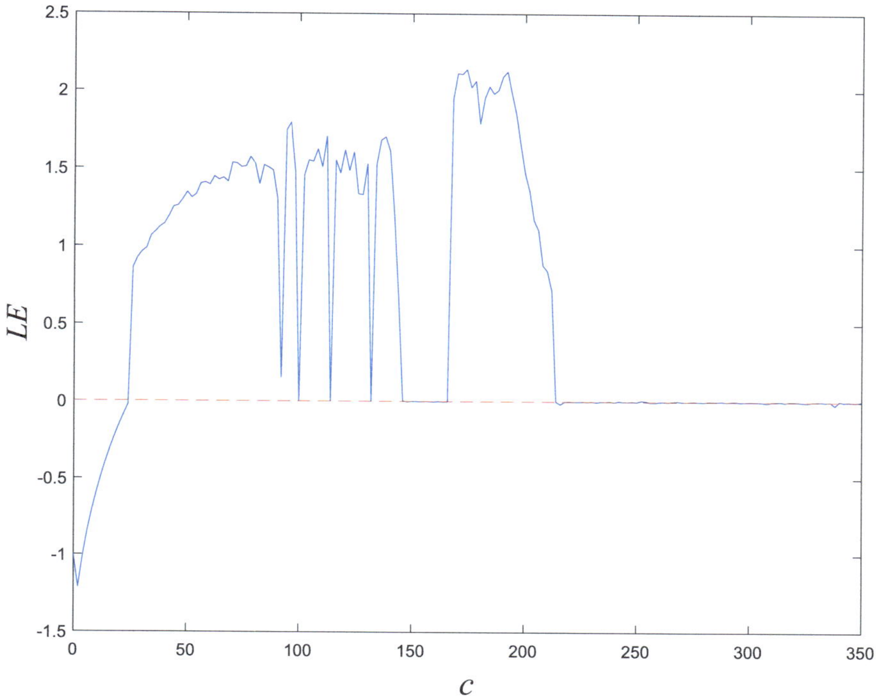

Fig. 1.7 Maximum Lyapunov exponent of the Lorenz system (1.31) with $a = 10$, $b = 8/3$ and $c \in [0, 350]$

$$J_{(x',y',z')} = \begin{bmatrix} -a & a & 0 \\ c + k_1 - z' & k_2 - 1 & k_3 - x' \\ y' & x' & -b \end{bmatrix}. \tag{1.36}$$

Clearly, at the equilibrium point $(x', y', z') = (0, 0, 0)$, the parameter k_3 does not affect the eigenvalues. This implies that k_3 does not contribute to the Lyapunov exponents of the controlled Lorenz system; therefore, it may be set to zero.

In order to obtain chaotic behaviors from the controlled system, the Shilnikov criterion (Sect. 1.2.4) suggests a simple option, among several possibilities: choose $k_1 = a$ and $k_2 = 1 + c$, yielding the controller

$$u = -ax + (1 + c)y.$$

This controller leads the controlled, currently non-chaotic Lorenz equation (1.32) and Eq. (1.33), to become a chaotic system [25]:

$$\begin{cases} \dot{x} = a\,(y - x)\,, \\ \dot{y} = (c - a)\,x - xz + cy, \\ \dot{z} = xy - bz, \end{cases} \tag{1.37}$$

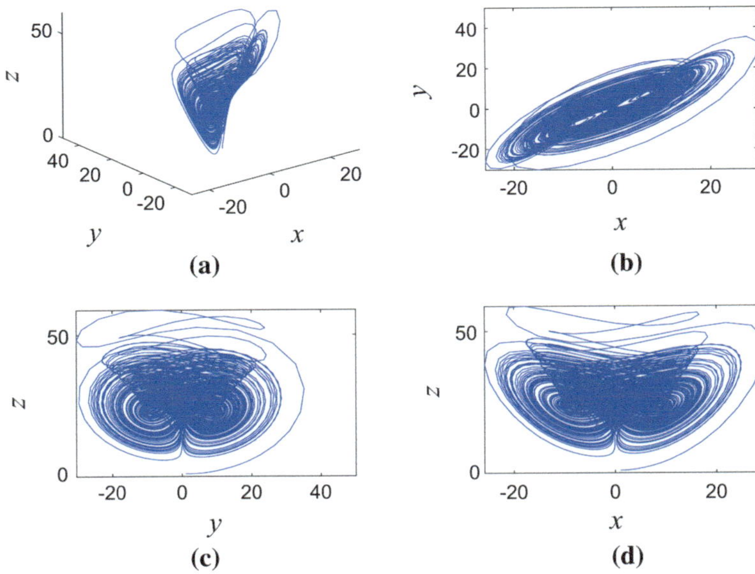

Fig. 1.8 Chen attractor [28]. **a** 3-D space. **b** x-y plane. **c** y-z plane. **d** x-z plane

where parameters $a = 35$, $b = 3$, and $c = 28$.

The system (1.37) was then referred to as the Chen system by others [26, 27].

The Chen system appears to be quite similar to the Lorenz system. However, the Chen system is topologically non-equivalent to the Lorenz system, namely there does not exist any homeomorphism that can transform one system to another, and has much more complex dynamical behaviors with many essentially different dynamical properties [28].

Figure 1.8 shows the phase portrait of the Chen system with the attractor clearly distinctive from the Lorenz attractor.

Further, it is easy to find the three equilibrium points of the Chen system:

$$A\,(0,0,0)\,,\quad B\left(3\sqrt{7},3\sqrt{7},21\right),\quad C\left(-3\sqrt{7},-3\sqrt{7},21\right).$$

The corresponding Jacobian eigenvalues at point A are $\lambda_1 = -30.8359$, $\lambda_2 = -23.8359$, $\lambda_3 = -3$; those at point B are $\lambda_1 = -18.4280$, $\lambda_2 = 4.2140 - 14.8846i$, $\lambda_3 = 4.2140 + 14.8846i$; those at point C are $\lambda_1 = -18.4280$, $\lambda_2 = 4.2140 - 14.8846i$, $\lambda_3 = 4.2140 + 14.8846i$.

Figure 1.9 illustrates the Poincaré map across several sections, showing multiple sheets of the Chen attractors.

Figure 1.10 shows the maximum Lyapunov exponent of the Chen system with $a \in [30, 40]$, $b = 3$, and $c = 28$.

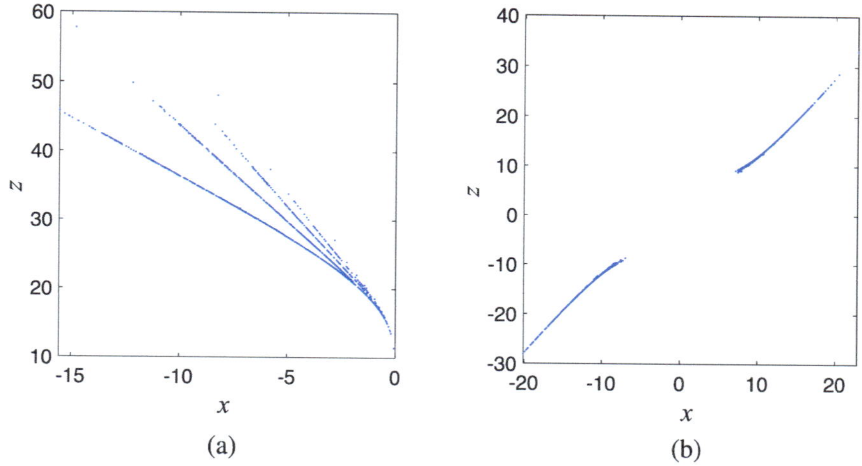

Fig. 1.9 Poincaré map of the Chen system [28]. **a** $y = 0$. **b** $z = 20$

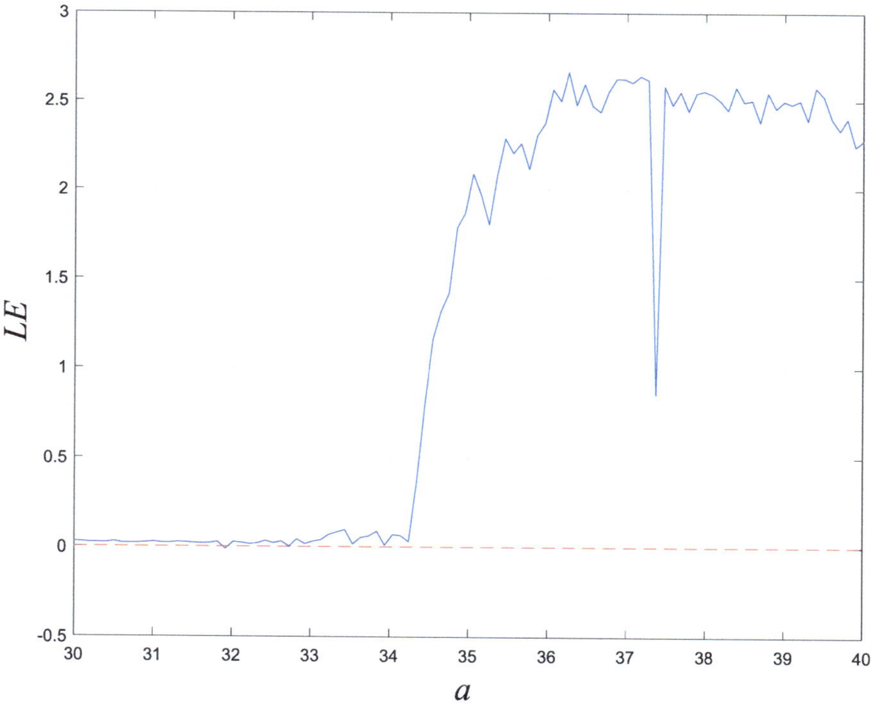

Fig. 1.10 Maximum Lyapunov exponent of the Chen system with $a \in [30, 40]$, $b = 3$ and $c = 28$ [28]

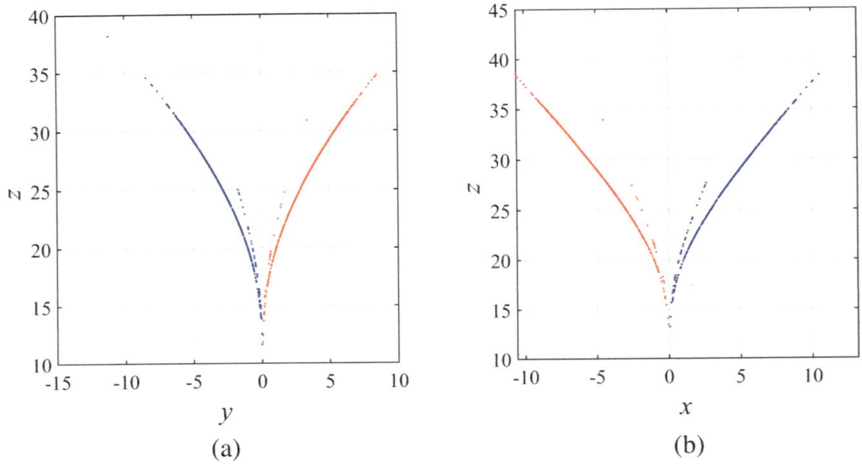

Fig. 1.11 Poincaré map of the Lü system [29, 30]. **a** $x = 0$. **b** $y = 0$

1.3.4 Lü System

Following the controlled-Lorenz system approach developed by Chen in [25], Lü and Chen [29, 30] furthermore discovered an even simpler 3-D autonomous chaotic system. Characterized as a structured root-critical system, it has since been referred to by others as the Lü system [31]. Notably, this system serves as a transitional bridge between the Lorenz system and the Chen system, representing a transition from one to the other.

Specifically, the Lü system is described by

$$\begin{cases} \dot{x} = a\left(y - x\right), \\ \dot{y} = -xz + cy, \\ \dot{z} = xy - bz, \end{cases} \tag{1.38}$$

where, similarly, $a,\ b,\ c$ are constant parameters. When $a = 36$ and $b = 3$ are fixed while c varies, the attractor generated by this system exhibits different characteristics: it resembles the Lorenz attractor for $12.7 < c < 17.0$, transits through an intermediate shape for $18.0 < c < 22.0$, and then forms a Chen-like attractor for $23.0 < c < 28.5$.

Figure 1.11 shows the Poincaré map of the Lü system across several sections, showing multiple sheets of the Lü attractors.

Figure 1.12 shows the Lü attractor generated by the system (1.38).

Figure 1.13 shows the maximum Lyapunov exponents of the Lü system with $a = 36$ and $b = 3$, $c \in [12, 30]$.

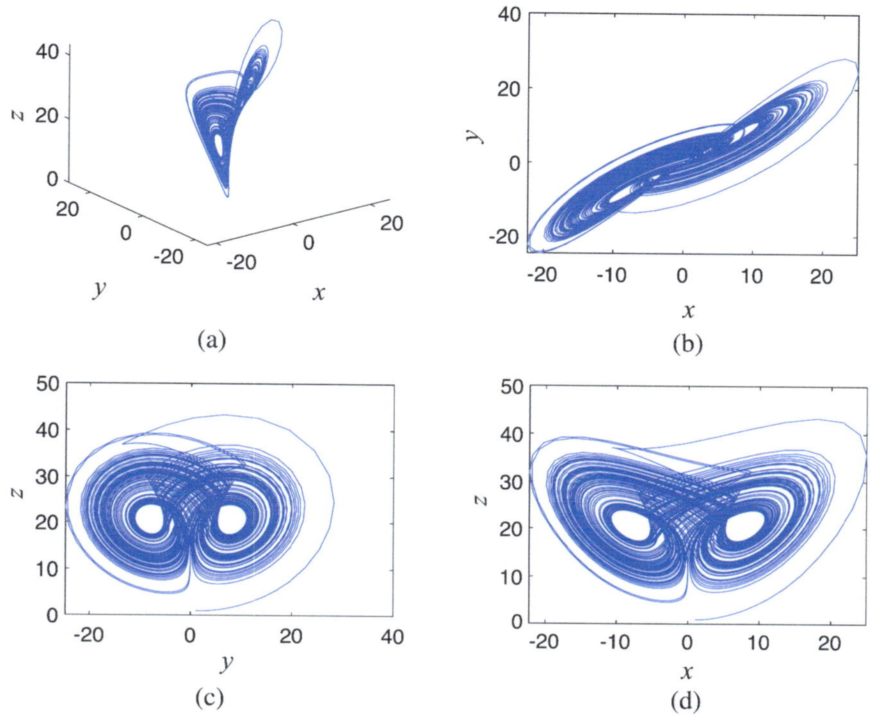

Fig. 1.12 Lü attractor [29, 30]. **a** 3-D space. **b** x-y plane. **c** y-z plane. **d** x-z plane

1.3.5 Unified System

Building on the concept of the generalized Lorenz canonical form [32], as summarized by Chen in [28], to reveal and describe the relationships between the Lorenz and Chen systems, particularly the transition from one to the other, the following unified system was constructed in [33]:

$$\begin{cases} \dot{x} = (25\alpha + 10)(y - x), \\ \dot{y} = (28 - 35\alpha)x - xz + (29\alpha - 1)y, \\ \dot{z} = xy - \frac{\alpha+8}{3}z, \end{cases} \tag{1.39}$$

where $\alpha \in [0, 1]$.

According to a classification established by Vanecek and Celikovsky in [34], the linear part of the Lorenz-Chen-Lü systems, denoted $A = [a_{ij}]_{3\times3}$, has a critical value determined by $a_{12}a_{21}$. Based on the classification:

1. When $0 \leq \alpha < 0.8$, system (1.39) belongs to the generalized Lorenz system, since for these values of α, one has $a_{12}a_{21} > 0$.
2. When $\alpha = 0.8$, system (1.39) becomes the Lü system with $a_{12}a_{21} = 0$.

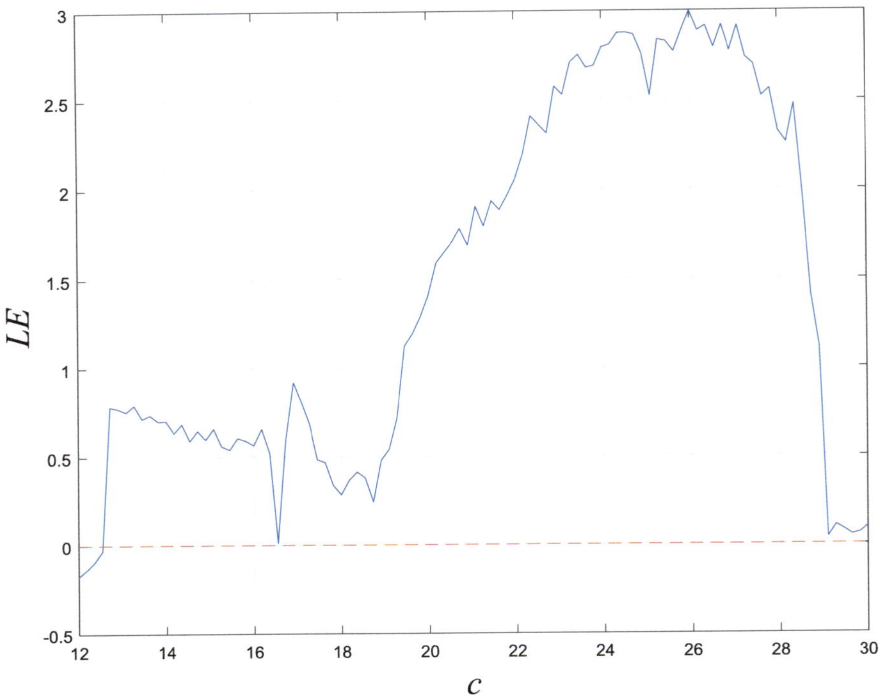

Fig. 1.13 Maximum Lyapunov exponent of the Lü system with $a = 36$ and $b = 3$, $c \in [12, 30]$ [29, 30]

3. When $0.8 < \alpha \leq 1$, it belongs to the generalized Chen system formulated in [32], since for these values of α, one has $a_{12}a_{21} < 0$.

   Some special features of System (1.39) are summarized as follows:

1. System (1.39) is a chaotic system when $\alpha \in [0, 1]$.
2. System (1.39) is very simple and has only one key parameter, α.
3. System (1.39) connects the Lorenz and Chen systems, realizing the entire transition spectrum between them.
4. The control of the parameter α in system (1.39) reveals the evolution of dynamical behaviors from the Lorenz attractor to the Chen attractor.
5. The periodic windows of system (1.39) explain the fundamental reason why the Lorenz and Chen attractors have similar but distinctive topological structures.

   The system (1.39) has three equilibrium points:

$$S_0\,(0, 0, 0)\,, \quad S_{\pm}\left(\pm\sqrt{(8+\alpha)\,(9-2\alpha)}, \pm\sqrt{(8+\alpha)\,(9-2\alpha)}, 28-6\alpha\right).$$

Among them, the two equilibria S_- and S_+ are symmetrically positioned with respect to the z-axis. All the three equilibrium points of system (1.39) remain unstable for all $\alpha \in [0, 1]$ [33].

1.4 Multiscroll Chaotic Attractors

1.4.1 An Overview

Chaos theory has been intensively investigated within the science, mathematics and engineering communities for over half a century [35–37]. In particular, Chua's circuit, a paradigm of chaos and a bridge between electronic circuits and the mathematical chaos theory, has been widely studied and used as a platform for various engineering applications [22]. In recent years, extended from Chua's circuit [23, 38, 39], theoretical design and hardware implementation of different chaotic oscillators have attracted considerable attention, aiming at real-world applications of many chaos-based technologies and information systems. It stimulates the current research interest in creating complex multiscroll chaotic attractors using simple electronic circuits and devices.

Over the past three decades, basic theoretical design approaches have been developed utilizing piecewise linear functions, convolutional neural networks, nonlinear modulating functions, fundamental circuits, switching manifolds, multifolded torus forming, and other techniques and methods [40–42]. The dynamical mechanisms in generating multiscroll chaotic attractors from these approaches are now well understood, and new research continues in the design and applications of various multidirectional multiscroll chaotic or hyperchaotic attractors.

In general, multiscroll chaotic attractors are generated, displayed and verified through numerical simulations. However, advances have been made in theoretical proofs and analog circuit realizations. Despite many physical challenges and limitations, it is now possible to implement up to 1-D 14-scroll, 2-D 14×10-grid scroll, and 3-D $10 \times 10 \times 10$-grid (a total of 1000) scroll chaotic attractors by using electronic circuits. Success notwithstanding, producing more scrolls via hardware implementation remains a technical challenge, and rigorous theoretical proofs of the existence of multiscroll chaotic attractors with multidirectional orientations are very difficult due to their complex dynamical behaviors and lack of powerful mathematical tools, especially for switching systems. This remains an important and stimulating topic for future research.

Multiscroll chaotic attractors are expected to have numerous practical applications in technical fields such as digital and secure communications, synchronous prediction, random bit generation, and information systems, among others [43]. Realizing the promising potential of multiscroll chaotic attractors for engineering applications requires further exploration with greater efforts.

The field of multiscroll chaotic attractor generation has witnessed significant advancements over the past few decades. These developments include not only the progress of in-depth theoretical studies but also the recognition of numerous novel real-world applications. This promising field deserves continued attention from the engineering, physics, applied mathematics, and electronics communities.

1.4.2 Developments

After decades of rapid growth, the generation of multiscroll chaotic attractors has developed into a relatively mature state forming a self-contained research domain [44–48]. This section provides a brief overview of the evolution of the investigation in multiscroll chaotic attractors [49–51].

Looking back to the past, Suykens and Vandewalle were the first to introduce a family of n-double scroll chaotic attractors through using a quasi-linear function [52, 53]. They also designed a simple recurrent neural network model [54], capable of producing a chaotic attractor similar to the double-scroll attractor of Chua's circuit. Arena et al. [55–57] realized the generalized n-double-scroll Chua's circuit by using a three-state-controlled Cellular Neural Network (CNN) with a generalized piecewise linear (PWL) output function. Then, Suykens et al. [58] proposed a method for generating a more complete family of n-scroll chaotic attractors instead of n-double-scroll ones. Around the same time, Suykens and Chua [59] investigated n-double-scroll attractors produced by hypercubes in 1-D CNNs. Later, Aziz-Alaoui [60, 61] presented a PWL function approach for creating multispiral chaotic attractors from both autonomous and non-autonomous differential equations. Yalcin et al. [62] introduced a technique for generating a family of n-double-scroll hyperchaotic attractors and they also constructed a simple circuit model for generating n-scroll chaotic attractors [63].

The main design idea in most of the aforementioned methodologies is the same–to add some additional breakpoints into the PWL function of the nonlinear resistor in Chua's circuit, or other nonlinear circuits [64]. More recently, Zhang et al. [65] proposed a multi-segment memristor model based on a multi-segment linear function within the Sprott-B system. 1-D memristive multiscroll chaotic attractors, 2-D memristive multiscroll chaotic attractors, and even higher-D memristive multiscroll chaotic attractors are produced separately, each introducing a different number of memristors. Yu et al. [66] proposed a piecewise Lorenz system, equipped with a staircase function and an evenly symmetric PWL function, which is autonomous and can generate various grid multiwing butterfly chaotic attractors without requiring any external forcing. Escalante et al. [67] presented a new class of PWL dynamical system without equilibria, which exhibits a 3-D grid multiscroll chaotic attractor, whereas the number of scrolls of the generated attractor can be easily changed by altering the number of the linear parts of three piecewise constant functions. A particular system with a 3-D grid multiscroll attractor, arranged of $3 \times 3 \times 3$ configuration, was demonstrated.

From a similar but different approach, Tang et al. [68] introduced a sine function to generate n-scroll chaotic attractors using a simple circuit. Similarly, Ozguz et al. [69, 70] suggested a nonlinear transconductor method for creating n-scroll chaotic attractors. Lü et al. [33, 71–73] presented a switching manifold approach for generating chaotic attractors with multiple-merged basins of attraction. Cafagna and Grassi [74, 75] developed a coupling Chua's circuit method using sine nonlinearity instead of PWL nonlinearity to create 1-D n-scroll, 2-D $n \times m$-grid scroll, and 3-D

$n \times m \times l$-grid scroll hyperchaotic attractors. Yu et al. [76] proposed a family of n-scroll hyperchaotic chaotic attractors in a 4-D system, and they also introduced a nonlinear modulating function approach for generating n-scroll chaotic attractors based on a general jerk circuit [77–79], which can arbitrarily design and generate different swings, widths, slopes, breakpoints, equilibriums, shapes, and even some general phase portraits. Further, they constructed a family of multifolded torus chaotic attractors [80–82]. Moreover, Ahmad [83] created n-scroll chaotic attractors from fractional-order systems. Sun et al. [42] designed autonomous memristor chaotic systems with infinitely many chaotic attractors, with 3-D, 4-D, and 5-D memristor chaotic systems in standard forms with sine functions and tangent functions.

Regarding electronic circuits, it is well known that the step-function circuit, hysteresis-function circuit and saturate-function circuit are the three types of basic circuits, which have been intensively investigated and applied [84–89]. Yalcin et al. [90] proposed a method using the step-function circuit to generate a family of scroll-grid chaotic attractors, including 1-D n-scroll, 2-D $n \times m$-grid scroll, and 3-D $n \times m \times l$-grid scroll chaotic attractors. Elwakil and Kennedy [91, 92] investigated some hysteresis-based chaotic oscillators. Lü et al. [93] introduced a hysteresis-based circuit series for creating 1-D n-scroll, 2-D $n \times m$-grid scroll and 3-D $n \times m \times l$-grid scroll chaotic attractors. They also initiated a saturated circuit series approach for generating 1-D n-scroll, 2-D $n \times m$-grid scroll, and 3-D $n \times m \times l$-grid scroll attractors [94]. Ma et al. [47] proposed generating multiscroll chaotic attractors by utilizing a step function. Based on the multilevel-logic pulse control method, Anam et al. [49] designed simplified mirror-symmetric multidirectional multiscroll chaotic attractors using multilevel pulse excitation sources. Hong et al. [50] introduced a systematic method for generating multidirectional multi-double-scroll chaotic attractors without altering the original nonlinear functions. By employing the multi-pulse control technique in a 3-D double-scroll chaos system, they were able to generate a new family of double-scroll grid attractors, including 1-D double-scroll, 2-D double-scroll, and 3-D double-scroll chaotic attractors.

Most of the aforementioned multiscroll chaotic attractors were verified only through numerical simulations [95–99]. However, it is widely recognized among electronic engineers that physically realizing them using analog circuits is significantly challenging [30, 56, 100–102]. Despite this difficulty, great efforts have been made by many researchers. In this endeavor, Arena et al. [57] experimentally verified some n-double-scroll chaotic attractors using a state-controlled CNN-based circuit. Yalcin et al. [103] experimentally confirmed the 3- and 5-scroll chaotic attractors in a generalized Chua's circuit. They also physically realized a 6-scroll chaotic attractor in a generalized Chua's circuit using a rescaling breakpoints technique [104]. Elwakil and Kennedy [105–108] presented a systematic circuit design method for the realization of a class of hysteresis-based chaotic oscillators. Tang et al. [68] designed an electronic circuit to experimentally produce 6-, 7-, 8-, 9-scroll chaotic attractors in a modified Chua's circuit with sine nonlinearity instead of PWL functions. It is interesting to note that, until 2001, the largest number of scrolls that were experimentally confirmed was limited to 9. Hardware implementation of 10- or more-scroll attractors requires a very large dynamic range in hardware, necessitating a very high voltage supply with appropriate differential amplifiers, or suitable scaling of voltages [64]. However, physical constraints often limit or even prohibit such circuit realizations. Zhong et al. [102] were the first to experimentally produce and observe a 10-scroll chaotic attractor clearly visible on an oscilloscope, using a systematic chaos generation method. Then, Yu et al. [77, 78] constructed a novel block circuit to realize a 12-scroll chaotic attractor from a generalized jerk circuit. They also physically realized a 9-folded torus chaotic attractor [79, 82] and presented an autonomous system-based approach for creating compound chaotic attractors from a class of generalized Lorenz systems through switching control [109].

It is worth noting that physically realizing multiple nonlinear resistors with many segments simultaneously in an electronic circuit is very difficult [110–115]. To achieve this, the electronic device must have an extremely wide dynamic range, and the slopes and breakpoints of the segments must be adjustable–both easily and independently–to achieve the desired effects simultaneously [116–122]. Therefore, the circuit implementation for generating chaotic attractors of 1-D n-scroll with $n \geq 10$, 2-D $n \times m$-grid scroll with $n, m \geq 10$, and 3-D $n \times m \times l$-grid scroll with $n, m, l \geq 10$, is technically challenging [93, 102, 123]. In realizing such multidirectional multiscroll chaotic attractors, Yalcin et al. [90] experimentally generated and observed 2-D 3×3-grid scroll and 3-D $2 \times 2 \times 2$-grid scroll chaotic attractors. Along this line, Lü et al. [124] were able to construct a block circuit to experimentally verify 2-D 3×11-grid scroll and 3-D $3 \times 3 \times 11$-grid scroll hysteresis chaotic attractors. They were even able to physically realize 1-D 14-scroll, 2-D 14×10-grid scroll, and 3-D $10 \times 10 \times 10$-grid (a total of 1000) scroll chaotic attractors [124]. This marked the first experimental generation and observation of 2-D $n \times m$-grid scroll (with $n, m \geq 10$) and 3-D $n \times m \times l$-grid scroll (with $n, m, l \geq 10$) chaotic attractors observable on an oscilloscope, thereby opening a new direction for the collaborative control of scroll networks. Furthermore, they provided a theoretical design principle for hardware implementation of a large number of multiscroll chaotic attractors with multidirectional orientations [93, 123, 124]. Moreover, Ai et al. [125] presented three transformation approaches based on the 1-D multiscroll Chua's circuit for generating grid multiscroll chaotic attractors, generating $(2n + 2) \times (2m + 2)$ scroll and $(2n + 3) \times (2m + 1)$ scroll chaotic attractors.

From a theoretical point of view, although all the aforementioned multiscroll chaotic attractors could be confirmed via numerical simulation as well as circuit implementation [29, 126–128], it is more desirable to theoretically prove the existence of these visible multiscroll chaotic attractors [129–133]. However, this is a very difficult task, due to the complex dynamical behaviors of multiscroll chaotic attractors [134–136]. Nevertheless, Lü et al. [88, 94] theoretically verified the chaotic behaviors of saturated multiscroll chaotic attractors using a 2-D Poincaré return map. Encouragingly, this constructive proof approach can be easily extended to proving the existence of many other complex multiscroll chaotic attractors.

1.4.3 Potential Applications

Multiscroll chaotic attractors have many practical applications in, for example, broadband signal generation, CNNs, digital and secure communications, as well as efficient liquid mixing. To increase the complexity of chaotic dynamics for higher security [137], multiscroll chaotic attractors could be used for communication instead of the general topologically simple chaotic attractors. Suykens and Vandewalle [138] used the 5-scroll chaotic attractor in their Belgium K. U. Leuven time-series prediction competition in 1998. Eguchi et al. [139] introduced a new digital chaotic circuit for generating multiscroll chaotic attractors, with potential applications in digital devices. Yalcin et al. [140] experimentally verified a nonlinear H_∞ synchronization scheme with 5-scroll chaotic attractors. Tang et al. [141] proposed a secure digital communication system that can resist intrusion of eavesdroppers during transmissions of signals and data, for which multiscroll attractors show better performances. Yalcin et al. [142–144] designed a "true random bit generator" based on a double-scroll chaotic attractor. Fontes et al. [145] proposed a system based on master-slave chaotic synchronization with synchronization conditions derived for a general higher-dimensional chaos generator map, and its performance was evaluated in terms of bit error rate. They also discussed the chaos-based cryptography from a perspective of both cryptography and chaos theory [146]. Gao et al. [147] designed a video encryption scheme based on temporal action segmentation, which involves the identification and extraction of pivotal frames from a video dataset, followed by the encryption of these significant key frames. In addition, many examples have demonstrated promising applications of multiscroll chaotic attractors in a variety of engineering scenarios [2, 148–151].

As a key step in the development of chaos-based technologies, the ongoing research on multiple convolutional chaotic attractors is expected to soon yield exciting and promising features for practical applications.

The potential applications of multiscroll chaotic attractors are diverse, including but not limited to the following [145, 152]:

1. In signal processing, multiscroll chaotic attractors can generate complex signals, such as broadband signal generators, useful for applications in radars, radio communications and other fields that require spectrum expansion.
2. Multiscroll chaotic attractors can enhance control systems' performances, such as nonlinear and synchronization control, improving control systems' stability and robustness.
3. Multiscroll chaotic attractors have been demonstrated to produce high-quality random number sequences, which are imperative in cryptography, simulations and modeling. These sequences exhibit a high degree of randomness and unpredictability, satisfying the requirements of various applications that demand high randomness and complexity.
4. In the domain of information security, multiscroll chaotic attractors have been employed to engineer secure information systems, including identity authentication systems, and to access control systems. These systems benefit from the

chaotic attractors' intricate dynamic properties, which fortify the security and resilience of information systems against cyberattacks.
5. Multiscroll chaotic attractors have been employed in the domain of biomedical engineering for signal processing and pattern recognition. For instance, in the context of electroencephalogram (EEG) signal processing and electrocardiogram (ECG) analysis, chaotic systems have been shown to offer significant assistance to medical diagnosis and treatment, where multiscroll chaotic attractors may enhance our understanding of some complex pathological symptoms.

In summary, multiscroll chaotic attractors hold significant potential for applications across various fields, including communications, signal processing, control systems, random number generation, and information security [25, 153–156], among others. As our understanding of their properties and features deepens and as the technological advancements progress, the scope of their applications is expected to expand with more diversify [157–161]. To fully harness the promising potential of multiscroll chaotic attractors for engineering applications, continued research work and dedicated effort are truly essential.

1.5 Software

This section introduces some useful software for analyzing multiscroll chaotic attractors.

1.5.1 MATLAB

MATLAB [162] is a multi-paradigm numerical computing environment and a fourth-generation programming language developed by MathWorks. It is a proprietary programming language that facilitates matrix manipulation, plotting of functions and data, algorithm implementation, user-interface creation, and interfacing with programs written in other languages such as C, C++, Java, Fortran, and Python.

Primarily intended for numerical computing, MATLAB also offers an optional toolbox that utilizes the MuPAD symbolic engine, providing access to symbolic computing capabilities. Additionally, the Simulink package enhances MATLAB by offering graphical multi-domain simulation and model-based design for dynamic and embedded systems.

The features of MATLAB bear a strong similarity to those of C++. It includes a specialized MATLAB editor for writing programs, and executing a MATLAB program is quite straightforward. The following highlights some of the main features of MATLAB:

1. Computation involving matrices and vectors is both efficient and accurate.

2. Creating graphics is user-friendly, making it popular in scientific and engineering fields.
3. It is an object-oriented programming language with various tools designed to improve program performances and enhance graphics.
4. It includes file input/output (I/O) functions, allowing users to develop applications using the MATLAB programming language.

MATLAB has proven to be highly useful across a wide range of fields, with strong capabilities in vector computation, numerical matrix operations and algorithm manipulation. One of its primary uses is solving complex systems of equations. Simulation is another major function of MATLAB, on numerical tasks including imaging, figure visualization and data analysis.

Users can easily develop specialized toolboxes to address specific needs arising from scientific research. Most simulation figures of multiscroll chaotic attractors and bifurcation diagrams demonstrated in this book were generated by MATLAB.

1.5.2 Mathematica

Mathematica [163] is a powerful computational software widely used in chaos theory and dynamical systems analysis. It provides robust tools for symbolic computation, numerical simulation and visualization of various chaotic attractors, including

1. Differential Equation Solver: Used for solving chaotic system equations numerically.
2. 2-D and 3-D Plotting: For visualizing various chaotic attractors, including multi-scroll attractors.
3. Symbolic Computation: Enables parameter analysis and real-time system behavioral exploration.
4. Bifurcation Analysis: While Mathematica lacks a built-in bifurcation tool, it allows the user to custom bifurcation diagrams using iteration and plotting functions.

1.6 Organization of the Book

This book offers a comprehensive overview of various methods for generating multiscroll chaotic attractors.

It begins with a review of the multi-breakpoint piecewise linear function approach for generating such attractors (Chap. 2) and subsequently explores nonlinear modulation function methods (Chap. 3). The discussion progresses to various techniques for generating multidirectional multiscroll chaotic attractors using foundational circuit designs (Chap. 4).

Additionally, the book introduces fundamental methods for generating multi-scroll chaotic attractors from simple linear systems using switching-manifold control (Chap. 5) and presents transformation methods for creating grid multiscroll chaotic attractors (Chap. 6). It also examines the generation of multiscroll attractors in fractional-order systems (Chap. 7) and investigates a modified folded torus circuit for creating multi-folded torus chaotic attractors (Chap. 8). Several typical approaches for generating hyperchaotic multiscroll attractors are then discussed (Chap. 9), followed by an introduction to nonautonomous chaotic systems that can produce multiscroll chaotic attractors (Chap. 10).

The book also reviews some other effective techniques for generating multiscroll chaotic attractors (Chap. 11). Finally, it provides a brief overview of the digital implementation, control, synchronization, and potential engineering applications of multiscroll chaotic attractors (Chap. 12).

References

1. Poincaré, H., Maitland, F.: Science and Method. Courier Corporation (2003)
2. Lorenz, E.N.: Deterministic nonperiodic flow. J. Atmos. Sci. **20**, 130–41 (1963)
3. Stewart, I.: The Lorenz attractor exists. Nature **406**, 948–949 (2000)
4. Ruelle, D., Takens, F.: On the nature of turbulence. Commun. Math. Phys. **20**, 167–192 (1971); **23**, 343–344 (1971)
5. Gleick, J.: Solving the Mathematical Riddle of Chaos. New York Times (10 June 1084)
6. Li, T., Yorke, J.: Period three implies chaos. Am. Math. Mon. **82**, 985–992 (1975)
7. Shaw, R.: Strange attractors, chaotic behavior, and information flow. Z. Naturforsch. **36**, 80–112 (1981)
8. Grassberger, P., Procaccia, I.: Measuring the strangeness of strange attractors. Phys. D: Nonlinear Phenom. **9**, 189–208 (1983)
9. Grassberger, P., Procaccia, I.: Characterization of strange attractors. Phys. Rev. Lett. **50**, 346 (1983)
10. Eckmann, J.P., Ruelle, D.: Ergodic theory of chaos and strange attractors. Rev. Mod. Phys. **57**, 617 (1985)
11. Tucker, W.: Computing accurate Poincaré maps. Phys. D **171**, 127–137 (2002)
12. Takens, F.: Detecting strange attractors in turbulence. In: Dynamical Systems and Turbulence, Warwick 1980: Proceedings of a symposium held at the University of Warwick 1979/80. Springer, Berlin, pp. 366–381 (2006)
13. Mandelbrot, B.B.: The Fractal Geometry of Nature. Revised and enlarged edition. W.H. Freeman and Co., New York (1983)
14. Kaplan, J.L., Yorke, J.A.: Chaotic behavior of multidimensional difference equations. In: Functional Differential Equations and Approximation of Fixed Points, Proceedings, Bonn, July, Berlin, pp. 204–227. Springer, Heidelberg (1978)
15. Wolf, A., Swift, J.B., Swinney, H.L., et al.: Determining Lyapunov exponents from a time series. Phys. D: Nonlinear Phenom. **16**, 285–317 (1985)
16. Eckmann, J.P., Kamphorst, S.O., Ruelle, D., et al.: Liapunov exponents from time series. Phys. Rev. A **34**, 4971 (1986)
17. Sano, M., Sawada, Y.: Measurement of the Lyapunov spectrum from a chaotic time series. Phys. Rev. Lett. **55**, 1082 (1985)
18. Barna, G., Tsuda, I.: A new method for computing Lyapunov exponents. Phys. Lett. A **175**, 421–427 (1993)

19. Nicolis, G.: Introduction to Nonlinear Science. Cambridge University Press, UK (1995)
20. Afraimovich, V.S., Bykov, V.V., Shilnikov, L.P.: Attractive non-rough limit sets of Lorenz-attractor type. Trudy Moskovskoe Matematicheskoe Obshchestvo **44**, 150–212 (1982)
21. Afraimovich, V.S., Bykov, V.V., Shilnikov, L.P.: On the origin and structure of the Lorenz attractor. Akademiia Nauk SSSR Doklady **234**, 336–339 (1977)
22. Chua, L.O., Komuro, M., Matsumoto, T.: The double scroll family. IEEE Trans. Circuits Syst. **33**, 1072–1118 (1986)
23. Chua, L.O.: CNN: A Paradigm for Complexity. World Scientific, Singapore (1998)
24. Matsumoto, T., Chua, L., Komuro, M.: The double scroll. IEEE Trans. Circuits Syst.-I **32**, 797–818 (1985)
25. Chen, G., Ueta, T.: Yet another chaotic attractor. Int. J. Bifur. Chaos **9**, 1465–1466 (1999)
26. Ueta, T., Chen, G.: Bifurcation analysis of Chen's equation. Int. J. Bifur. Chaos **10**, 1917–1931 (2000)
27. Wang, X.F.: Chen's attractor: a new chaotic attractor. (Chinese) Control Theory Appl. **16**, 719 (1999)
28. Chen, G.: Generalized Lorenz systems family (2006). arXiv:2006.04066
29. Lü, J., Chen, G., Zhang, S.: Dynamical analysis of a new chaotic attractor. Int. J. Bifur. Chaos **12**, 1001–1015 (2002)
30. Lü, J., Chen, G.: A new chaotic attractor coined. Int. J. Bifur. Chaos **12**, 659–661 (2002)
31. Yu, Y., Zhang, S.: Controlling uncertain Lü system using backstepping design. Chaos Solit. Fract. **15**, 897–902 (2003)
32. Čelikovský, S., Chen, G.: On a generalized Lorenz canonical form of chaotic systems. Int. J. Bifur. Chaos **12**, 1789–1812 (2002)
33. Lü, J., Chen, G., Cheng, D.Z., Čelikovský, S.: Bridge the gap between the Lorenz system and the Chen system. Int. J. Bifur. Chaos **12**, 2917–2926 (2002)
34. Vanecek, A., Celikovsky, S.: Control Systems: From Linear Analysis to Synthesis of Chaos. Prentice Hall, London (1996)
35. Chen, G., Dong, X.: From Chaos to Order: Methodologies. Perspectives and Applications, World Scientific, Singapore (1998)
36. Chen, G., Lü, J.: Dynamics of the Lorenz System Family: Analysis, Control and Synchronization. (in Chinese). Science Press, Beijing (2003)
37. Chen, G., Yu, X.: Chaos Control: Theory and Applications. Springer, Berlin (2003)
38. Kennedy, M.P.: Three steps to chaos, Part II: a Chua's circuit primer. IEEE Trans. Circuits Syst. **40**, 657–674 (1993)
39. Yu. S., Tang, W.K.S., Chen. G.: Generation of $n \times m$-scroll attractors under a Chua's circuit framework. Int. J. Bifur. Chaos **17**, 3951–3964 (2007)
40. Lü, J., Chen, G.: Generating multiscroll chaotic attractors: theories, methods and applications. Int. J. Bifur. Chaos **16**, 775–858 (2006)
41. Lai, Q., Kuate, P.D.K., Liu, F., Iu, H.H.C.: An extremely simple chaotic system with infinitely many coexisting attractors. IEEE Trans. Circuits Syst.-II **67**, 1129–1133 (2019)
42. Sun, J., Zhao, X., Fang, J., Wang, Y.: Autonomous memristor chaotic systems of infinite chaotic attractors and circuitry realization. Nonlinear Dyn. **94**, 2879–2887 (2018)
43. Chen, Z., Wen, G., Zhou, H., Chen, J.: Generation of grid multi-scroll chaotic attractors via hyperbolic tangent function series. Optik **130**, 594–600 (2017)
44. Carbajal-Gómez, V.H., Sánchez-López, C.: Determining accurate Lyapunov exponents of a multiscroll chaotic attractor based on SNFS. Nonlinear Dyn. **98**, 2389–2402 (2019)
45. Gao, Y., Li, Q., Li, X., Qian, G.: Construction of multi-scroll chaotic attractors with exponential function. In: IEEE International Conference on Digital Signal Processing, pp. 542–544 (2016)
46. Han, F., Lü, J., Yu, X., et al: Generating multi-scroll chaotic attractors via a linear second-order hysteresis system. In: Dynamics of Continuous, Discrete and Impulsive Systems. Series B, Applications and Algorithms, vol. 12, pp. 95–110 (2005)
47. Ma, Y., Jiang, H., Ju, L.: Circuit implementation of multiscroll chaotic attractors generated using step function. Int. J. Bifur. Chaos **25**, 1550160 (2015)

48. Yu, S., Tang, W.K.S.: Generation of $n \times m$-scroll attractors in a two-port RCL network with hysteresis circuits. Chaos Solit. Fract. **39**, 821–830 (2009)
49. Azam, A., Aqeel, M., Sunny, D.A.: Generation of multidirectional mirror symmetric multi-scroll chaotic attractors (MSMCA) in double wing satellite chaotic system. Chaos Solit. Fract. **155**, 111715 (2022)
50. Hong, Q., Xie, Q., Xiao, P.: A novel approach for generating multidirection multi-double-scroll attractors. Nonlinear Dyn. **87**, 1015–1030 (2017)
51. Suykens, J.A.K., Vandewalle, J.: Quasilinear approach to nonlinear systems and the design of n-double scroll ($n = 1, 2, 3, 4, \ldots$). IEE Proc. G **138**, 595–603 (1991)
52. Suykens, J.A.K., Vandewalle, J.: Between n-double sinks and n-double scrolls ($n = 1, 2, 3, 4, \ldots$). In: International Symposium on Nonlinear Theory and Its Applications (NOLTA), December 5–10, Hawaii, USA, pp. 829–834 (1993)
53. Suykens, J.A.K., Vandewalle, J.: Generation of n-double scrolls ($n = 1, 2, 3, 4, \ldots$). IEEE Trans. Circuits Syst.-I **40**, 861–867 (1993)
54. Suykens, J.A.K., Vandewalle, J.: Learning a simple recurrent neural state space model to behave like Chua's double scroll. IEEE Trans. Circuits Syst.-I **42**, 499–502 (1995)
55. Arena, P., Baglio, S., Fortuna, L., Manganaro, G.: Chua's circuit can be generated by CNN cells. IEEE Trans. Circuits Syst.-I **42**, 123–125 (1995)
56. Arena, P., Baglio, S., Fortuna, L., Manganaro, G.: Generation of n-double scrolls via cellular neural networks. Int. J. Circuit Th. Appl. **24**, 241–252 (1996)
57. Arena, P., Baglio, S., Fortuna, L., Manganaro, G.: State controlled CNN: a new strategy for generating high complex dynamics. IEICE Trans. Fund. **E79-A**, 1647–1657 (1996)
58. Suykens, J.A.K., Huang, A., Chua, L.O.: A family of n-scroll attractors from a generalized Chua's circuit. Int. J. Electron. Commun. **51**, 131–138 (1997)
59. Suykens, J.A.K., Chua, L.O.: n-Double scroll hypercubes in 1-D CNNs. Int. J. Bifur. Chaos **7**, 1873–1885 (1997)
60. Aziz-Alaoui, M.A.: Differential equations with multispiral attractors. Int. J. Bifur. Chaos **9**, 1009–1039 (1999)
61. Aziz-Alaoui, M.A.: Multispiral chaos. In: 2nd International Conference Control of Oscillations and Chaos, vol. 1, July 5–7. St. Petersberg, Russsia, pp. 88–91 (1999)
62. Yalcin, M.E., Suykens, J., Vandewalle, J.: Hyperchaotic n-scroll attractors. Nonlinear Dyn. Electron. Syst. 25–28 (2000)
63. Yalcin, M.E., Ozoguz, S., Suykens, J.A.K., Vandewalle, J.: n-scroll chaos generators: a simple circuit model. Electron. Lett. **37**, 147–148 (2001)
64. Yalcin, M.E.: Cellular Neural Networks, Multi-scroll Chaos and Synchronization: Theory. Applications and Implementations, World Scientific, Singapore (2005)
65. Zhang, J., Zuo, J., Wang, M., Guo, Y., Xie, Q., Hou, J.: Design and application of multi-scroll chaotic attractors based on a novel multi-segmented memristor. Chaos Solit. Fract. **181**, 114676 (2024)
66. Yu, S., Lü, J., Chen, G., Yu, X.: Design and implementation of grid multiwing butterfly chaotic attractors from a piecewise Lorenz system. IEEE Trans. Circuits Syst.-II **57**, 803–807 (2010)
67. Escalante-González, R.J., Campos-Cantón, E.: A class of piecewise linear systems without equilibria with 3-D grid multiscroll chaotic attractors. IEEE Trans. Circuits Syst.-II **66**, 1456–1460 (2018)
68. Tang, W.K.S., Zhong, G.Q., Chen, G., Man, K.F.: Generation of n-scroll attractors via sine function. IEEE Trans. Circuits Syst.-I **48**, 1369–1372 (2001)
69. Ozoguz, S., Elwakil, A.S., Salama, K.N.: n-scroll chaos generator using nonlinear transconductor. Electron. Lett. **38**, 685–686 (2002)
70. Salama, K.N., Ozoguz, S., Elwakil, A.S.: Generation of n-scroll chaos using nonlinear Transconductors. In: IEEE International Symposium on Circuits and Systems, vol. 3, May 25–28, Bangkok, Thailand, pp. 176–179 (2003)
71. Lü, J., Yu, X., Chen, G.: Generating chaotic attractors with multiple merged basins of attraction: a switching piecewise-linear control approach. IEEE Trans. Circuits Syst.-I **50**, 198–207 (2003)

72. Lü, J., Yu, X., Chen, G.: Switching control for multi-scroll chaos generation: an overview. In: International Conference on Physics and Control, vol. 2, August 20–22, pp. 420–428 (2003)
73. Lü, J., Zhou, T., Chen, G., Yang, X.: Generating chaos with a switching piecewise-linear controller. Chaos **12**, 344–349 (2002)
74. Cafagna, D., Grassi, G.: Hyperchaotic coupled Chua circuits: an approach for generating new $n \times m$-scroll attractors. Int. J. Bifur. Chaos **13**, 2537–2550 (2003)
75. Cafagna, D., Grassi, G.: New 3-D scroll attractors in hyperchaotic Chua's circuit forming a ring. Int. J. Bifur. Chaos **13**, 2889–2903 (2003)
76. Yu, S., Lin, Q.H., Qiu, S.: Simulation investigation on multi-scroll chaotic and hyperchaotic attractors for four-dimensional systems. Acta Phys. Sinica **52**, 25–33 (2003)
77. Yu, S., Lü, J., Leung, H., Chen, G.: N-scroll chaotic attractors from a general jerk circuit. In: IEEE International Symposium on Circuits and Systems, May 23–26, Kobe, Japan, pp. 1473–1476 (2005)
78. Yu, S., Lü, J., Leung, H., Chen, G.: Design and implementation of n-scroll chaotic attractors from a general jerk circuit. IEEE Trans. Circuits Syst.-I **52**, 1459–1476 (2005)
79. Yu, S., Qiu, S., Lin, Q.: New results of study on generating multiple-scroll chaotic attractors. Sci. in China, Ser. F: Inf. Sci. **46**, 104–115 (2003)
80. Yu, S., Lü, J., Chen, G.: Multifolded torus chaotic attractors: theoretical design and circuit implementation. Chaos **17**, 013118 (2007)
81. Yu, S., Lü, J., Leung, H., Chen, G.: Generating multi-folded torus chaotic attractors. In: International Conference on Physics and Control, August 24–26, pp. 249–253 (2005)
82. Yu, S., Lin, Q., Qiu, S.: A family of multiple-folded torus chaotic attractors. Acta Phys. Sinica **53**, 2084–2088 (2004)
83. Ahmad, W.M.: Generation and control of multi-scroll chaotic attractors in fractional order systems. Chaos Solit. Fract. **25**, 727–735 (2005)
84. Fujiwara, T., Horio, Y., Aihara, K.: An integrated multi-scroll circuit with floating-gate MOSFETs. In: IEEE International Symposium on Circuits and Systems, vol. 3, May 25–28, Bangkok, Thailand, pp. 180–183 (2003)
85. Han, F., Yu, X., Chen, G., et al.: Generation of multi-scroll chaos using second-order linear systems with hysteresis. In: IEEE International Symposium on Circuits and Systems, vol. 3, May 25–28, Bangkok, Thailand, pp. 84–87 (2003)
86. Han, F., Yu, X., Wang, Y., et al.: n-scroll chaotic oscillators by second-order systems and double-hysteresis blocks. Electron. Lett. **39**, 1636–1638 (2003)
87. Han, F.: Multiscroll Chaos Generation via Linear Systems and Hysteresis Function Series, Ph.D. Thesis, RMIT University, Melbourne, Australia (2004)
88. Lü, J., Chen, G., Yu, X., Leung, H.: Generating multi-scroll chaotic attractors via switching control. In: 5th Asian Control Conference, vol. 3, July 20–23, Melbourne, Australia, pp. 1753–1761 (2004)
89. Nakagawa, S., Saito, T.: An RC OTA hysteresis chaos generator. IEEE Trans. Circuits Syst.-I **43**, 1019–1021 (1996)
90. Yalcin, M.E., Suykens, J.A.K., Vandewalle, J., Ozoguz, S.: Families of scroll grid attractors. Int. J. Bifur. Chaos **12**, 23–41 (2002)
91. Elwakil, A.S., Kennedy, M.P.: Chaotic oscillators derived from Saito's double-screw hysteresis oscillator. IEICE Trans. Fund. **E82**-A, 1769–1775 (1999)
92. Elwakil, A.S., Kennedy, M.P.: Construction of classes of circuit-independent chaotic oscillators using passive-only nonlinear devices. IEEE Trans. Circuits Syst.-I **48**, 289–307 (2005)
93. Lü, J., Yu, S., Leung, H., Chen, G.: Experimental verification of 3-D hysteresis multi-scroll chaotic attractors. In: IEEE International Symposium on Circuits and Systems, vol. 4, May 23–26, Kobe, Japan, pp. 3391–3394 (2005)
94. Lü, J., Chen, G., Yu, X., Leung, H.: Design and analysis of multiscroll chaotic attractors from saturated function series. IEEE Trans. Circuits Syst.-I **51**, 2476–2490 (2004)
95. Li, Q., Yang, X.S., Yang, F.: Multiple-scrolls chaotic attractor and circuit implementation. Electron. Lett. **39**, 1 (2003)

96. Saito, T., Nakagawa, S.: Chaos from a hysteresis and switched circuit. Phil. Trans. R. Soc. Lond. **A353**, 47–57 (1995)
97. Saito, T.: An approach toward higher dimensional hysteresis chaos generators. IEEE Trans. Circuits Syst.-I **37**, 399-409 (1990)
98. Storace, M., Parodi, M.: Simple realisation of hysteresis chaos generator. Electron. Lett. **34**, 10–11 (1998)
99. Wang, C.H., Xu, H., Yu, F.: A novel approach for constructing high-order Chua's circuit with multi-directional multi-scroll chaotic attractors. Int. J. Bifur. Chaos **23**, 1350022 (2013)
100. Yang, X.S., Li, Q.: Generate n-scroll attractor in linear system by scalar output feedback. Chaos Solit. Fract. **18**, 25–29 (2003)
101. Zhang, C., Yu, S.: On constructing complex grid multi-wing hyperchaotic system: theoretical design and circuit implementation. Int. J. Circuit Theory and Appl. **41**, 221–237 (2013)
102. Zhong, G., Man, K., Chen, G.: A systematic approach to generating n-scroll attractors. Int. J. Bifur. Chaos **12**, 2907–2915 (2002)
103. Yalcin, M.E., Suykens, J.A.K., Vandewalle, J.: Experimental confirmation of 3- and 5-scroll attractors from a generalized Chua's circuit. IEEE Trans. Circuits Syst.-I **47**, 425–429 (2000)
104. Yalcin, M.E., Suykens, J.A.K., Vandewalle, J.: On the realization of n-scroll attractors. IEEE International Symposium on Circuits and Systems, vol. 5, May 30–June 2, Orlando, Florida, USA, pp. 483–486 (1999)
105. Elwakil, A.S., Kennedy, M.P.: Improved implementation of Chua's chaotic oscillator using current feedback op amp. IEEE Trans. Circuits Syst.-I **47**, 76–79 (2000)
106. Elwakil, A.S., Kennedy, M.P.: Systematic realization of a class of hysteresis chaotic oscillators. Int. J. Circuit Th. Appl. **28**, 319–334 (2000)
107. Elwakil, A.S., Salama, K.N., Kennedy, M.P.: A system for chaos generation and its implementation in monolithic form. In: IEEE International Symposium on Circuits and Systems, vol. 5, May 28–31, Geneva, Switzerland, pp. 217–220 (2000)
108. Elwakil, A.S.: Chaotic oscillator configuration using a frequency dependent negative resistor. Int. J. Circuit Th. Appl. **28**, 69–76 (2000)
109. Yu, S., Lü, J., Chen, G.: Anti-Control Methods and Applications in Dynamical Systems (in Chinese). Science Press, Beijing (2013)
110. Danca, M.F., Feckan, M., Kuznetsov, N., Chen, G.: Looking more closely at the Rabinovich-Fabrikant system. Int. J. Bifur. Chaos **26**, 1650038 (2016)
111. Danca, M.F.: Hidden chaotic attractors in fractional-order systems. Nonlinear Dyn. **89**, 577–586 (2017)
112. Elwakil, A.S., Ozoguz, S., Kennedy, M.P.: Creation of a complex butterfly attractor using a novel Lorenz-type system. IEEE Trans. Circuits Syst.-I **49**, 527–530 (2002)
113. Pikovskii, A.S., Rabinovich, M.I., Trakhtengerts, V.Y.: Onset of stochasticity in decay confinement of parametric instability. Sov. Phys. -JETP. **47**, 715–719 (1978)
114. Vaidyanathan, S., Volos, C.: Analysis and adaptive control of a novel 3-D conservative no-equilibrium chaotic system. Arch. Control Sci. **25**, 333–353 (2015)
115. Yan, D., Ji'e, M., Wang, L., et al.: Generating novel multi-scroll chaotic attractors via fractal transformation. Nonlinear Dyn. **107**, 3919–3944 (2022)
116. Elwakil, A.S.: Nonautonomous pulse-driven chaotic oscillator based on Chua's circuit. Microelectron. J. **33**, 479–486 (2002)
117. Hong, Q., Xie, Q., Shen, Y., Wang, X.: Generating multi-double-scroll attractors via non-autonomous approach. Chaos **26**, 083110 (2016)
118. Lai, Q., Nestor, T., Kengne, J., Zhao, X.: Coexisting attractors and circuit implementation of a new 4-D chaotic system with two equilibria. Chaos Solit. Fract. **107**, 92–102 (2018)
119. Matsumoto, T., Chua, L., Kobayashi, K.: Hyper chaos: Laboratory experiment and numerical confirmation. IEEE Trans. Circuits Syst. **33**, 1143–1147 (1986)
120. Matsumoto, T., Chua, L., Tokunaga, R.: Chaos via torus breakdown. IEEE Trans. Circuits Syst. **34**, 240–253 (1987)
121. Pan, W., Chen, L., Zhang, W.: Design of a class of fractional-order hyperchaotic multi-directional multi-scroll attractors. Math. Methods Appl. Sci. **44**, 2416–2430 (2021)

122. Yin, Y.: Synchronization of chaos in a modified Chua's circuit using continuous control. Int. J. Bifur. Chaos **6**, 2101–2117 (1996)
123. Lü, J., Yu, S., Leung, H.: Design of 3-D multi-scroll chaotic attractors via basic circuits. In: 4th International Conference on Engineering Applications and Computational Algorithms, vol. 8, pp. 324–328 (2005)
124. Lü, J., Yu, S., Leung, H., Chen, G.: Experimental verification of multidirectional multiscroll chaotic attractors. IEEE Trans. Circuits Syst.-I **53**, 149–165 (2006)
125. Ai, X., Sun, K., He, S., Wang, H.: Design of grid multiscroll chaotic attractors via transformations. Int. J. Bifur. Chaos **25**, 1530027 (2015)
126. Lü, J., Chen, G., Cheng, D.: A new chaotic system and beyond: the generalized Lorenz-like system. Int. J. Bifur. Chaos **14**, 1507–1537 (2004)
127. Lü, J., Chen, G., Yu, Y.: Asymptotic analysis of a modified Lorenz system. Chinese Phys. Lett. **19**, 1260 (2002)
128. Miranda, R., Stone, E.: The proto-Lorenz system. Phys. Lett. A **178**, 105–113 (1993)
129. Bao, H., Ding, R., Chen, B., Xu, Q., Bao, B.: Two-dimensional non-autonomous neuron model with parameter-controlled multi-scroll chaotic attractors. Chaos Solit. Fract. **169**, 113228 (2023)
130. Cheng, G., Li, D., Yao, Y., Gui, R.: Multi-scroll chaotic attractors with multi-wing via oscillatory potential wells. Chaos Solit. Fract. **174**, 113837 (2023)
131. Lü, J., Murali, K., Sinha, S., et al.: Generating multi-scroll chaotic attractors by thresholding. Phys. Lett. A **372**, 3234–3239 (2008)
132. Li, C., Chen, G., Liao, X., Yu, J.: Hopf bifurcation and chaos in tabu learning neuron models. Int. J. Bifur. Chaos **15**, 2633–2642 (2005)
133. Liu, W., Chen, G.: Can a three-dimensional smooth autonomous quadratic chaotic system generate a single four-scroll attractor? Int. J. Bifur. Chaos **14**, 1395–1403 (2004)
134. Chen, Z., Wen, G., Zhou, H., Chen, J.: A new $m \times n$-grid double-scroll chaotic attractors from Rucklidge chaotic system. Optik **136**, 27–35 (2017)
135. Rucklidge, A.M.: Chaos in models of double convection. J. Fluid Mech. **237**, 209–229 (1992)
136. Silva, C.P.: Shil'nikov's theorem–a tutorial. IEEE Trans. Circuits Syst.-I **40**, 675–682 (1993)
137. Wada, M., Nishio, Y., Ushida, A.: Chaotic itinerancy phenomena on coupled n-double scrolls chaotic circuits. In: IEEE International Symposium on Circuits and Systems, May 30–June 10, Orlando, USA, vol. 5, pp. 487–490 (1999)
138. Suykens, J.A.K., Vandewalle, J.: The K. U. Leuven time series prediction competition. In: Nonlinear Modeling: Advanced Black-Box Techniques, pp. 241–253. Kluwer, Boston, USA (1998)
139. Eguchi, K., Inoue, T., Tsuneda, A.: Synthesis and analysis of a digital chaos circuit generating multiple-scroll strange attractors. IEICE Trans. Fund. **E82-A**, 965–972 (1999)
140. Yalcin, M.E., Suykens, J.A.K., Vandewalle, J.: Experimental confirmation of a nonlinear $H\infty$ synchronization scheme with 5-scroll attractors. In: International Workshop on Nonlinear Dynamics of Electronic Systems, July 15–17, Ronne, Denmark, pp. 257–260 (1999)
141. Tang, W.K.S., Man, K.F., Chen, G.: Digitized n-scroll attractors model for secure communications. In: IEEE International Symposium Circuits Systems, vol. 3, May 6–9, Sydney, Australia, pp. 787–790 (2001)
142. Yalcin, M.E., Ozoguz, S., Suykens, J.A.K., Vandewalle, J.: Scroll maps from n-scroll attractors. In: 10th International Workshop on Nonlinear Dynamics of Electronic Systems, June 21–23, Izmir, Turkey, pp. 45–48 (2002)
143. Yalcin, M.E., Suykens, J.A.K., Vandewalle, J.: A double scroll based true random bit generator. In: IEEE International Symposium on Circuits and Systems, vol. 4, May 23–26, Vancouver, Canada, pp. 581–584 (2004)
144. Yalcin, M.E., Suykens, J.A.K., Vandewalle, J.: True random bit generation from a double-scroll Attractor. IEEE Trans. Circuits Syst.-I **51**, 1395–1404 (2004)
145. Fontes, R.T., Eisencraft, M.: A digital bandlimited chaos-based communication system? Commun. Nonlinear Sci. Numer. Simul. **37**, 374–385 (2016)

146. Kocarev, L.: Chaos-based cryptography: a brief overview. IEEE Circuits Syst. Mag. **1**, 6–21 (2001)
147. Gao, S., Iu, H.H.C., Mou, J., et al.: Temporal action segmentation for video encryption. Chaos Solit. Fract. **183**, 114958 (2024)
148. Lin, H., Wang, C., Sun, Y., Wang, T.: Generating n-scroll chaotic attractors from a memristor-based magnetized Hopfield neural network. IEEE Trans. Circuits Syst.-II **70**, 311–315 (2022)
149. Yu, S., Ma, Z., Qiu, S., et al.: Generation and synchronization of n-scroll chaotic and hyperchaotic attractors in fourth-order systems. Chinese Phys. B **13**, 317 (2004)
150. Zhang, C., Yu, S., Chen, G.: Design and implementation of compound chaotic attractors. Int. J. Bifur. Chaos **22**, 1250120 (2012)
151. Zou, Y., Luo, X., Jiang, P., et al.: Controlling the chaotic n-scroll Chua's circuit. Int. J. Bifur. Chaos **13**, 2709–2714 (2003)
152. Tse, C.K., Lau, F.C.: Chaos-Based Digital Communication Systems: Operating Principles. Analysis Methods and Performance Evaluation, Springer, Berlin (2003)
153. Shimizu, T., Morioka, N.: On the bifurcation of a symmetric limit cycle to an asymmetric one in a simple model. Phys. Lett. A **76**, 201–204 (1980)
154. Yu, S., Tang, W.K.S., Lü, J., Chen, G.: Design and implementation of multi-wing butterfly chaotic attractors via Lorenz-type systems. Int. J. Bifur. Chaos **20**, 29–41 (2010)
155. Yu, S., Tang, W.K.S., Lü, J., Chen, G.: Generating $2n$-wing attractors from Lorenz-like systems. Int. J. Circuit Th. Appl. **38**, 243–258 (2010)
156. Zhou, T., Chen, G.: Classification of chaos in 3-D autonomous quadratic systems-I: basic framework and methods. Int. J. Bifur. Chaos **16**, 2459–2479 (2006)
157. Doungmo Goufo, E.F.: Implementation of multi-folded torus attractors via a piecewise system with a piecewise linear odd function. Fractals **30**, 2240232 (2022)
158. Ozoguz, S., Elwakil, A.S., Kennedy, M.P.: Experimental verification of the butterfly attractor in a modified Lorenz system. Int. J. Bifur. Chaos **12**, 1627–1632 (2002)
159. Sprott, J.C.: Some simple chaotic flows. Phys. Rev. E **50**, R647 (1994)
160. Yu, S., Lü, J., Yu, X., Chen, G.: Design and implementation of grid multiwing hyperchaotic Lorenz system family via switching control and constructing super-heteroclinic loops. IEEE Trans. Circuits Syst.-I **59**, 1015–1028 (2012)
161. Yu, X., Lü, J., Chen, G.: Theoretical design and circuit implementation of multidirectional multitorus chaotic attractors. IEEE Trans. Circuits Syst.-I **54**, 2087–2098 (2007)
162. Singh, Y.K., Chaudhuri, B.B.: MATLAB Programming. PHI Learning Pvt. Ltd. (2007)
163. Wolfram, S.: The Mathematica Book. Wolfram Research, Inc. (2003)

Chapter 2
Generating n-Scroll Chaotic Attractors with PWL Functions

Abstract This chapter discusses the design and implementation of n-scroll chaotic attractors using piecewise linear (PWL) functions. It outlines some methods for generating n-scroll chaotic attractors, $n \times m$-scroll attractors, and grid multiwing butterfly attractors, building on extensions of Chua's circuit and other simple nonlinear systems. The chapter also introduces some techniques, such as modifications of nonlinear functions, integration of memristors, and generalizations of basic circuits. To illustrate the feasibility and scalability of these designs for practical applications in mulitscroll attractors analysis, numerical simulations and circuit implementations are presented.

2.1 Designing n-Double-Scroll Chaotic Attractors with Quasi-Linear Functions

This section provides an overview of the quasi-linear function method for generating n-double-scroll chaotic attractors.

The method is based on Chua's circuit [1–3], a quasi-linear function is used as a qualitative method that provides a global perspective on the dynamic behavior of a system, serving as a complement to the traditional linearization techniques. Suykens and Vandewalle [4, 5] took this approach to study system (1.27), proceeding as follows.

First, the system is rewritten in the following form:

$$\begin{pmatrix} \dot{x} \\ \dot{y} \\ \dot{z} \end{pmatrix} = \begin{pmatrix} -a - k\,(x) & a & 0 \\ b & -b & 1 \\ 0 & -c & 0 \end{pmatrix} \begin{pmatrix} x \\ y \\ z \end{pmatrix} = A \begin{pmatrix} x \\ y \\ z \end{pmatrix}, \tag{2.1}$$

where $x = V_{C1}$, $y = V_{C2}$, $z = i_L$, $a = 1/RC_1$, $b = 1/RC_2$, $c = 1/L$, and $k\,(x) = g\,(x)/C_1 x$, which is referred to as a quasi-linear function.

The characteristic equation of the matrix A in the above system is

J. Lü et al., *Chaotic Systems with Multiscroll Attractors*,
https://doi.org/10.1007/978-3-032-09854-2_2

$$1+\frac{\lambda^2+b\lambda+c}{\lambda^3+(a+b)\lambda^2+c\lambda+ca}k(x)=0, \tag{2.2}$$

which can be used to plot the root-locus for the function $k(x)$.

To produce a 2-double-scroll chaotic attractor from system (2.1), Suykens and Vandewalle [4, 5] modified the function $k(x)$ as follows:

$$k_2(x)=\begin{cases}-\frac{8}{7}a, & \text{if } 0\le x\le \delta_1,\\ \frac{2a(x-\delta_1)}{7(\delta_2-\delta_1)}-\frac{8}{7}a, & \text{if } \delta_1\le x\le \delta_2,\\ \frac{2a(x-\delta_2)}{7(\delta_3-\delta_2)}-\frac{6}{7}a, & \text{if } \delta_2\le x\le \delta_3,\\ -\frac{8}{7}a, & \text{if } \delta_3\le x\le \delta_4,\\ \frac{a(-0.5x-0.15|x+\delta_4|+0.15|x-\delta_4|)}{0.7x}, & \text{if } x\ge \delta_4,\end{cases} \tag{2.3}$$

where a, δ_1, δ_2, δ_3, and δ_4 are parameters determined in the design.

The nonlinear functions $k_2(x)$ and $g_2(x)$ used to generate the 2-double-scroll attractor are illustrated in the two subfigures of Fig. 2.1, respectively.

With $a=7$, $\delta_1=5$, $\delta_2=7.5$, $\delta_3=10$, and $\delta_4=12$, system (2.1) produces a 2-double-scroll chaotic attractor, as shown in Fig. 2.2a.

To generate n-double-scroll ($n\ge 2$) chaotic attractors, Suykens and Vandewalle [4, 5] further extended system (2.1) as follows:

$$\begin{cases}\dot{x}=(-a-k_n(x))x+ay,\\ \dot{y}=bx-by+z,\\ \dot{z}=-cy,\end{cases} \tag{2.4}$$

where

$$k_n(x)=\begin{cases}\alpha_1, & \text{if } 0\le|x|\le\delta_1,\\ \alpha_2\frac{|x|-\delta_{1+3(i-2)}}{\delta_{2+3(i-2)}-\delta_{1+3(i-2)}}+\alpha_1, & \text{if } \delta_{1+3(i-2)}\le|x|\le\delta_{2+3(i-2)},\\ -\alpha_2\frac{|x|-\delta_{2+3(i-2)}}{\delta_{3(i-1)}-\delta_{2+3(i-2)}}+\alpha_1+\alpha_2, & \text{if } \delta_{2+3(i-2)}\le|x|\le\delta_{3(i-1)},\\ \alpha_1, & \text{if } \delta_{3(i-1)}\le|x|\le\delta_{1+3(i-1)},\\ \frac{\alpha_3}{|x|}\begin{pmatrix}\beta_1|x|+\beta_2\left||x|+\delta_{1+3(n-1)}\right|\\ +\beta_3\left||x|-\delta_{1+3(n-1)}\right|\end{pmatrix}, & \text{if } |x|\ge\delta_{1+3(n-1)},\end{cases}$$

and $2\le i\le n$.

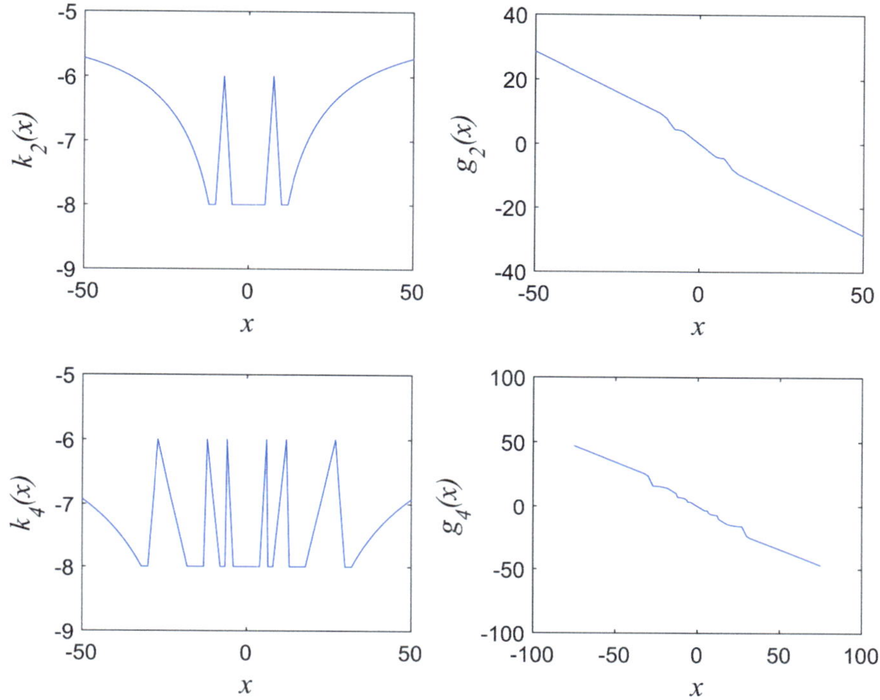

Fig. 2.1 Functions $k_n(x)$ and $g_n(x)$ used to generate the n-double scroll chaotic attractors ($n = 2, 4$) [4, 5]

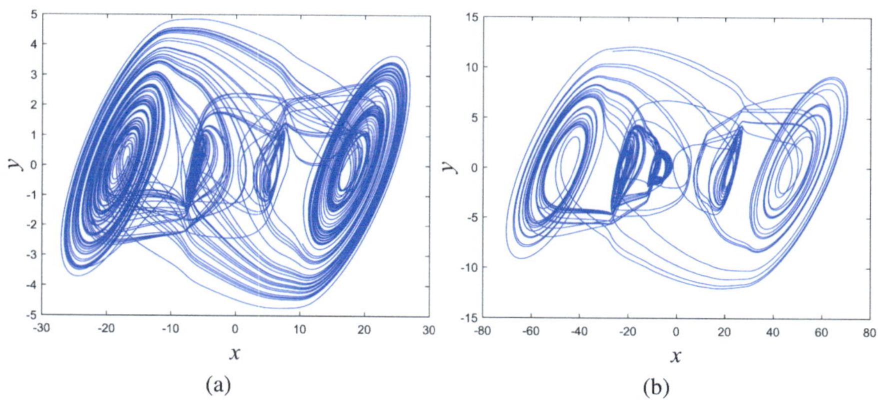

Fig. 2.2 The n-Double-scroll attractors generated using the quasi-linear function [4, 5]. **a** 2-double-scroll attractor. **b** 4-double-scroll attractor

Let $a = 7$ and $(\delta_{2+3(i-2)} + \delta_{1+3(i-2)})/(\delta_{2+3(i-2)} - \delta_{1+3(i-2)}) = 5$, for $2 \leq i \leq n$. Then, system (2.4) can produce n-double-scroll chaotic attractors. For instance, with parameters $a = 7$, $b = 0.7$, $c = 7$, $\delta_1 = 4$, $\delta_2 = 6$, $\delta_3 = 6.5$, $\delta_4 = 8$, $\delta_5 = 12$, $\delta_6 = 13$, $\delta_7 = 18$, $\delta_8 = 27$, $\delta_9 = 30$, and $\delta_{10} = 32$, the system generates a 4-double-scroll chaotic attractor, as shown in Fig. 2.2b. The nonlinear functions $k_4(x)$ and $g_4(x)$ that generate this 4-double-scroll attractor are shown in the lower two subfigures of Fig. 2.1.

Note, however, that since the dimension of those scrolls near the origin will become very small, such scrolls are physically invisible for larger values of n.

2.2 n-Scroll Chaotic Attractors from the Generalized Chua's Circuit

This section introduces the generalized Chua's circuit and uses it to generate n-scroll chaotic attractors.

Suykens et al. [6] proposed the generalized Chua's circuit, described by

$$\begin{cases} \dot{x} = \alpha\left(y - h\left(x\right)\right), \\ \dot{y} = x - y + z, \\ \dot{z} = -\beta y, \end{cases} \tag{2.5}$$

where n is a positive integer, and

$$h\left(x\right) = m_{2n-1}x + \frac{1}{2}\sum_{i=1}^{2n-1}\left(m_{i-1} - m_i\right)\left(|x + c_i| - |x - c_i|\right).$$

Here, system (2.5) is determined by the parameters set $\{\alpha, \beta, m, c\}$, where $m = m_0, \ldots, m_{2n-1}$, and $c = c_0, c_1, \ldots, c_{2n-1}$. When $\alpha = 9$, $\beta = 14.286$, $m = \{-1/7, 2/7, -4/7, 2/7, -4/7, 2/7\}$, and $c = \{1, 2.15, 3.6, 8.2, 13\}$, system (2.5) has a 3-double-scroll chaotic attractor, as shown in Fig. 2.3a. When $\alpha = 9$, $\beta = 14.286$, $m = \{0.9/7, -3/7, 3.5/7, -2.4/7, -1.68/7, 2.52/7\}$, and $c = (1, 2.15, 3.6, 6.2, 9, 14, 25)$, the system has a 7-scroll chaotic attractor, as shown in Fig. 2.3b.

Remark 2.1 For n-double-scroll attractors, the components of the vector m alternate in sign, with m_0 having negative sign. In contrast, for n-scroll attractors with an odd number of scrolls, the components of m alternate in sign with m_0 having positive sign.

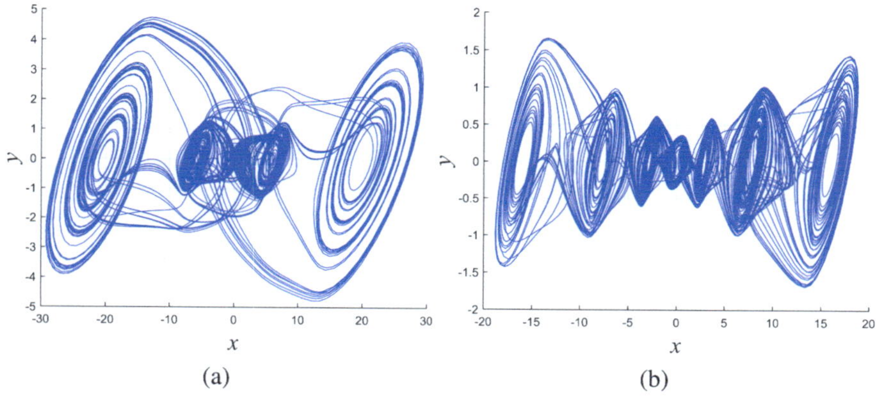

Fig. 2.3 n-Scroll chaotic attractors generated from the generalized Chua's circuit [6]. **a** 3-Double-scroll attractor. **b** 7-Scroll attractor

2.3 n-Scroll Chaotic Attractors from the Unfolded Chua's Circuit

This section presents a family of n-scroll chaotic attractors derived from the unfolded Chua's circuit. Figure 2.4 shows the unfolded Chua's circuit, described by

$$\begin{cases} \dfrac{dV_{C1}}{d\tau} = \dfrac{1}{C_1}\left(G\left(V_{C2} - V_{C1}\right) - f\left(V_{C1}\right)\right), \\ \dfrac{dV_{C2}}{d\tau} = \dfrac{1}{C_2}\left(G\left(V_{C1} - V_{C2}\right) + i_3\right), \\ \dfrac{di_3}{d\tau} = -\dfrac{1}{L}\left(V_{C2} + R_0 i_3\right), \end{cases} \tag{2.6}$$

where

$$f\left(V_{C1}\right) = G_b V_{C1} + 0.5\left(G_a - G_b\right)\left(\left|V_{C1} + E\right| - \left|V_{C1} - E\right|\right).$$

Let $x = V_{C1} = G_b/E, y = V_{C2} = G_b/E, z = i_3/EG, t = \tau G/C_2, m_0 = G_a/G, m_1 = G_b/G, \alpha = C_2/C_1, \beta = C_2/LG^2$, and $\gamma = C_2 R_0/GL$. Then, the corresponding dimensionless state equations of the circuit are obtained as

Fig. 2.4 The unfolded Chua's circuit [7]

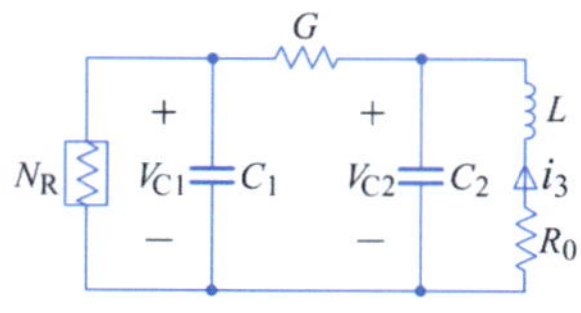

$$\begin{cases} \dfrac{dx}{dt} = \alpha \left(y - x - f_2\left(x \right) \right), \\ \dfrac{dy}{dt} = x - y + z, \\ \dfrac{dz}{dt} = -\beta y - \gamma z, \end{cases} \tag{2.7}$$

where

$$f_2\left(x \right) = m_1 x + \frac{1}{2}\left(m_0 - m_1 \right)\left[|x+1| - |x-1| \right].$$

When $\alpha = 9.365$, $\beta = 11.7$, $\gamma = 0.04$, $m_0 = -8/7$, and $m_1 = -5/7$, system (2.7) has a chaotic attractor, as shown in Fig. 2.5.

To create n-scroll chaotic attractors, Aziz-Alaoui [7] modified the unfolded Chua's circuit (2.7) as follows:

$$\begin{cases} \dfrac{dx}{dt} = \alpha \left(y - x - f_N\left(x \right) \right), \\ \dfrac{dy}{dt} = x - y + z, \\ \dfrac{dz}{dt} = -\beta y - \gamma z, \end{cases} \tag{2.8}$$

where

$$f_N\left(x \right) = \begin{cases} m_k x + \operatorname{sgn}\left(x \right) \xi_k, & \text{if } \; s_{k-1} \le |x| \le s_k, k \in I_{N-2}, \\ m_{N-1} x + \operatorname{sgn}\left(x \right) \xi_{N-1}, & \text{if } \; |x| \ge s_{N-2}. \end{cases} \tag{2.9}$$

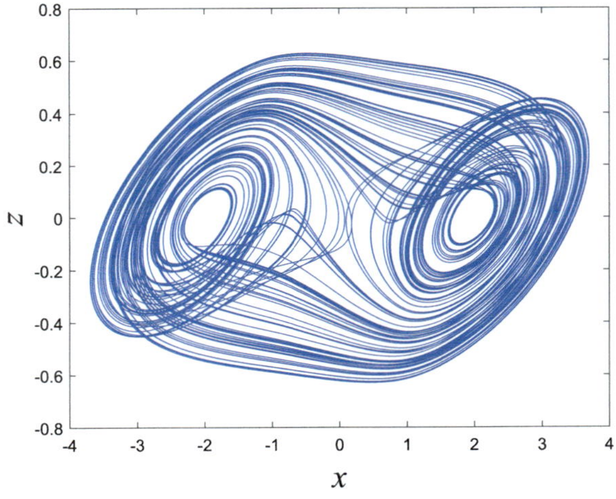

Fig. 2.5 The double-scroll attractor of the unfolded Chua's circuit [7]

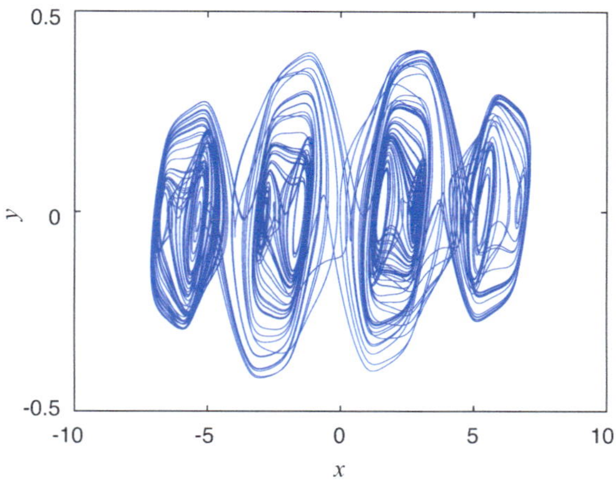

Fig. 2.6 12-Scroll chaotic attractor generated from the unfolded Chua's circuit [7]

Let $N \geq 2$, with sets $I_N = \{0, \dots, N\}$ and $I_N^* = \{1, \dots, N\}$. Define $\{m_k\}_{k \in I_{N-1}}$ and $\{\xi_k\}_{k \in I_{N-1}}$ as two finite real sequences, and let $\{s_k\}_{k \in I_{N-2}}$ be a strictly increasing finite sequence of positive real numbers. Additionally, let $s_{-1} = 0, s_0 = 1, s_N = +\infty$, and $\xi_0 = 0$. The parameters m_k, $k \in I_{N-1}$, represent the slopes of f_N across each linear segment $[-s_0, s_0]$ and $\left[-s_{k-1}, s_k\right]$, $k \in I_{N-1}^*$.

Remark 2.2 The function f_N is continuous if the parameters $(\xi_k)_{k \in I_{N-1}}$ satisfy the following condition [7]:

$$\zeta_{k+1} = (m_k - m_{k+1})\, s_k + \zeta_k,$$

for $\xi_0 \in \mathbf{R}$, $k \in I_{N-2}^*$.

Therefore, $\xi_k = \xi_0 + \sum_{j=1}^{k} \left(m_{j-1} - m_j\right) s_{j-1}$, for $\xi_0 \in \mathbf{R}$, $k \in I_{N-2}^*$.

System (2.8) is defined by the parameter set $\mathbf{B}_N^C = \{\alpha, \beta, \gamma\} \cup \mathbf{B}_N \subset \mathbf{R}^{2N+3}$, where $\mathbf{B}_N = \left\{(s_k)_{k \in I_{N-2}}, (m_k)_{k \in I_{N-1}}\right\}$. To obtain an even number of scrolls, the following parameter set may be used: $\alpha = 9.265$, $\beta = 11.79$, $\gamma = 0.04$, $m_{2j} = m_0 = -8/7$, and $m_{2j+1} = m_1 = -5/7$, $j = 1, 2, \dots$. Conversely, to obtain an odd number of scrolls, one may use $\alpha = 10.40$, $\beta = 12.5709$, $\gamma = 0.005$, $m_{2j} = m_0 = -5/7$, and $m_{2j+1} = m_1 = -8/7$, $j = 1, 2, \dots$.

With

$$\mathbf{B}_{12}^C = \left\{ \begin{array}{l} \alpha = 9.365, \beta = 11.79, \gamma = 0.04, s_0 = 1, s_1 = 1.8, \\ s_2 = 2.7, s_3 = 3.36, s_4 = 5.0, s_5 = 5.7, s_6 = 6.6, \\ s_7 = 7.15, s_8 = 9.0, s_9 = 9.95, s_{10} = 10.75, \end{array} \right\},$$

system (2.8) generates a 12-scroll chaotic attractor, as shown in Fig. 2.6.

With

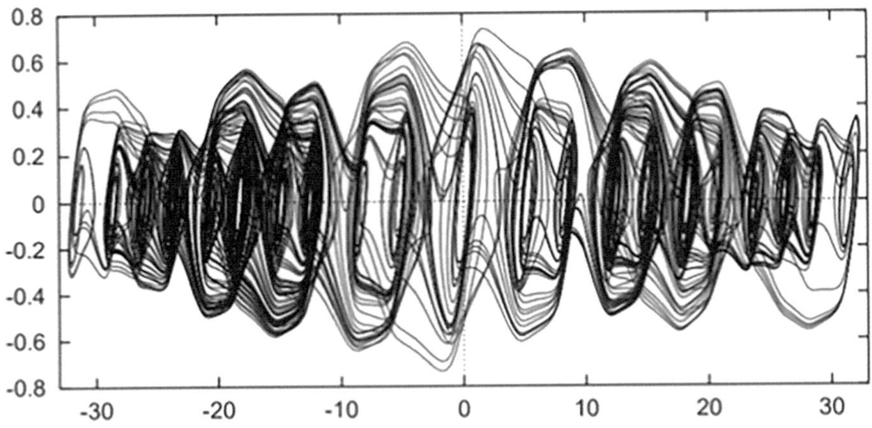

Fig. 2.7 21-Scroll chaotic attractor generated from the unfolded Chua's circuit. Reprinted from [7], with permission from World Scientific Pub. Co.

$$\mathbf{B}_{21}^{C} = \left\{ \begin{array}{l} \alpha = 10.40, \beta = 12.5709, \gamma = 0.005, s_0 = 1, s_1 = 4.5, s_2 = 5.8, \\ s_3 = 7.8, s_4 = 9.0, s_5 = 11.8, s_6 = 12.9, s_7 = 14.5, s_8 = 15.43, \\ s_9 = 17.7, s_{10} = 18.7, s_{11} = 20.2, s_{12} = 21.5, s_{13} = 23.3, \\ s_{14} = 24.2, s_{15} = 25.8, s_{16} = 26.5, s_{17} = 28.2, s_{18} = 29.12, \\ s_{19} = 31.0, \end{array} \right\},$$

system (2.8) generates a 21-scroll chaotic attractor, as shown in Fig. 2.7.

2.4 Multiscroll Chaotic Attractors from the Modified Brockett System

This section introduces the modified Brockett system capable of generating n-scroll chaotic attractors. The Brockett system is described by [7, 8]

$$\begin{cases} \dot{x} = y, \\ \dot{y} = z, \\ \dot{z} = -\beta y - \gamma z + g(x), \end{cases} \tag{2.10}$$

where

$$g(x) = \begin{cases} -Kx, & \text{if } |x| < 1, \\ 2Kx - 3k\operatorname{sgn}(x), & \text{if } 1 < |x| < 3, \\ 3K\operatorname{sgn}(x), & \text{if } |x| > 3. \end{cases}$$

With $K = -1.8$, system (2.10) generates a double-scroll chaotic attractor. Since the region $|x| > 3$ does not influence the dynamics of system (2.9) [9], $g(x)$ can be replaced by

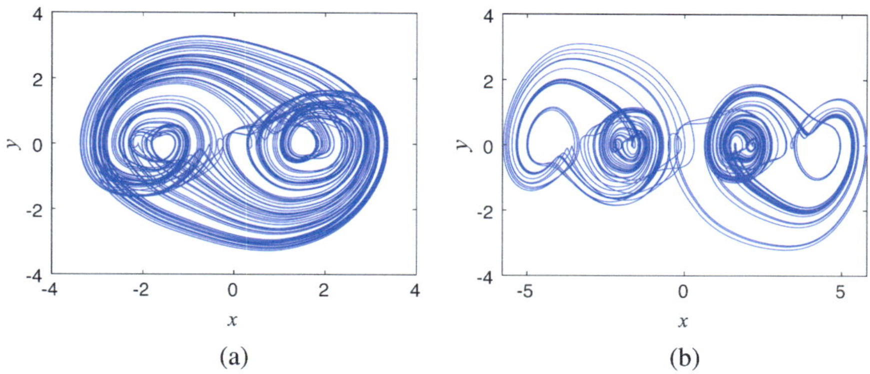

Fig. 2.8 n-Scroll chaotic attractors generated from the modified Brockett system [7, 8]. **a** 4-Scroll chaotic attractor. **b** 6-Scroll chaotic attractor

$$f_2(x) = m_1 x + \frac{1}{2}(m_0 - m_1)[|x+1| - |x-1|]. \tag{2.11}$$

When $m_0 = -K = 1.8$ and $m_1 = 2K$, the system (2.10) with Eq. (2.11) generates a similar double-scroll chaotic attractor.

To generate n-scroll chaotic attractors, Aziz-Alaoui [7] extended system (2.10) to derive the following modified Brockett system:

$$\begin{cases} \dot{x} = y, \\ \dot{y} = z, \\ \dot{z} = -\beta y - \gamma z + f_N(x), \end{cases} \tag{2.12}$$

where $f_N(x)$ is given in Eq. (2.9).

Define the parameters set: $\mathbf{B}_N^B = \{\beta, \gamma\} \cup \mathbf{B}_N \subset \mathbf{R}^{2N+1}$, where

$$\mathbf{B}_N = \left\{(s_k)_{k\in I_{N-2}}, (m_k)_{k\in I_{N-1}}\right\} \subset \mathbf{R}^{2N-1},$$

let $m_0 = m_{2i} = -K$ and $m_1 = m_{2i+1} = 2K$, $j = 1, 2, \ldots$.

Figure 2.8a displays a 4-scroll chaotic attractor [7], where $s_0 = 1.0$, $K = -1.99$, and $B_4^B = \{\beta = 1.06, \gamma = 0.827, s_1 = 1.65, s_2 = 2.1\}$. Figure 2.8b shows a 6-scroll chaotic attractor [7], where $s_0 = 1.0$, $K = -1.99$, and $B_6^B = B_4^B \cup \{s_3 = 2.5, s_4 = 4.1\}$.

2.5 Multiscroll Chaotic Attractors from the Sprott-B System

Introducing memristors into traditional chaotic systems facilitates the generation of multiscroll chaotic attractors, significantly expanding the possibilities for information processing and encryption applications of chaotic systems.

Zhang et al. [10] proposed a multi-segment memristor model constructed using a PWL function. By incorporating this model into the Sprott-B system, memristive multiscroll chaotic attractors (MMSCAs) can be generated in one, two or three dimensions. The dimensionality and complexity of such attractors, namely 1-D, 2-D, and 3-D MMSCAs, can be determined by the number of memristors utilized in the system.

Sprott [11] proposed several simple chaotic systems, with the Sprott-B system expressed as follows:

$$\begin{cases} \dfrac{dx}{dt} = yz, \\ \dfrac{dy}{dt} = x - y, \\ \dfrac{dz}{dt} = 1 - xy, \end{cases} \tag{2.13}$$

where x, y, and z are the state variables.

Numerical simulation results of the 3-D phase diagram for this system are presented in Fig. 2.9. The figure depicts a three-dimensional phase portrait of a chaotic attractor, plotted in the x-y-z coordinates. The trajectory appears to be a complex, having an intertwined structure of chaotic dynamics, with multiple loops and folds. This specific attractor exhibits a butterfly-like or double-scroll topology, commonly observed from such as the Lorenz system. The axes x, y, and z represent the state variables of the dynamical system, with the attractor confined within a bounded region in 3-D space.

To incorporate memristors into the Sprott-B system, Zhang et al. [10] developed a new type of multi-terminal memristor. Based on a PWL function, the multi-segment memristor model is described by

$$\begin{cases} i = W(\phi)\, v = k\phi v, \\ \dfrac{d\phi}{dt} = cv + dg(\phi), \end{cases} \tag{2.14}$$

where $W(\phi)$ represents the inductance of the memristor, v is the port voltage of the memristor, ϕ denotes the magnetic flux, c and d are expansion control parameters, and $g(\phi)$ is odd-symmetric about the coordinate origin.

By incorporating the memristor model (2.14) into system (2.13), the 1-D MMSCAs model is given by

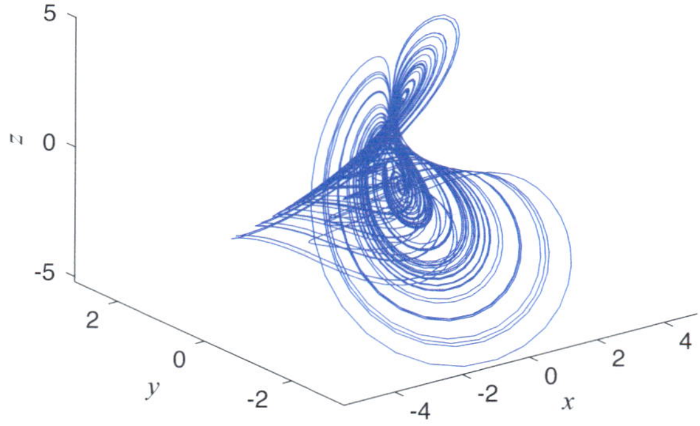

Fig. 2.9 Phase diagram of the Sprott-B system [11]

$$\begin{cases} \dfrac{dx}{dt} = ayz + kx\phi_1 \sin(2\pi f t), \\ \dfrac{dy}{dt} = x - y, \\ \dfrac{dz}{dt} = 1 - bxy, \\ \dfrac{d\phi}{dt} = cx + dg(\phi_1), \end{cases} \tag{2.15}$$

where x, y, z, and ϕ_1 are state variables, and a, b, c, d, f, k are control parameters with $\sin(2\pi f t)$ representing a nonlinear time variable. Here, set the values $a = 2$, $b = 1, c = 2, d = -4, f = 2, k = 0.1$, and the initial conditions $(x_0, y_0, z_0, \phi_{10}) = (0.1, 0.1, 0.1, 0.1)$.

When N_1 takes different values, the numerical simulation results of the 1-D MMSCA in the x-ϕ_1 plane are shown in Fig. 2.10. Figure 2.10 demonstrates that the chaotic attractors of system (2.13) undergo topological replication along the ϕ_1 direction due to the influence of the memristor, resulting in connected MMSCAs.

When N_1 is odd, the number of chaotic attractors generated is even, while when N_1 is even, the number of attractors is odd. As a result, system (2.15) can generate any number of chaotic attractors along the ϕ_1 direction. The total number of chaotic attractors produced by the system is $N_1 + 1$, and the system has $2(N_1 + 1)$ vortices.

By introducing another memristor model (2.14) into the 1-D MMSCAs, the corresponding 2-D MMSCAs model can be obtained, as follows:

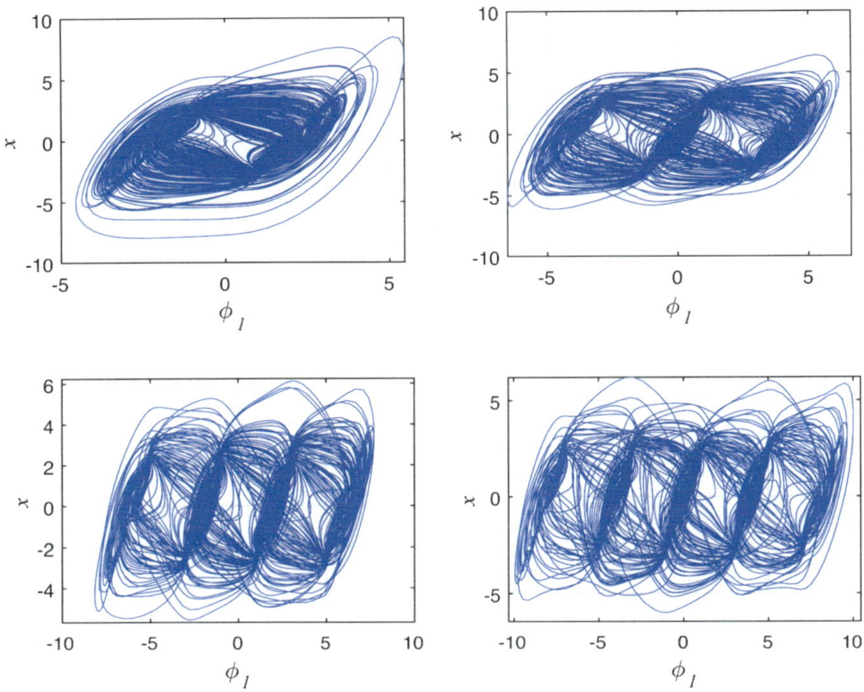

Fig. 2.10 Phase diagrams of the 1-D MMSCAs [10]. **a** $N_1 = 1$. **b** $N_1 = 2$. **c** $N_1 = 3$. **d** $N_1 = 4$

$$\begin{cases} \dfrac{dx}{dt} = ayz + kx\phi_1 \sin(2\pi f t), \\ \dfrac{dy}{dt} = x - y + ky\phi_2 \sin(2\pi f t), \\ \dfrac{dz}{dt} = 1 - bxy, \\ \dfrac{d\phi_1}{dt} = cx + dg(\phi_1), \\ \dfrac{d\phi_2}{dt} = cy + dg(\phi_2). \end{cases} \tag{2.16}$$

With the control parameters a, b, c, d, f, and k being constant, and the initial values being $(x_0, y_0, z_0, \phi_{10}, \phi_{20}) = (0.1, 0.1, 0.1, 0.1, 0.1)$, the numerical simulation results of the 2D-MMSCAs in the ϕ_1-ϕ_2 plane for different values of N_1 and N_2 are shown in Fig. 2.11.

Figure 2.11 illustrates that the chaotic attractors of system (2.13) first replicate topologically along the ϕ_1 direction and then along the ϕ_2 direction, forming a connected mesh-like structure of multiscroll chaotic attractors. The number of scrolls

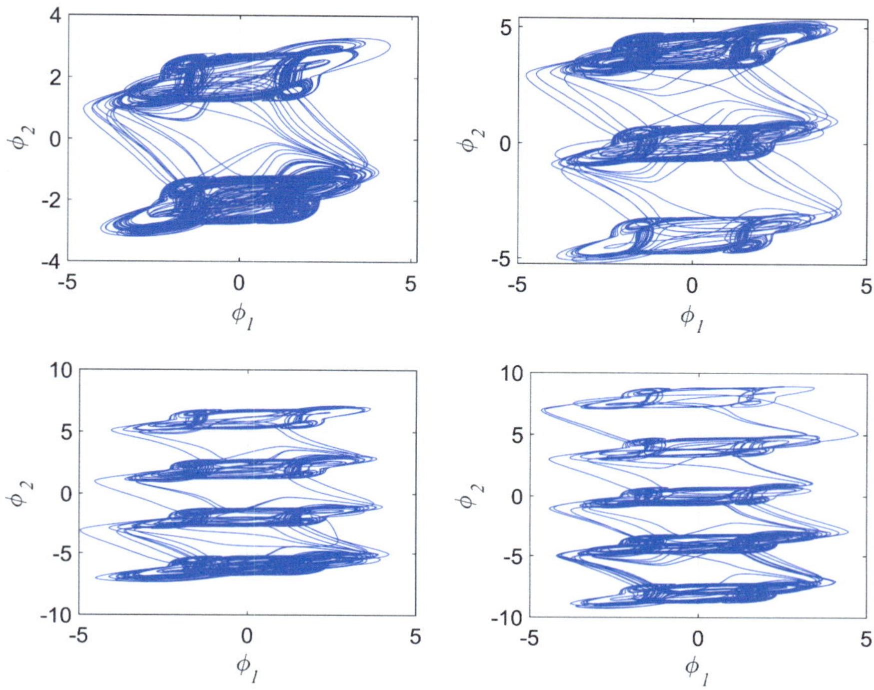

Fig. 2.11 Phase diagrams of the 2-D MMSCAs [10]. **a** $N_1 = 1$ and $N_2 = 1$. **b** $N_1 = 1$ and $N_2 = 2$. **c** $N_1 = 1$ and $N_3 = 3$. **d** $N_1 = 1$ and $N_3 = 4$

generated is $N_2 + 1$, and the total number of chaotic attractors is $(N_1 + 1)(N_2 + 1)$, with the total number of vortices being $2(N_1 + 1)(N_2 + 1)$.

Figure 2.11 presents the 2-D phase portraits of the chaotic attractors in the (ϕ_1, ϕ_2) plane, illustrating the evolution of the system dynamics led by varying parameters or initial conditions: Fig. 2.11a depicts a chaotic attractor with a double-scroll structure, where the trajectories oscillate between two distinct regions in the phase space. Figure 2.11b shows a more complex attractor with four scrolls aligned vertically, indicating a higher degree of chaotic behavior. Figure 2.11c and d demonstrate an even more intricate attractor with six vertically distributed scrolls, indicating the increased system complexity. The series of phase portraits highlights the progression in scroll quantity and distribution as system parameters are varied and adjusted, demonstrating the flexibility of the system in generating multiscroll chaotic attractors.

Next, by introducing another memristor model (2.14) into the 2-D MMSCAs, the mathematical model of 3-D MMSCAs is obtained as follows:

$$\begin{cases} \dfrac{dx}{dt} = ayz + kx\phi_1 \sin(2\pi f t), \\ \dfrac{dy}{dt} = x - y + ky\phi_2 \sin(2\pi f t), \\ \dfrac{dz}{dt} = 1 - bxy + kz\phi_3 \sin(2\pi f t), \\ \dfrac{d\phi_1}{dt} = cx + dg(\phi_1), \\ \dfrac{d\phi_2}{dt} = cy + dg(\phi_2), \\ \dfrac{d\phi_3}{dt} = cz + dg(\phi_3). \end{cases} \tag{2.17}$$

Keeping the control parameters a, b, c, d, f, and k fixed, and setting the initial values $(x_0, y_0, z_0, \phi_{10}, \phi_{20}, \phi_{30}) = (0.1, 0.1, 0.1, 0.1, 0.1, 0.1)$, the 3-D MMSCAs phase portraits in the ϕ_1-ϕ_2-ϕ_3 space are obtained, as shown in Fig. 2.12, for various values of N_1, N_2, and N_3.

Figure 2.12 demonstrates that the chaotic attractors of system (2.13) first replicate topologically along the ϕ_1 direction, then along the ϕ_2 direction, and finally along the ϕ_3 direction, resulting in a 3-D MMSCAs. The number of scrolls generated is $(N_2 + 1)(N_3 + 1)$, and the total number of chaotic attractors is $(N_1 + 1)(N_2 + 1)(N_3 + 1)$, while the number of vortices is $2(N_1 + 1)(N_2 + 1)(N_3 + 1)$.

Figure 2.12 presents the 3-D phase portraits of the multiscroll chaotic attractors. The state variables ϕ_1, ϕ_2, and ϕ_3 represent the system's dimensions. Figure 2.12a shows a configuration with a specific scroll pattern distributed along the coordinate axes, forming structured chaotic dynamics. Figure 2.12b presents an attractor with increased complexity and a denser distribution of scrolls, indicating a more intricate chaotic behavior. Figure 2.12c and d display a symmetrical and evenly distributed multiscroll attractor, showcasing the system's capability to produce highly ordered yet chaotic dynamics. Each subfigure highlights the variation of the chaotic behavior due to parameter or initial condition changes in the system.

A time scale transformation is applied to the equations of the 3-D MMSCA system by introducing $t = \tau_0 t$, where τ_0 is the time scale transformation factor. A larger value of τ_0 results in a faster change in the time domain, generating denser phase trajectories. The transformed equations are as follows:

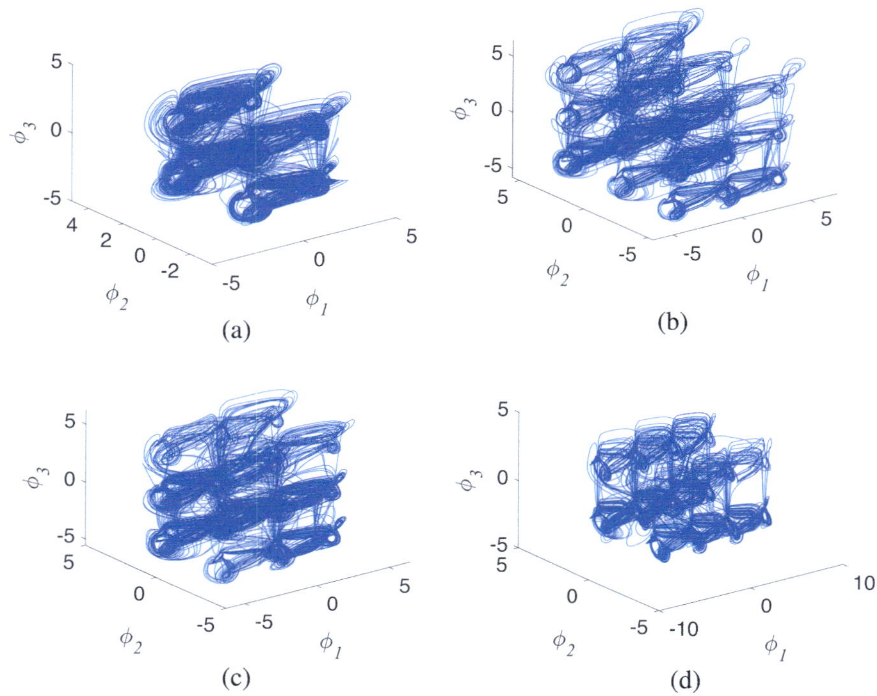

Fig. 2.12 Phase diagrams of the 3-D MMSCAs [10]. **a** $N_1 = 1$, $N_2 = 1$, and $N_3 = 1$. **b** $N_1 = 2$, $N_2 = 2$, and $N_3 = 2$. **c** $N_1 = 2$, $N_2 = 1$, and $N_3 = 1$. **d** $N_1 = 3$, $N_2 = 1$, and $N_3 = 1$

$$\begin{cases} \dfrac{dx}{dt} = \tau_0 ayz + \tau_0 kx\phi_1 \sin(2\pi ft), \\ \dfrac{dy}{dt} = \tau_0 x - \tau_0 y + \tau_0 ky\phi_2 \sin(2\pi ft), \\ \dfrac{dz}{dt} = \tau_0 - \tau_0 bxy + \tau_0 kz\phi_3 \sin(2\pi ft), \\ \dfrac{d\phi_1}{dt} = \tau_0 cx + \tau_0 dg(\phi_1), \\ \dfrac{d\phi_2}{dt} = \tau_0 cy + \tau_0 dg(\phi_2), \\ \dfrac{d\phi_3}{dt} = \tau_0 cz + \tau_0 dg(\phi_3). \end{cases} \tag{2.18}$$

By setting $\tau_0 = 1000$ and using the control parameters $a = 2$, $b = 1$, $c = 2$, $d = -4$, $f = 2$, and $k = 0.1$, one obtains

$$
\begin{cases}
\dfrac{dx}{dt} = -2000ayz + 100x\varphi_1 \sin(2\pi f t)\,, \\
\dfrac{dy}{dt} = -1000(-x) - 1000y + 100y\varphi_2 \sin(2\pi f t)\,, \\
\dfrac{dz}{dt} = 1000 - 1000bxy + 100z\varphi_3 \sin(2\pi f t)\,, \\
\dfrac{d\varphi_1}{dt} = 2000x - 4000g(\varphi_1)\,, \\
\dfrac{d\varphi_2}{dt} = 2000y - 4000g(\varphi_2)\,, \\
\dfrac{d\varphi_3}{dt} = 2000z - 4000g(\varphi_3)\,,
\end{cases}
\tag{2.19}
$$

and

$$
\begin{cases}
\dfrac{dx}{dt} = -\dfrac{1}{10R_1C_1}y(-z) + \dfrac{1}{10R_bC_1}x\phi_1 \sin(2\pi f t)\,, \\
\dfrac{dy}{dt} = -\dfrac{1}{10R_4C_2}(-x) - \dfrac{1}{R_5C_2}y + \dfrac{1}{100R_bC_2}y\phi_2 \sin(2\pi f t)\,, \\
\dfrac{dz}{dt} = \dfrac{1}{10R_8C_3} - \dfrac{1}{10R_9C_3}xy + \dfrac{1}{100R_bC_3}z\phi_3 \sin(2\pi f t)\,, \\
\dfrac{d\phi_1}{dt} = \dfrac{1}{R_cC_1}x - \dfrac{1}{R_dC_1}g(\phi_1)\,, \\
\dfrac{d\phi_2}{dt} = \dfrac{1}{R_cC_2}y - \dfrac{1}{R_dC_2}g(\phi_2)\,, \\
\dfrac{d\phi_3}{dt} = \dfrac{1}{R_cC_3}z - \dfrac{1}{R_dC_3}g(\phi_3)\,.
\end{cases}
\tag{2.20}
$$

Based on Eq. (2.19), the circuit design schematic for the 3-D MMSCAs can be sketched, as shown in Fig. 2.13. In the diagram, the function generator is used to implement the nonlinear time variable $\sin(2\pi f t)$. When only switch S_1 is closed, it produces 1-D MMSCAs. When both switches S_1 and S_2 are closed, the system generates 2-D MMSCAs. When switches S_1, S_2, and S_3 are all closed, they yield 3-D MMSCAs.

Equation (2.20) is derived from Fig. 2.14, and a comparison of Eqs. (2.19) and (2.20) yields

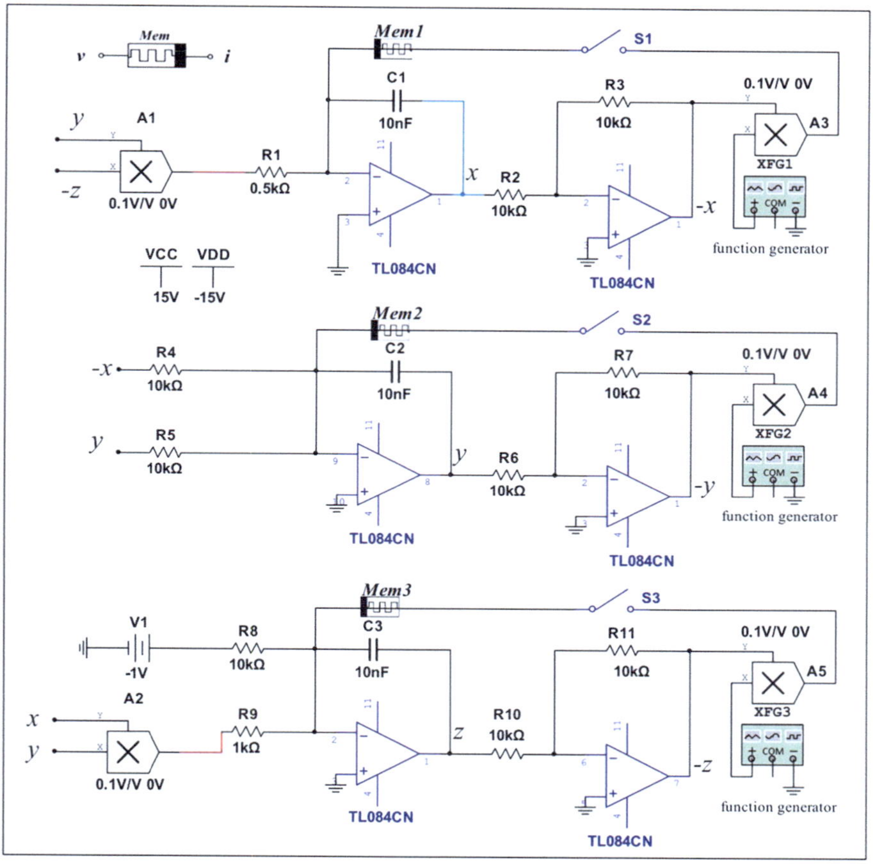

Fig. 2.13 Circuit design for system (2.16). Reprinted from [10], with permission from Elsevier

$$
\begin{cases}
2000 = \dfrac{1}{10R_1C_1} = \dfrac{1}{R_cC_1} = \dfrac{1}{R_cC_2} = \dfrac{1}{R_cC_3}, \\
100 = \dfrac{1}{100R_kC_1} = \dfrac{1}{100R_kC_2} = \dfrac{1}{100R_kC_3}, \\
1000 = \dfrac{1}{R_4C_2} = \dfrac{1}{R_5C_2} = \dfrac{1}{10R_8C_3} = \dfrac{1}{10R_9C_3}, \\
4000 = \dfrac{1}{R_dC_1} = \dfrac{1}{R_dC_2} = \dfrac{1}{R_dC_3}.
\end{cases}
\tag{2.21}
$$

By setting $C_1 = C_2 = C_3 = 10\,\text{nF}$, the following component values are obtained from Eq. (2.21): $R_1 = 5\,\text{k}\Omega$, $R_4 = R_5 = R_8 = 100\,\text{k}\Omega$, $R_9 = R_k = 10\,\text{k}\Omega$, $R_c = 50\,\text{k}\Omega$, and $R_d = 25\,\text{k}\Omega$.

NI Multisim 14.0 is used for circuit simulation, with the operational amplifier model TL084CN and the multiplier model AD633. The power supply voltage for both the operational amplifier and the multiplier is set to ± 15 V. The function generator is configured with a frequency of $f = 10\,\text{kHz}$ and an amplitude of $A = 1\,\text{V}$.

When only switch S_1 is closed and N_1 is set to 2, 3, and 5, the simulated results of the 1-D MMSCA circuit are presented in Fig. 2.14a. This figure reveals a chaotic attractor characterized by a complex intertwined trajectory, highlighting the chaotic nature of the dynamic system in the 2-D plane. When both switches S_1 and S_2 are closed, but N_1 and N_2 are varied, the simulated results of the 2-D MMSCA circuit are shown in Fig. 2.14b. This attractor displays an organized multiloop structure with scroll-like features, demonstrating the system's capacity to generate intricate patterns.

Additionally, when switches S_1, S_2, and S_3 are all closed, while N_1, N_2, and N_3 take different values, the simulated results for the 3-D MMSCA circuit are depicted in Fig. 2.14c. The attractor illustrates a denser and more expansive trajectory, forming a web-like structure that signifies higher complexity and potential multistability. Furthermore, when $N_1 = 3$, $N_2 = 3$, and $N_3 = 3$, the simulated results of the 2-D MMSCA circuit are shown in Fig. 2.14d. This figure demonstrates a multiscroll attractor with distinct layered patterns, showcasing the system's capability to produce repeated and well-organized scroll structures.

Based on Fig. 2.15, four identical circuits are constructed and circuit simulations are performed using NI Multisim 14.0 software in [12]. The initial conditions for $(E_{x_i}, E_{y_i}, E_{z_i}, E_{\phi_{1i}})$ are set to $(1, 1, 1, 1)$, $(1, 1, -1, -1)$, $(1, -1, -1, 1)$, and $(-1, -1, 1, 1)$, where $i = 1, 2, 3, 4$. By closing only switch S_1, the simulation results for the coexisting attractors of 1-D MMSCAs are presented in Fig. 2.16a. Similarly, two identical circuits are built based on Fig. 17, with initial conditions set to $(E_{x_i}, E_{y_i}, E_{z_i}, E_{\phi_{1i}}, E_{\phi_{2i}}) = (1, 1, 1, 1, 1)$ and $(1, 1, -1, -1, -1)$, where $i = 1, 2$. Closing switches S_1 and S_2, the simulation results with the coexisting attractors of 2-D MMSCAs are shown in Fig. 2.16b. For 3-D MMSCAs, the settings for initial conditions are $(E_{x_i}, E_{y_i}, E_{z_i}, E_{\phi_{1i}}, E_{\phi_{2i}}, E_{\phi_{3i}}) = (1, 1, 1, 1, 1, 1)$ and $(1, 1, -1, -1, -1, -1)$, $i = 1, 2$. By closing switches S_1, S_2, and S_3, the simulation results are shown in Fig. 2.16c and d.

These results reveal that, under different initial conditions, the system can generate chaotic attractors with a unique stretching structure. The consistency between the circuit implementation and numerical simulation results confirms the existence of coexisting attractors and multistability within the system. This agreement verifies the reliability of the modeling and simulation methods, providing a basis for further investigation of these types of chaotic systems.

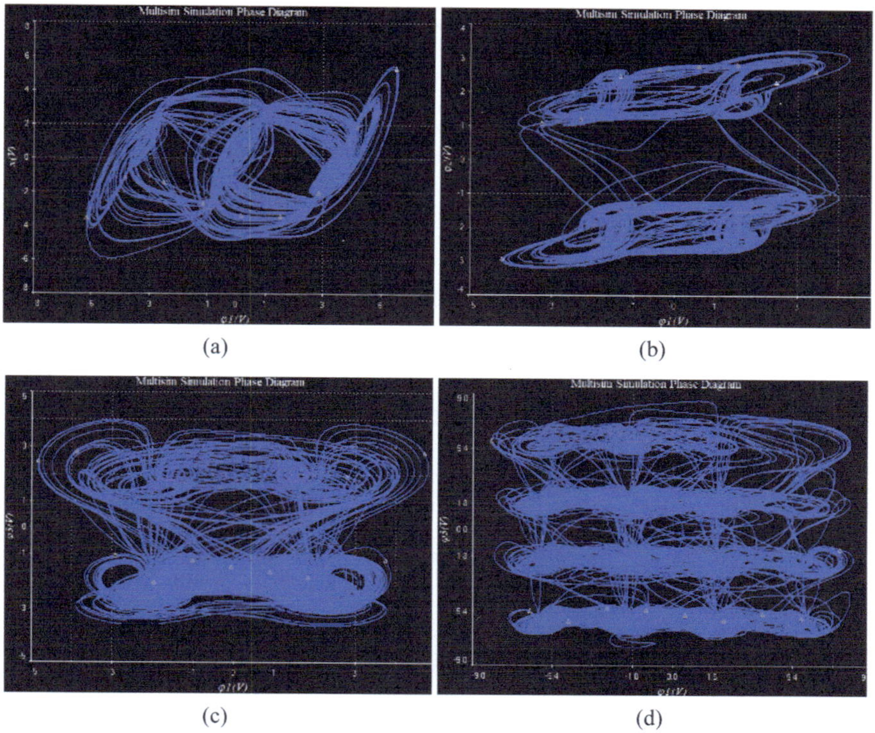

Fig. 2.14 Multisim simulation results of 1-D MMSCAs, 2-D MMSCAs, and 3-D MMSCAs. **a** 1-D MMSCAs with $N_1 = 2$. **b** 2-D MMSCAs with $N_1 = 1$ and $N_2 = 1$. **c** 3-D MMSCAs with $N_1 = 1$, $N_2 = 1$, and $N_3 = 1$. **d** 2-D MMSCAs with $N_1 = 3$, $N_2 = 3$, and $N_3 = 3$. Reprinted from [10], with permission from Elsevier

2.6 Grid Multiwing Butterfly Chaotic Attractors from a PWL System

The intrinsic dynamics of the Lorenz system are confined to the positive half-space with respect to the vertical axis, limited by a threshold effect. To address this limitation, Yu et al. [13] proposed a piecewise Lorenz system, which incorporates both a staircase function and an even-symmetric PWL function.

The piecewise Lorenz system considered here is described by [14]

$$\begin{cases} \dfrac{dx}{dt} = \alpha\,(y - x)\,, \\ \dfrac{dy}{dt} = -\operatorname{sgn}(x) \cdot z, \\ \dfrac{dz}{dt} = -1 + f\,(x)\,, \end{cases} \tag{2.22}$$

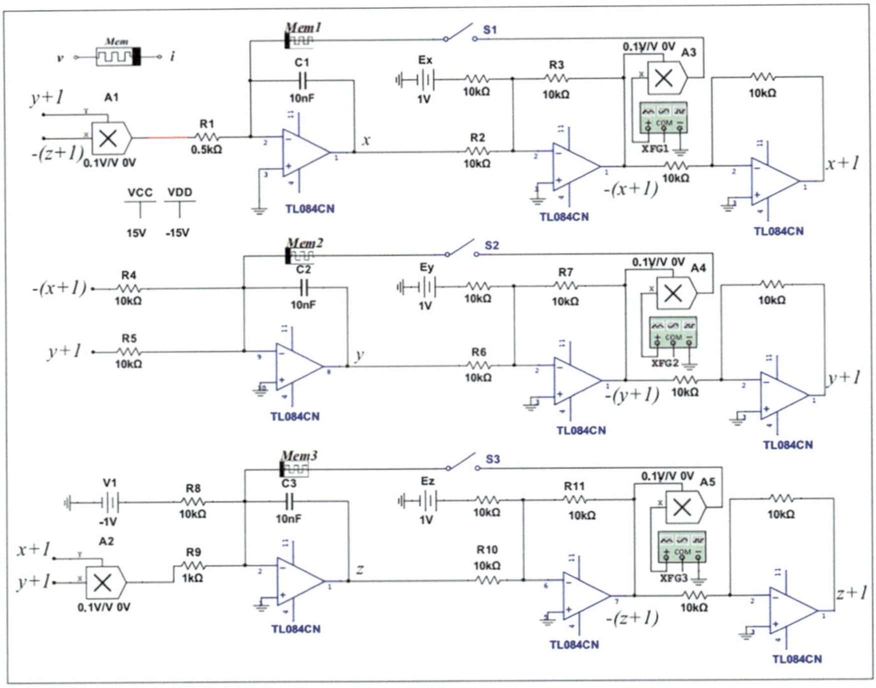

Fig. 2.15 Circuit design for coexisting attractor. Reprinted from [12], with permission from Elsevier

where sgn(x) is the standard sign function and $f(x) = |x|$ is the absolute value function. System (2.22) with $\alpha = 0.5$ has a typical double-wing butterfly chaotic attractor.

Similar to most Lorenz-like systems, the chaotic dynamics of system (2.22) are generally confined to the positive half-space with respect to the z-axis, or its mirrored counterpart, due to a limiting threshold effect. To overcome this threshold limitation, the following modification to the system is considered:

$$\begin{cases} \dfrac{dx}{dt} = \alpha\,(y - x)\,, \\ \dfrac{dy}{dt} = -\text{sgn}(x)\cdot\text{sgn}(z)\,[z - F\,(z)]\,, \\ \dfrac{dz}{dt} = \text{sgn}(z)\,[-1 + f\,(z)]\,, \end{cases} \tag{2.23}$$

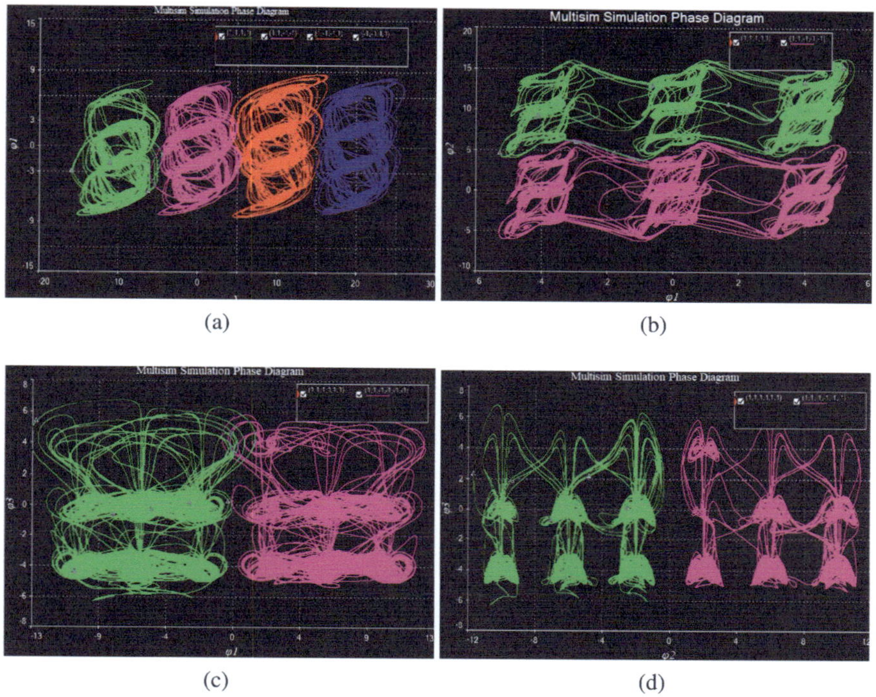

Fig. 2.16 Multisim simulation results of 1-D MMSCAs, 2-D MMSCAs, and 3-D MMSCAs. **a** 1-D MMSCAs with $N_1 = 2$. **b** 2-D MMSCAs with $N_1 = 1$ and $N_2 = 1$. **c** 3-D MMSCAs with $N_1 = 1$, $N_2 = 1$, and $N_3 = 1$. **d** 2-D MMSCAs with $N_1 = 3$, $N_2 = 3$, and $N_3 = 3$. Reprinted from [12], with permission from Elsevier

where

$$\begin{cases} f(x) = |kx| + \sum\limits_{m=1}^{M} L_m \left[\operatorname{sgn}(x + E_m) - \operatorname{sgn}(x - E_m) - 2\right], \\ F(z) = F_0 \operatorname{sgn}(z) + \sum\limits_{n=1}^{N} F_n \left[\operatorname{sgn}(z + W_n) - \operatorname{sgn}(z - W_n)\right]. \end{cases}$$

In this system, α, k, L_m, E_m, F_0, F_n, and W_n, with $1 \leqslant m \leqslant M$ and $1 \leqslant n \leqslant N$, are system parameters, and $M, N = 0, 1, 2, \ldots$. For simplicity, set $F_n = F_0$ for $1 \leqslant n \leqslant N$. Here, $f(x)$ is a standard sawtooth wave function that controls the number and size of the wings in the x-axis direction, while $F(z)$ is a stair-wave function that determines the number and size of the wings in the z-axis direction. In Eq. (2.23), the term $\operatorname{sgn}(z)[z - F(z)]$ is an even-symmetric function, which is key to overcome the threshold effect and to generate various multiwing butterfly chaotic attractors.

This system can generate a maximum of $(2M + 2) \times (2N + 2)$ grid multiwing chaotic attractors, $M, N = 0, 1, 2, \ldots$.

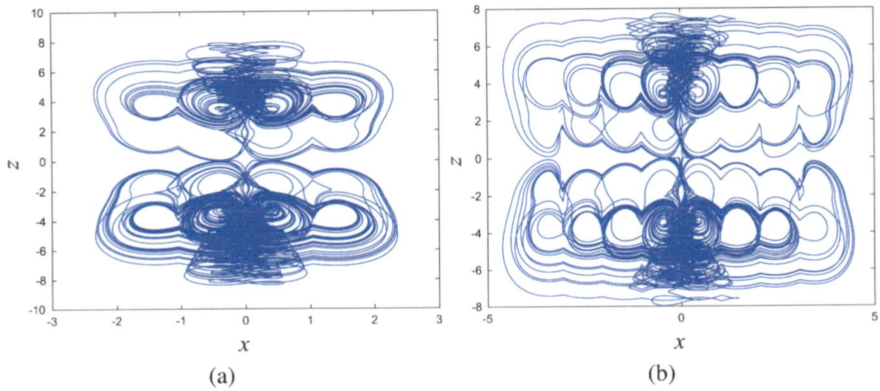

Fig. 2.17 Grid multiwing butterfly chaotic attractors [13]. **a** 4×4-grid multiwing. **b** 8×4-grid multiwing

When $M = N = 0$, the summation terms in the above formula vanish. Consequently, the recursive formulas for L_m, E_m, and W_n become

$$\begin{cases} L_m = \dfrac{A}{A_m}, \\ E_m = \dfrac{mA}{k}, \\ W_n = 2nF_n, \end{cases} \tag{2.24}$$

where the recursive parameters M, k, A, A_i, L_i, E_i, and W_i are given in Tables I and II in [13], which are not detailed here.

Similarly, based on Eq. (2.24) and following the procedure shown in [13], all parameters can be calculated, with the results presented in Tables I and II in [13], which are not shown here. When $N = 1$ and $M = 1, 3$, the system generates 4×4- and 8×4-grid multiwing butterfly chaotic attractors, as shown in Fig. 2.17a and b, respectively.

In Fig. 2.17a, the phase portrait demonstrates a symmetric chaotic attractor with a densely intertwined structure. The attractor features multiple lobes that appear to form two mirrored patterns around the origin along the z-axis. The attractor exhibits a high level of complexity and dynamic variability. In Fig. 2.17b, the phase portrait shows a more expansive chaotic attractor compared to that in (a), with additional lobes and layers extending further along the x- and z-axes. The attractor maintains symmetry and displays a more intricate scroll-like structure, demonstrating the system's capability to generate multiscroll patterns. Both figures highlight the richness and complexity of the chaotic dynamics in this system. It is worth noting that the system becomes increasingly sensitive to parameter mismatches as M and N increase.

Based on Eqs. (2.23) and (2.24) and the results presented in Tables I and II in [13], a module-based circuit diagram can be designed to implement the system with

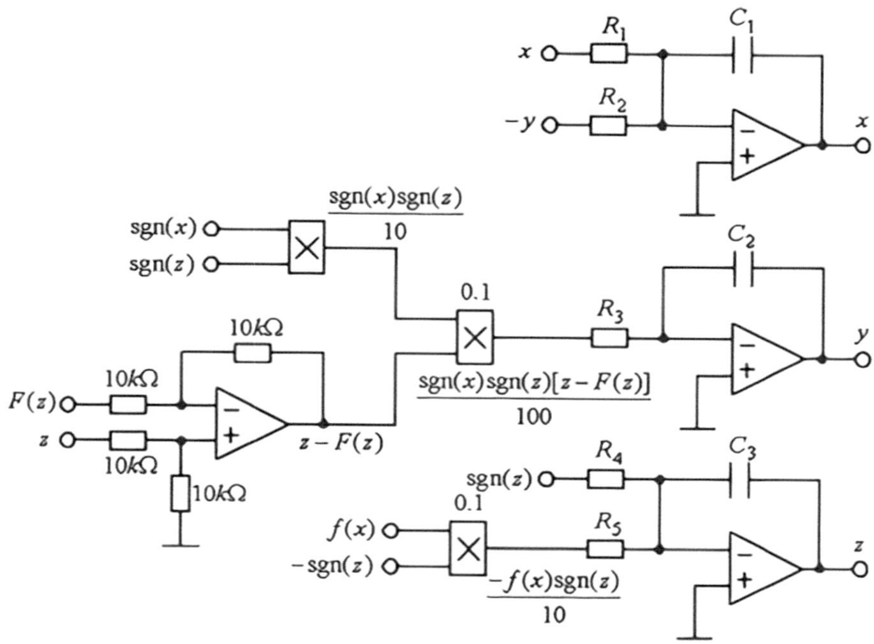

Fig. 2.18 Module-based circuit diagram for implementing the grid multiwing Lorenz system. Reprinted from [13], with permission from the IEEE

the simulated grid multiwing butterfly chaotic attractors. Figure 2.18 shows the main components of this circuit diagram for the grid multiwing Lorenz system. For simplicity, the subcircuit diagrams for the nonlinear functions sgn(·), −sgn(·), $f(x)$, $F(z)$, and their respective circuit parameters are not included here. Figure 2.19 presents the experimental observations of the grid multiwing butterfly chaotic attractors on the x-y plane, with $x = 0.6$ V/div and $y = 2.6$ V/div in Fig. 2.19a, and $x = 1.25$ V/div and $y = 2.6$ V/div in Fig. 2.19b.

2.7 Circuit Realization of n-Scroll Chaotic Attractors

This section discusses the circuit realization for n-scroll chaotic attractors.

2.7.1 State-Controlled CNN Realization

Arena et al. [9, 15] experimentally confirmed an n-double scroll chaotic attractor by a state-controlled CNN-based circuit.

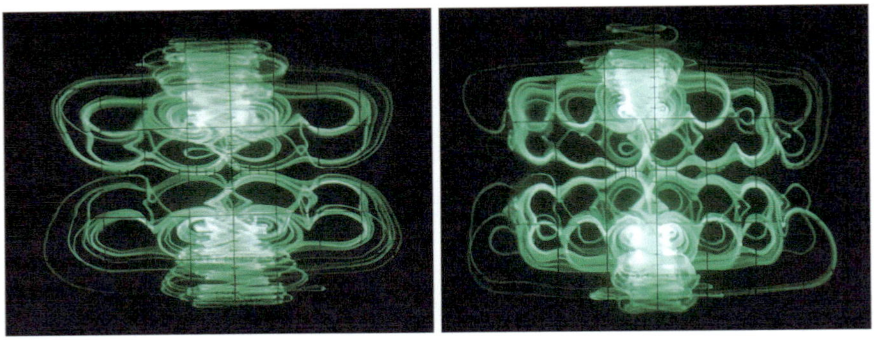

(a) (b)

Fig. 2.19 Experimental observations of the grid multiwing butterfly chaotic attractors on the x-y plane. **a** 4×4-grid multiwing. **b** 8×4-grid multiwing. Reprinted from [13], with permission from the IEEE

Consider the generalized Chua's circuit

$$\begin{cases} \dfrac{dx}{dt} = \alpha \left(y - h\left(x \right) \right), \\ \dfrac{dy}{dt} = x - y + z, \\ \dfrac{dz}{dt} = -\beta y - \gamma z, \end{cases} \tag{2.25}$$

where $h(x)$ is a PWL function defined by

$$h(x) = m_{2n-1}x + \sum_{k=1}^{2n-1} \frac{m_{k-1} - m_k}{2} \left(|x + b_k| - |x - b_k| \right),$$

where α, β, γ, the $2n$ coefficients m_k, and the $2n - 1$ breakpoints b_k are all system parameters.

It can be verified [9, 15] that system (2.25) is equivalent to a single-layer three-cell state-controlled CNN with parameters $C_{1,2} = \alpha$, $C_{1,1} = 1 - \alpha m_{2n-1}$, $A_{1,1} = 1$, $C_{3,3} = 1 - \gamma$, $C_{2,1} = C_{2,3} = 1$, $C_{3,2} = -\beta$, and the following nonlinear output function:

$$y_j = \frac{1}{2} \sum_{k=1}^{2n-1} n_k \left(\left| x_j + b_k \right| - \left| x_j - b_k \right| \right),$$

where $n_k = \alpha(m_k - m_{k-1})$, $k = 1, \ldots, 2n - 1$.

Figure 2.20 shows a 2-double-scroll chaotic attractor generated from the above state-controlled CNN realization [9, 15].

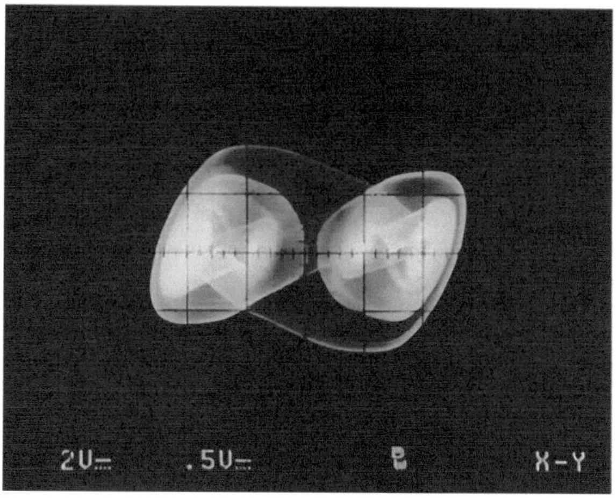

Fig. 2.20 Observed 2-double-scroll chaotic attractor from the state-controlled CNN realization. (h-axis: x_1; v-axis: x_2). Reprinted from [16], with permission from World Scientific Pub. Co.

2.7.2 Circuit Realization of 3 and 5-Scroll Chaotic Attractors

Yalcin et al. [17] experimentally verified 3- and 5-scroll chaotic attractors in the generalized Chua's circuit (2.5).

First, recall the circuit implementation of Chua's circuit. Figure 2.21 illustrates the voltage controlled voltage source (VCVS) realization of Chua's circuit. In this type of implementation, the corresponding state equations are given by

$$\begin{cases} \dfrac{dx}{dt} = \alpha\left(-\left(1+\delta\right)x + y + f\left(x\right)\right), \\ \dfrac{dy}{dt} = x - y + z, \\ \dfrac{dz}{dt} = -\beta y, \end{cases} \tag{2.26}$$

with $f(x) = -h(x) + (1+\delta)x$ and $\delta = 1$, where $h(x)$ is defined in Eq. (2.5).

Yalcin et al. [17] designed the circuit diagram shown in Fig. 2.22 to physically realize a 5-scroll chaotic attractor. The breakpoints are adjusted using the gains of inverting and non-inverting amplifiers as follows:

$$c_1 = \frac{E_{\text{sat}}}{1+\frac{R_{82}}{R_{81}}},\ c_2 = \frac{E_{\text{sat}}}{1+\frac{R_{62}}{R_{61}}},\ c_3 = \frac{E_{\text{sat}}}{1+\frac{R_{42}}{R_{41}}},\ c_4 = \frac{E_{\text{sat}}}{1+\frac{R_{22}}{R_{21}}}.$$

Additionally, the slopes are adjusted using the gains of inverting and noninverting amplifiers, as well as voltage branches, as follows:

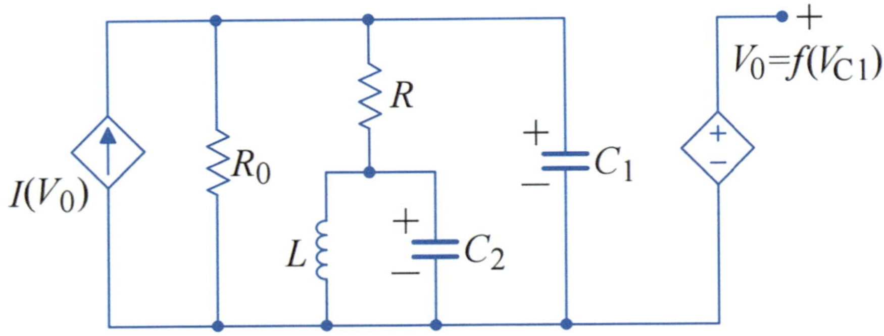

Fig. 2.21 VCVS implementation of Chua's circuit [17]

$$
\begin{aligned}
m_4 &= -\frac{1}{R_3}\frac{R_2}{R_1}\left(-\frac{R_{12}}{R_{11}}\right),\\
m_3 &= -\frac{1}{R_3}\frac{R_2}{R_1}\left(m_4 + \left(-\frac{R_{22}}{R_{21}}\right)\frac{R_{31}}{R_{31}+R_{32}}\right),\\
m_2 &= -\frac{1}{R_3}\frac{R_2}{R_1}\left(m_3 + \left(1+\frac{R_{42}}{R_{41}}\right)\frac{R_{52}}{R_{51}+R_{52}}\right),\\
m_1 &= -\frac{1}{R_3}\frac{R_2}{R_1}\left(m_2 + \left(-\frac{R_{62}}{R_{61}}\right)\frac{R_{72}}{R_{71}+R_{72}}\right),\\
m_0 &= -\frac{1}{R_3}\frac{R_2}{R_1}\left(m_1 + \left(1+\frac{R_{82}}{R_{81}}\right)\frac{R_{92}}{R_{91}+R_{92}}\right).
\end{aligned}
$$

By varying the value of R_{31} from 12280 Ω to 12740 Ω, a 3-scroll chaotic attractor can be generated, as shown in Fig. 2.23a. When all parameters are set according to Table 1 in [17], a 5-scroll chaotic attractor can be generated, as shown in Fig. 2.23b.

Figure 2.23a shows a 3-scroll chaotic attractor with the resistance value $R_{31} =$ 12.280 Ω. The phase portrait is shown on the h-v plane, where the h-axis is scaled to 0.5 V/div and the v-axis to 0.1 V/div. The three scrolls illustrate the system state transitions among three distinct chaotic regions.

Figure 2.23b displays a 6-scroll chaotic attractor with the resistance value $R_{31} =$ 12.740 Ω. This phase portrait is also represented on the h-v plane, with the h-axis scaled to 1.0 V/div and the v-axis to 0.2 V/div. The six scrolls reflect the increased complexity of the system, showing chaotic dynamics with more attractor regions.

These figures demonstrate how the number of scrolls and the chaotic behavior can be influenced by adjusting the system parameters, such as the resistance.

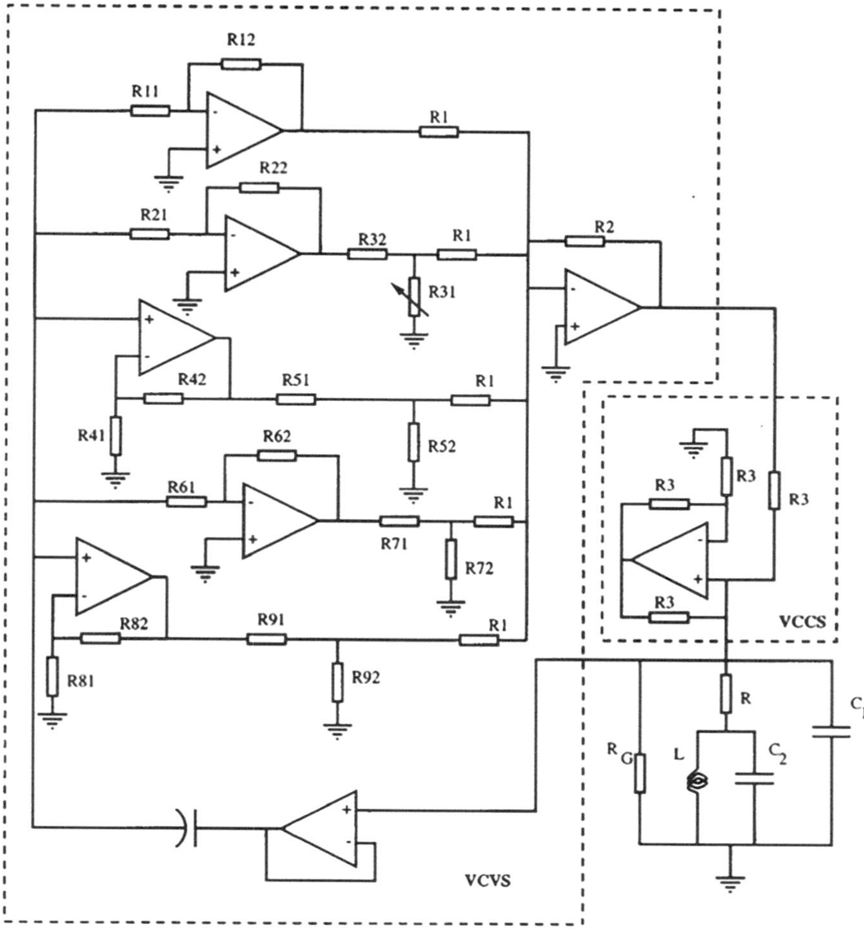

Fig. 2.22 Circuit implementation of 5-scroll attractor from a generalized Chua's circuit. Reprinted from [17], with permission from the IEEE

2.7.3 Realization of n-Scroll Chaotic Attractors

Using the scaling properties of the nonlinearity in the generalized Chua's circuit, Yalcin et al. [18] experimentally demonstrated a 6-scroll chaotic attractor.

According to Eq. (2.26), the equilibrium points of the generalized Chua's circuit are determined by the following condition:

$$\begin{cases} h(x) = 0, \\ x = -z, \\ y = 0, \end{cases} \tag{2.27}$$

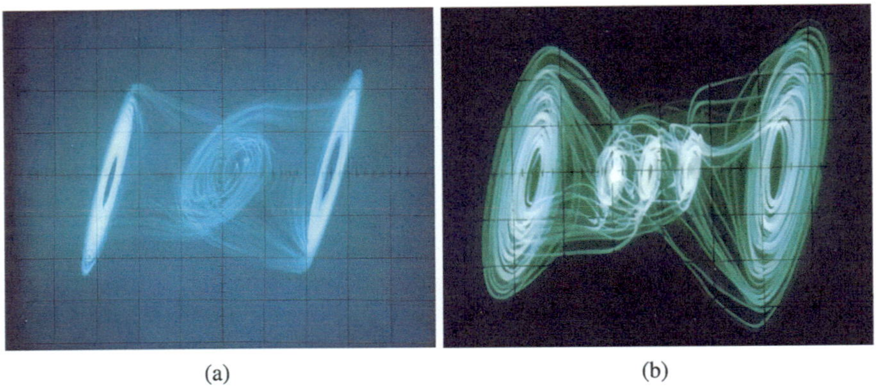

Fig. 2.23 Experimental observations of 3- and 5-scroll chaotic attractors. **a** 3-Scroll attractor with $R_{31} = 12.280\Omega$ (h-axis: 0.5 V/div; v-axis: 0.1 V/div). **b** 6-Scroll attractor with $R_{31} = 12.740\Omega$ (h-axis: 1 V/div; v-axis: 0.2 V/div. Reprinted from [16], with permission from World Scientific Pub. Co.

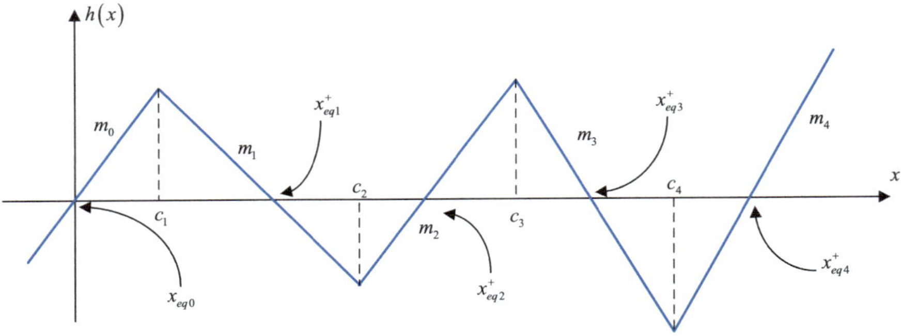

Fig. 2.24 PWL function of the generalized Chua's circuit [18]

with

$$h(x) = m_0 c_1 + \sum_{j=1}^{l-1} m_j (c_{j+1} - c_j) + \cdots + m_l (x_{eql} - c_l) = 0,$$

where $l = 1, \ldots, 2n - 1$. The equilibrium points x_{eql} are shown in Fig. 2.24.

The Jacobian matrices, evaluated at the first equilibrium point $O = (0, 0, 0)$ and at the i-th equilibrium point $eqi^{\pm} = (x^{\pm}_{eqi}, 0, x^{\pm}_{eqi})$, are obtained as follows:

$$\mathrm{J}\,(0) = \begin{pmatrix} 0 & \alpha & 0 \\ 1 & -1 & 1 \\ 0 & -\beta & 0 \end{pmatrix} + \begin{pmatrix} -\alpha m_0 & 0 & 0 \\ 0 & 0 & 0 \\ 0 & 0 & 0 \end{pmatrix},$$

and

$$J\left(eqi^{\pm}\right) = \begin{pmatrix} 0 & \alpha & 0 \\ 1 & -1 & 1 \\ 0 & -\beta & 0 \end{pmatrix} + \begin{pmatrix} -\alpha m_i & 0 & 0 \\ 0 & 0 & 0 \\ 0 & 0 & 0 \end{pmatrix},$$

where $l = 1, \ldots, 2n - 1$.

Yalcin et al. [18] applied a rescaling factor k to the breakpoints of the PWL function, ensuring that the slopes remain unchanged. After the rescaling, the PWL function takes the following form

$$h(x) = m_{2n-1}x + \frac{1}{2}\sum_{i=1}^{2n-1}(m_{i-1} - m_i)\left(\left|x + \frac{c_i}{k}\right| - \left|x - \frac{c_i}{k}\right|\right).$$

The new equilibrium points, $eqi^{\pm} = \left(x^{\pm}_{eqi}/k,\ 0,\ -x^{\pm}_{eqi}/k\right)$, are rescaled by the same factor k, and are governed by

$$m_0\frac{c_1}{k} + \sum_{j=1}^{l-1} m_j\left(\frac{c_{j+1}}{k} - \frac{c_j}{k}\right) + \cdots + m_j\left(\frac{x_{eq1}}{k} - \frac{c_j}{k}\right) = 0.$$

Figure 2.25 shows the circuit diagram for generating a 6-scroll chaotic attractor. After rescaling both current and time by a factor of 10,000, the component values are obtained as summarized in Table 1 in [18]. Figure 2.26 illustrates the observed 6-scroll chaotic attractor.

2.7.4 A Systematic Design Method for Generating n-Scroll Chaotic Attractors

Until the year 2000, the largest number of scrolls in a chaotic attractor that could be experimentally generated and observed was limited to 6 only. Implementing circuits with more than 6 scrolls required devices with a large dynamic range, which necessitates higher voltage supplies, suitable differential amplifiers, and an appropriate scaling of the voltages [19]. To that end, Zhong et al. [20] introduced a new circuit design technique that allows for the experimental verification of chaotic attractors with up to 10 scrolls.

Here, Chua's circuit (1.27) is revisited. By introducing additional breakpoints into the PWL function (1.28), the v-i characteristic of the nonlinear resistor is modified as follows:

$$f\left(V_{C1}\right) = m_{2n-1}V_{C1} + \frac{1}{2}\sum_{i=1}^{2n-1}(m_{i-1} - m_i) \times (|V_{C1} + b_i| - |V_{C1} - b_i|), \quad (2.28)$$

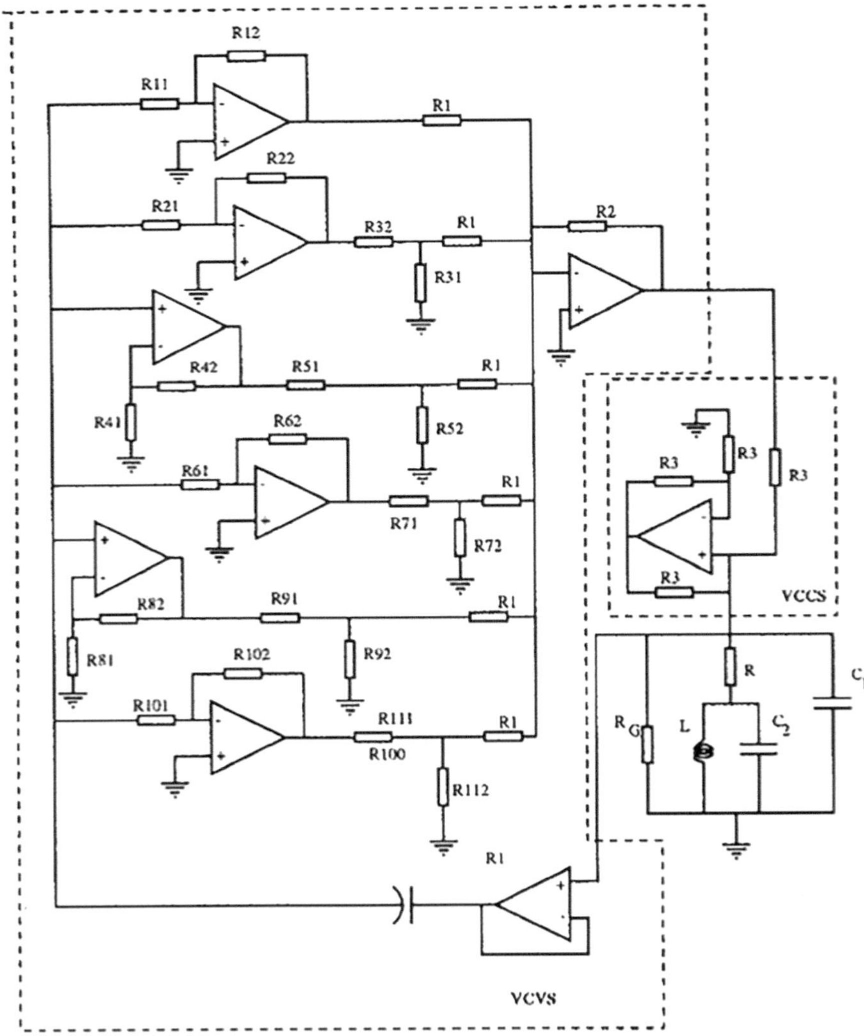

Fig. 2.25 Circuit diagram for generating a 6-scroll attractor from the generalized Chua's circuit. Reprinted from [18], with permission from the IEEE

where m_i and b_i represent the slopes and breakpoints of the i-th segment, respectively. Furthermore, chaotic attractors with an even number of $2n$ scrolls can be generated by incorporating the nonlinearity described in Eq. (2.28) into Chua's circuit. Similarly, chaotic attractors with an odd number of $2n - 1$ scrolls can be obtained with the following nonlinear resistor:

$$f(V_{C1}) = m_{2n-1}V_{C1} + \frac{1}{2}\sum_{i=2}^{2n-1}(m_{i-1} - m_i) \times (|V_{C1} + b_i| - |V_{C1} - b_i|). \quad (2.29)$$

In the diode of Chua's circuit depicted in Fig. 2.26, there are two fundamental circuit cells [20]. In cell I, the left operational amplifier operates in its linear region, while in cell II, the right operational amplifier spans its entire dynamic range, including both linear and saturated regions. The saturated point in this configuration defines the breakpoint of the PWL function. By combining cells I and II, nonlinear resistors with multiple segments can be constructed, using these two cells as the basic building blocks of the circuit.

Figure 2.27 shows the circuit diagram for a nonlinear resistor with 19 segments. It is evident that four pairs of cells, biased by a VCVS, are connected in parallel with Chua's diode. Additionally, two more breakpoints are introduced by connecting a pair of cells II with Chua's diode, allowing for the implementation of the nonlinear resistor described in Eq. (2.25) with $n = 5$. Each cell is offset by a VCVS to fine-tune the breakpoints. The measured v-i characteristic with 19 segments is shown in Fig. 2.29a.

Similarly, Fig. 2.28 presents the circuit diagram for a nonlinear resistor with 17 segments, obtained by removing resistor R_2 from the circuit shown in Fig. 2.27. This configuration enables the realization of the nonlinear resistor described by Eq. (2.29) with $n = 5$. The measured v-i characteristic with 17 segments is shown in Fig. 2.29b.

In Fig. 2.29a, the v-i curve exhibits a PWL characteristic with 19 distinct segments. Each segment represents a specific voltage-current relationship, with abrupt changes at the boundaries, indicating nonlinear transitions in the resistor's behavior.

In Fig. 2.29b, similar to Fig. 2.29a, the v-i curve also displays a PWL pattern but consists of 17 segments. The overall slope changes and transitions between segments highlight the resistor's nonlinear characteristics.

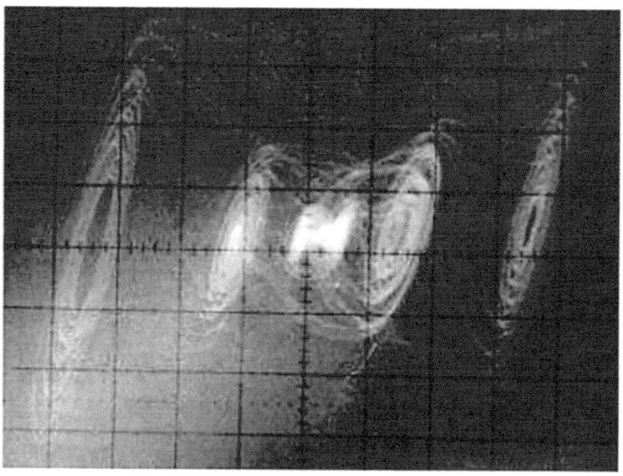

Fig. 2.26 Observed 6-scroll chaotic attractor (h-axis: 1 V/div; v-axis: 0.2 V/div). Reprinted from [16], with permission from World Scientific Pub. Co.

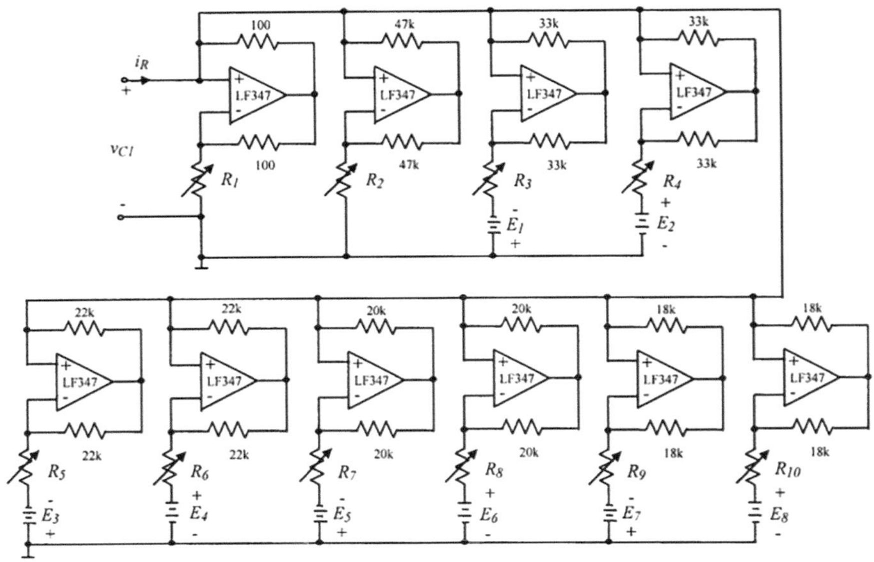

Fig. 2.27 Circuitry realizing the nonlinear resistor with 19 segments. Reprinted from [20], with permission from World Scientific Pub. Co.

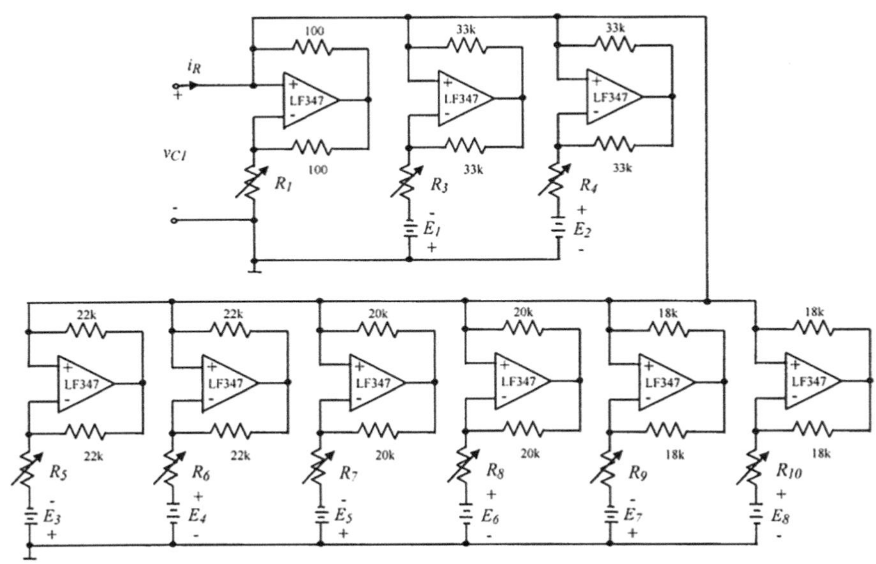

Fig. 2.28 Circuitry realizing the nonlinear resistor with 17 segments. Reprinted from [20], with permission from World Scientific Pub. Co.

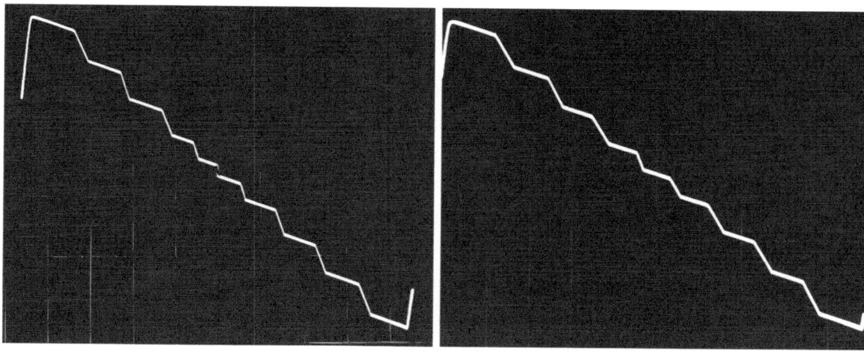

(a) (b)

Fig. 2.29 Measured v-i characteristic of the nonlinear resistor. **a** Measured v-i characteristic with 19 segments. **b** Measured v-i characteristic with 17 segments. Reprinted from [20], with permission from World Scientific Pub. Co.

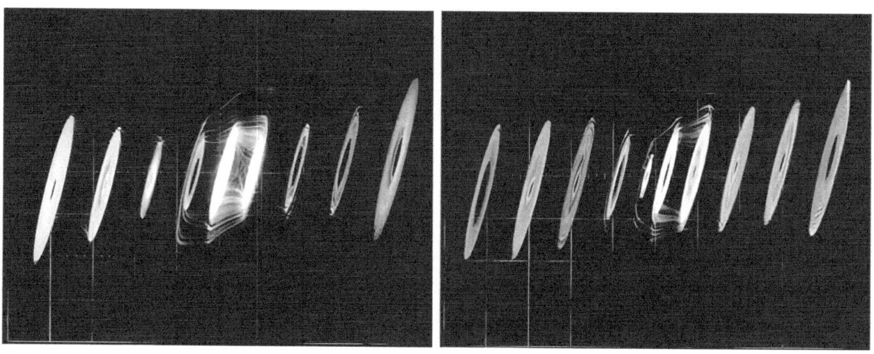

(a) (b)

Fig. 2.30 Phase portraits of n-scroll chaotic attractors on the v_{C1}-v_{C2} plane (h-axis: $v_{C1} =$ 2.5 V/div; v-axis: $v_{C2} = 0.5$ V/div. **a** 9-Scroll attractor. **b** 10-Scroll attractor. Reprinted from [20], with permission from World Scientific Pub. Co.

These measurements demonstrate the resistor's ability to generate complex nonlinear behaviors, critical for chaotic systems and applications that require segmented and adjustable resistance properties.

Zhong et al. [20] successfully experimentally verified chaotic attractors with 5, 6, 7, 8, 9, and 10 scrolls. Figure 2.30 display the experimentally observed 9-scroll and 10-scroll chaotic attractors, respectively. The system parameters for these experiments are provided in Table 1 of [20].

Figure 2.30 presents the oscilloscope visualization of chaotic attractors with distinct scroll dynamics.

In Fig. 2.30a, the attractor exhibits multiple scrolls arranged along one single axis, showcasing elliptical loops that are elongated and symmetrically distributed. The visualization highlights the spatial separation of the scrolls and the well-organized yet chaotic behaviors of the trajectories.

In Fig. 2.30b, similar to (a), the phase portrait shows a series of scrolls with slightly modified spacing and intensity, indicating allowable parameter variations in the chaotic system. The scrolls remain aligned along one principal axis, with dense trajectory loops forming each scroll of the attractor.

Both subfigures demonstrate the intricate and well-organized structures of chaotic attractors, demonstrating their sensitivity to system parameter variations and their capability of generating complex multi-scroll patterns. The gridlines provide a clear reference for the scrolls' spatial arrangement.

Remark 2.3 Generating attractors with a large number of scrolls is a technically challenging task due to the limitations imposed by the dynamic range of the physical devices available for experimentation. For this task, special design methods and technical skills beyond the conventional approaches are required, as can be seen from the above.

2.8 Generating $n \times m$-Scroll Attractors with the Modified Chua's Circuit

Yu et al. [21] demonstrates the generation of $n \times m$-scroll attractors under the Chua's circuit framework. By utilizing a sawtooth function $f_1(x)$ and a staircase function $f_2(y)$, it is possible to generate and observe various multiscroll attractors from a third-order circuit. This approach underscores the flexibility of Chua's circuit family in producing intricate dynamical behaviors through carefully chosen nonlinear circuit functions.

2.8.1 *Generating Attractors with the Modified Chua's Circuit*

Chua's circuit can be generalized in several ways: by incorporating a nonlinear function with both positive and negative slopes [16], adding more breakpoints to Chua's diode [20, 22], or using a sawtooth function [21] along with other non-piecewise linear (non-PWL) functions. In these cases, the generalized Chua's circuit is described in a uniform manner, as follows [21]:

$$\begin{cases} \dfrac{dx}{dt} = \alpha\left[y - f_1(x)\right], \\ \dfrac{dy}{dt} = x - y + z, \\ \dfrac{dz}{dt} = -\beta y, \end{cases} \tag{2.30}$$

where $\alpha > 0$, $\beta > 0$, and $f_1(x)$ to generate even scrolls is given by

$$f_1(x) = \xi \left(x - A_1 \left[-\mathrm{sgn}(x) + \sum_{i=0}^{N-1} (\mathrm{sgn}(x + 2iA_1) + \mathrm{sgn}(x - 2iA_1)) \right] \right), \tag{2.31}$$

or, to generate odd scrolls is given by

$$f_1(x) = \xi \left(x - A_1 \left[\sum_{i=0}^{N-1} (\mathrm{sgn}(x + (2i + 1)A_1) + \mathrm{sgn}(x - (2i + 1)A_1)) \right] \right), \tag{2.32}$$

with constants $\xi > 0$, $A_1 > 0$, $N \geqslant 0$, and

$$\mathrm{sgn}(x) = \begin{cases} 1, & \text{if } x > 0, \\ 0, & \text{if } x = 0, \\ -1, & \text{if } x < 0. \end{cases}$$

By utilizing Eq. (2.30) along with either Eq. (2.31) or Eq. (2.32), it is possible to generate attractors with either an even or an odd number of scrolls. For example, with parameters $\alpha = 10.0$, $\beta = 16.0$, $\xi = 0.25$, and $A_1 = 0.5$, 2-scroll and 8-scroll attractors can be produced by setting $N = 1$ and $N = 4$ in Eq. (2.31), respectively, as shown in Fig. 2.31.

Figure 2.31 displays two subfigures, illustrating phase portraits of chaotic attractors on the x-y plane. In Fig. 2.31a, the attractor exhibits a two-lobed structure with elliptical trajectories. The orbits have a dense and intertwined pattern, similarly to those in the classic chaotic systems with symmetry about the y-axis. The structure resembles a double-scroll attractor, characterized by a confined complex trajectory.

In Fig. 2.31b, the attractor is more expansive, consisting of multiple loops arranged in a scroll-like pattern along the x-axis. This multiscroll attractor features an organized and layered trajectory with increased scroll complexity compared to the one in the left subfigure. The trajectories extend further along the x-axis, showing the system's capacity to produce more intricate chaotic behaviors.

Both subfigures show the diversity and intricate dynamics of chaotic systems on the x-y plane in the phase space.

Similarly, 3-scroll and 9-scroll attractors can be generated by using $N = 1$ and $N = 4$ in Eq. (2.32), as demonstrated in Figs. 2.32. This figure showcases the phase portraits of the chaotic attractors generated using Eqs. (2.30) and (2.31), showing their scroll-like structures on the x-y plane.

In Fig. 2.32a, the attractor exhibits a three-lobed structure with elliptical and intertwined trajectories. The scrolls are symmetrically distributed along the x-axis, creating a balanced yet chaotic pattern. The dense and intricate loops illustrate the system's multistability and complex dynamics.

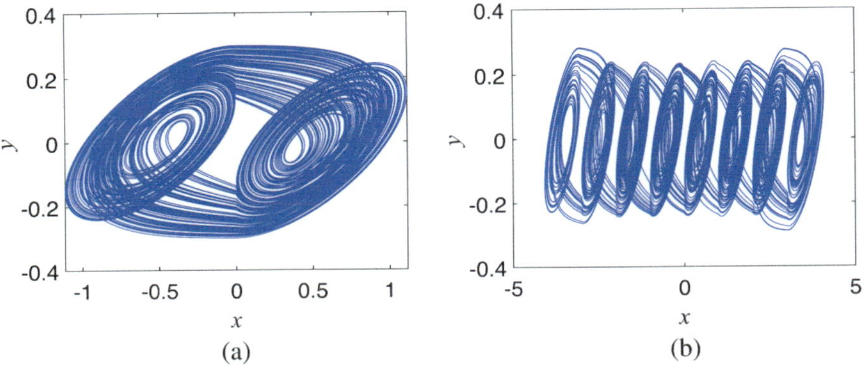

Fig. 2.31 Even number of scrolls generated by Eqs. (2.30) and (2.31) [21]. **a** 2-scroll ($N = 1$). **b** 8-scroll ($N = 4$)

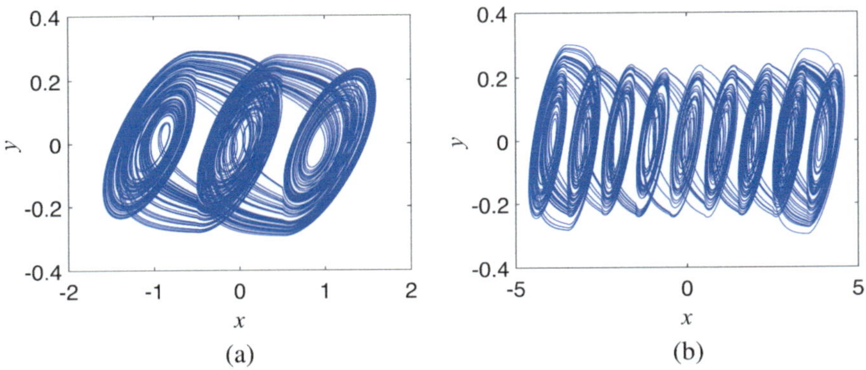

Fig. 2.32 Odd number of scrolls generated by Eqs. (2.30) and (2.31) [21]. **a** 3-scroll ($N = 1$). **b** 9-scroll ($N = 4$)

In Fig. 2.32b, the attractor shows a more complex structure, with 9 scrolls extending along the x-axis. The scrolls form distinct layers, demonstrating the system's ability to generate more elaborate and organized chaotic patterns. The increased number of scrolls enhances the attractor's complexity and spatial coverage. Both subfigures demonstrate the system's capability to produce multiscroll chaotic attractors with increasing complexity as N grows. The clear differentiation in scroll numbers reflects the system's parameter-dependent dynamics.

To further extend the number of scrolls in the y-direction, an additional nonlinear function $f_2(y)$ is incorporated into the first and third equations of Eq. (2.30), resulting in the following modified model:

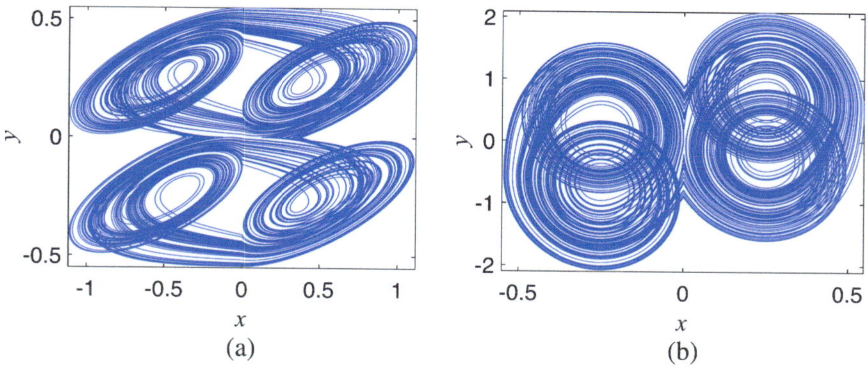

Fig. 2.33 2×2-scroll ($N = 1$, $M = 1$) generated from Eq. (2.33), with f_1 and f_2 defined in Eqs. (2.31) and (2.34), respectively [21]. **a** x-y phase portrait. **b** y-z phase portrait

$$\begin{cases} \dfrac{dx}{dt} = \alpha \left[y - f_2(y) - f_1(x) \right], \\ \dfrac{dy}{dt} = x - y + z, \\ \dfrac{dz}{dt} = -\beta \left[y - f_2(y) \right], \end{cases} \tag{2.33}$$

where $\alpha > 0$, $\beta > 0$, and $f_2(y)$ is given below, for $2M$-scrolls attractors in the y-direction and for $2M + 1$-scrolls attractors in the y-direction, respectively:

$$f_2(y) = A_2 \left[-\text{sgn}(y) + \sum_{j=0}^{M-1} (\text{sgn}(y + 2jA_2) + \text{sgn}(y - 2jA_2)) \right], \tag{2.34}$$

and

$$f_2(y) = A_2 \left[\sum_{j=0}^{M-1} (\text{sgn}(y + (2j+1)A_2) + \text{sgn}(y - (2j+1)A_2)) \right], \tag{2.35}$$

with $A_2 > 0$ and $M \geqslant 1$.

By keeping the system parameters fixed at $\alpha = 10$, $\beta = 16$, $\xi = 0.25$, $A_1 = 0.5$, and $A_2 = 0.25$, various $n \times m$-scroll attractors can be generated by adjusting the values of N and M. The resulting attractors are illustrated in Figs. 2.33 and 2.34.

Figure 2.33 depicts two distinct chaotic attractors on the x-y plane, showing their unique structures and dynamic behaviors.

In Fig. 2.33a, the attractor has a four-lobed structure, with trajectories forming compact elliptical loops. These loops are symmetrically arranged in pairs along the

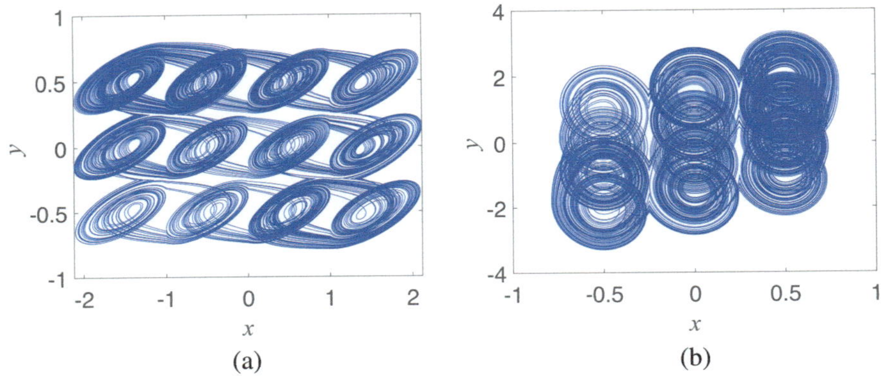

Fig. 2.34 4×3-scroll ($N = 2$, $M = 1$) generated from Eq. (2.33), with f_1 and f_2 defined in Eqs. (2.31) and (2.35), respectively [21]. **a** x-y phase portrait. **b** y-z phase portrait

x-axis and the y-axis, respectively, creating a pattern of four scrolls. The system's chaotic behavior is evident from the dense interwoven trajectories within each scroll.

In Fig. 2.33b, the attractor presents a two-lobed structure with larger and more expansive loops compared to (a). The scrolls are centered and more circular, forming a symmetric double-scroll attractor. The overlapping parts in the trajectories suggest intricate interactive dynamics in the system.

In Fig. 2.34a, the attractor exhibits a grid-like structure consisting of multiple scrolls arranged symmetrically in a 4×3 pattern. The trajectories form elliptical loops within each scroll, with clear separation and organization among the scrolls. This structured arrangement highlights the attractor's multiscroll nature and demonstrates the system's capacity for generating regular and repeated chaotic patterns.

In Fig. 2.34b, the attractor displays a clustered multiscroll structure, with the scrolls arranged more closely along the x-axis. Unlike the evenly spaced arrangement shown in Fig. 2.34a, the scrolls here are more compact and interconnected, forming a dense and layered pattern. The overlapping parts reflect a higher level of complexity and interaction among the trajectories.

Both subfigures demonstrate the diversity and intricacy of the chaotic attractors, showing how different parameters can influence the organization and density of the scrolls on the x-y plane in the phase space.

2.8.2 RLC Network Framework for Direct Realization

By introducing the transformations of $\tau = t/R_{C2}$, $\alpha = C_2/C_1$, $\beta = R^2C_2/L$, $x = V_{C1}$, $y = V_{C2}$, and $z = Ri_L$, the dimensionless state equations in Eq. (2.31) can be converted to their corresponding dimensional form, as follows:

$$\begin{cases} \dfrac{dV_{C1}}{dt} = \dfrac{1}{RC_2} \times \dfrac{C_2}{C_1} \left[V_{C2} - f_1\left(V_{C1}\right) - f_2\left(V_{C2}\right)\right], \\ \dfrac{dV_{C2}}{dt} = \dfrac{1}{RC_2} \left[V_{C1} - V_{C2} + R_i L\right], \\ \dfrac{dR_i L}{dt} = -\dfrac{1}{RC_2} \times \dfrac{R^2 C_2}{L} \left[V_{C2} - f_2\left(V_{C2}\right)\right], \end{cases} \tag{2.36}$$

or

$$\begin{cases} C_1 \dfrac{dV_{C_1}}{dt} = \dfrac{V_{C2} - V_{C_1}}{R} + iN, \\ C_2 \dfrac{dV_{C_2}}{dt} = \dfrac{V_{C1} - V_{C2}}{R} + iN, \\ L \dfrac{d\left(iL\right)}{dt} = -\left[V_{C2} - f_2\left(V_{C2}\right)\right], \end{cases} \tag{2.37}$$

where $f_2(V_{C2)}$ is a VCVS, and

$$i_N = \left(\frac{(V_{C1} - f_1(V_{C1}))}{R}\right) - \left(\frac{f_2(V_{C2})}{R}\right),$$

where i_N is generated by a voltage control current sources (VCCS).

For instance, by choosing the component values of $C_1 = 10\,\text{nF}$, $C_2 = 100\,\text{nF}$, $L = 18\,\text{mH}$, $R = 1.7\,\text{k}\Omega$, $R_1 = 10\,\text{k}\Omega$, $R_2 = 7.5\,\text{k}\Omega$, $R_F = 1\,\text{k}\Omega$, $R_X = 108\,\text{k}\Omega$, $R_Y = 54\,\text{k}\Omega$, and $V_{\text{sat}} = 13.5\,\text{V}$, with a supply voltage of $\pm 15\,\text{V}$ (which gives $\alpha = 10$, $\beta = 16$, $\xi = 0.25$, $A_1 = 0.5$, and $A_2 = 0.25$), it is possible to generate both n-scroll and $n \times m$-scroll attractors (with $m = 2, 3, 4$ and $n = 2, 3, 4$), by adjusting the switches and the reference voltages $D_{0,\pm 1}$ and $E_{0,\pm 1}$. Specific examples of these configurations are listed in Tables 1–3 in [21], with the corresponding $n \times m$-scroll attractors shown in Fig. 2.35.

Figure 2.35 presents two experimental x-y phase portraits obtained from an oscilloscope, illustrating multiscroll chaotic attractors with different configurations.

In Fig. 2.35a, the phase portrait shows a 2×2 scroll structure, forming four distinct elliptical scrolls symmetrically arranged on the x-y plane. Each scroll is well-organized, with minimal overlap between neighboring scrolls. The gridlines on the oscilloscope indicate a scale of 0.5 V/div along the x-axis and 0.25 V/div along the y-axis.

In Fig. 2.35b, the phase portrait depicts a more complex 4×4 scroll structure, consisting of 16 elliptical scrolls arranged in a regular grid pattern. The scrolls are closely spaced but remain clearly distinguishable. The oscilloscope's grid indicates a scale of 1.0 V/div along the x-axis and 0.5 V/div along the y-axis.

Both subfigures demonstrate the system's ability to generate organized multiscroll attractors on the x-y plane in the phase space, with increasing complexity and scroll density achieved by adjusting the system's parameters. The results highlight the practical realization of chaotic behaviors through circuit implementation.

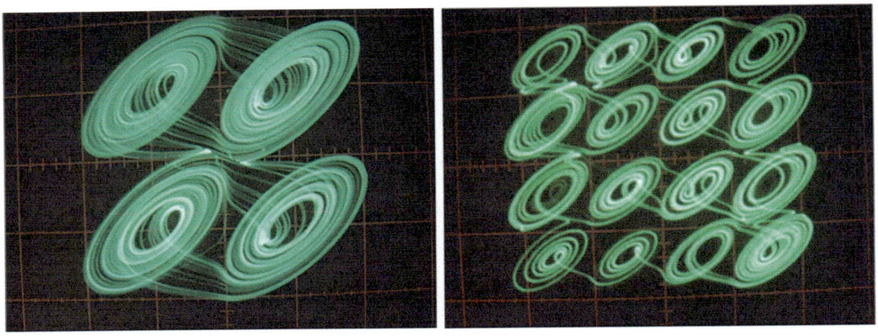

(a) (b)

Fig. 2.35 **a** x-y phase portrait 2×2 scrolls, X-scale: 0.5 V/div; Y-scale: 0.25V/div. **b** x-y phase portrait 4×4 scrolls, X-scale: 1.0 V/div; Y-scale: 0.5 V/div. Reprinted from [21], with permission from World Scientific Pub. Co.

From Figs. 2.36 and 2.37, one can observe that with the parameters $\alpha = 10$, $\beta = 16$, and $\xi = 0.25$, it is possible to generate n-scroll and $n \times m$-scroll attractors, where $m = 2$ to 6 and $n = 2$ to 8, depending on the settings of the switches and the reference voltages $D_{\pm i}$ and $E_{\pm j}$.

Some selected examples are listed in Tables 4–6 in [21]. For simplicity of presentation, only a few experimental snapshots of the results are shown in Figs. 2.38 and 2.39.

The dimensionless state equations can also be directly implemented using integrated circuits with operational amplifiers. The circuit schematic for this implementation is shown in Fig. 2.36, where the nonlinear functions $f_1(\cdot)$ and $f_2(\cdot)$ are realized using the circuits presented in Fig. 2.37. In this design, the summation terms in both nonlinear functions are generated by the outputs of the comparators.

Figure 2.38 presents two oscilloscope phase portraits, showing a 7×5 scroll chaotic attractor on different phase planes.

Figure 2.38a shows the x-y phase portrait, with a 7×5 scroll structure. The scrolls are organized into a grid with 7 columns along the x-axis and 5 rows along the y-axis. Each scroll forms a distinct elliptical shape, and the overall arrangement is symmetrical and evenly spaced. The scales for the oscilloscope are 1.0 V/div for the x-axis and 0.5 V/div for the y-axis.

Figure 2.38a shows the y-z phase portrait, with the same chaotic attractor from a different perspective. The scrolls appear more elongated and layered along the z-axis, reflecting a different structural view of the multiscroll system. Here, the scales are 0.5 V/div for the y-axis and 2.0 V/div for the z-axis.

Both subfigures demonstrate the intricate dynamics of the multiscroll chaotic systems, highlighting the attractor's organized yet complex behaviors on various phase planes. The grid overlay and voltage scales show the practical realization of such chaotic systems.

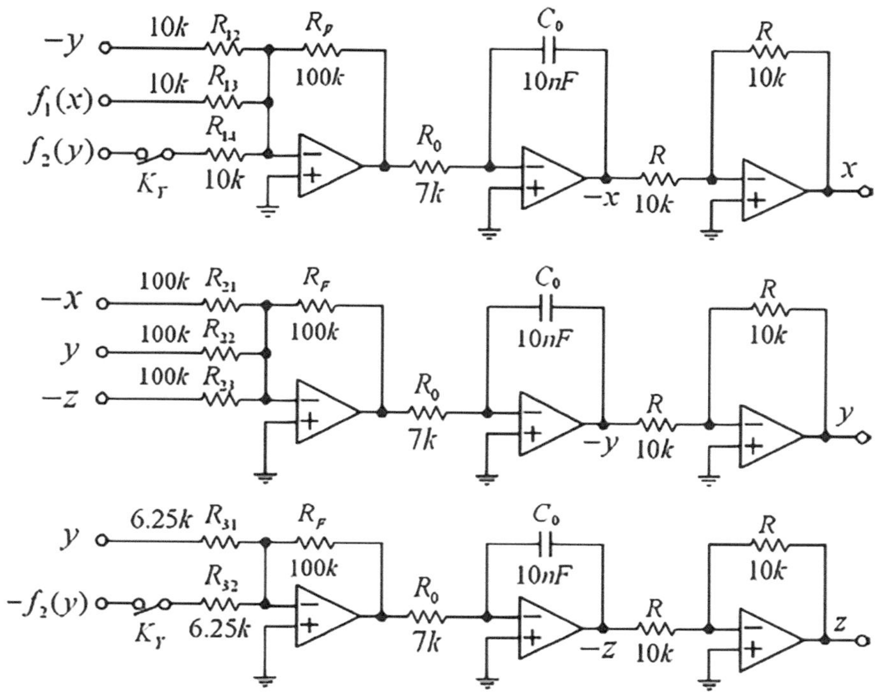

Fig. 2.36 Integrated circuit implementation. Reprinted from [21], with permission from World Scientific Pub. Co.

Figure 2.39 showcases the oscilloscope phase portraits of a chaotic attractor with a 9×6 scroll structure on different phase planes:

Figure 2.39a displays the x-y phase portrait, featuring a grid of 8 columns along the x-axis and 6 rows along the y-axis. Each scroll forms a well-organized elliptical loop, arranged symmetrically. The structure demonstrates the regularity and complexity of the chaotic attractor. The voltage scale is 1.0 V/div for the x-axis and 0.5 V/div for the y-axis.

Figure 2.39b shows the y-z phase portrait, where the scrolls appear layered and elongated along the z-axis. This perspective highlights the attractor's intricate multidimensional dynamics. The voltage scale is 0.5 V/div for the y-axis and 2.0 V/div for the z-axis.

Both subfigures illustrate the multiscroll chaotic attractor's complexity and organized distribution, showing its distinct geometric patterns across different phase planes.

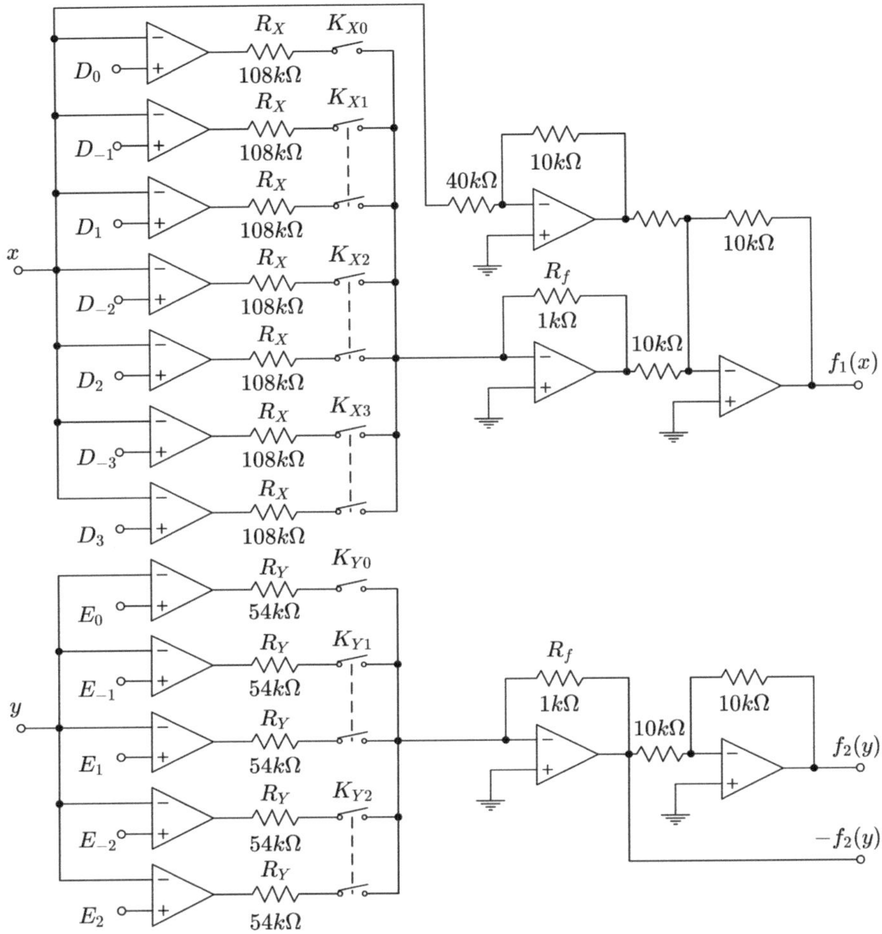

Fig. 2.37 Realization of functions f_1 and f_2. Reprinted from [21], with permission from World Scientific Pub. Co.

2.9 Conclusion

This chapter studies the design and realization of n-scroll chaotic attractors based on piecewise linear (PWL) functions. By building on traditional nonlinear systems such as Chua's circuit, unfolded Chua's circuit and the Brockett system, it presents scalable methods for generating complex multi-scroll chaotic attractors, including n-scroll, $n \times m$-scroll, and grid multiwing butterfly attractors.

The chapter introduces systematic methods for system and circuit designs, utilizing quasi-linear and PWL functions to regulate chaotic dynamics. The circuit implementations incorporate standard components such as operational amplifiers, memristors and function generators, verified through both simulations and experi-

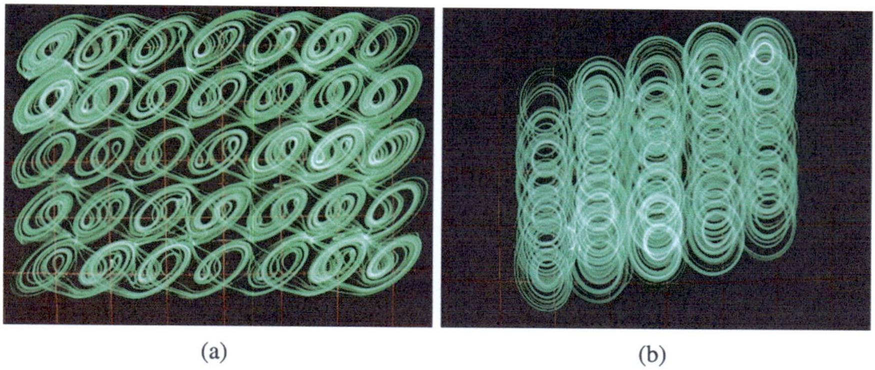

Fig. 2.38 7 × 5 scrolls. X-scale: 1.0 V/div; Y-scale: 0.5 V/div; Z-scale: 2.0 V/div. **a** x-y phase portrait. **b** y-z phase portrait. Reprinted from [21], with permission from World Scientific Pub. Co.

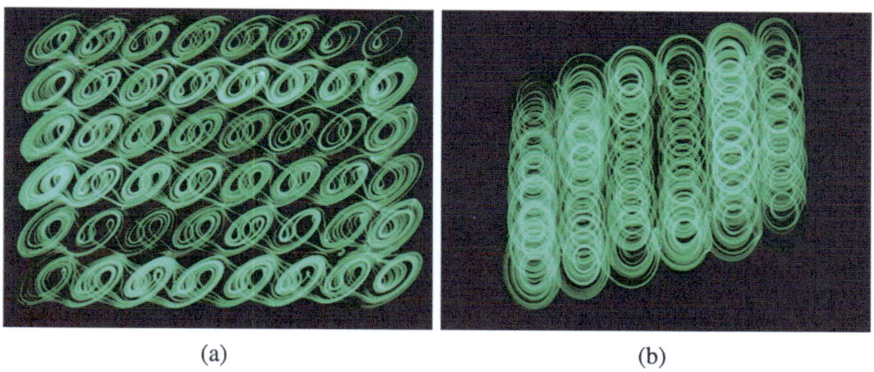

Fig. 2.39 8 × 6 scrolls. X-scale: 1.0 V/div; Y-scale: 0.5 V/div; Z-scale: 2.0 V/div. **a** x-y phase portrait. **b** y-z phase portrait. Reprinted from [21], with permission from World Scientific Pub. Co.

ments. Furthermore, the integration of multi-terminal memristors enhances the realization of 1-D, 2-D, and 3-D multi-scroll chaotic attractors, significantly improving the design flexibility.

These advancements offer valuable insights for practical applications in secure communications, signal generation and encryption technologies. The presented methods establish a scalable and efficient framework for designing chaotic systems, paving the way for further developments and innovations in the fields of nonlinear dynamical systems and chaos-based technologies.

References

1. Chua, L.O., Komuro, M., Matsumoto, T.: The double scroll family. IEEE Trans. Circuits Syst. **33**, 1072–1118 (1986)
2. Chua, L.O.: CNN: A Paradigm for Complexity. World Scientific, Singapore (1998)
3. Matsumoto, T., Chua, L., Komuro, M.: The double scroll. IEEE Trans. Circuits Syst.-I **32**, 797–818 (1985)
4. Suykens, J. A. K., Vandewalle, J.: Generation of n-double scrolls ($n = 1, 2, 3, 4, \ldots$). IEEE Trans. Circuits Syst.-I **40**, 861–867 (1993)
5. Suykens, J. A. K., Vandewalle, J.: Quasilinear approach to nonlinear systems and the design of n-double scroll ($n = 1, 2, 3, 4, \ldots$). IEE Proc. G **138**, 595–603 (1991)
6. Suykens, J.A.K., Huang, A., Chua, L.O.: A family of n-scroll attractors from a generalized Chua's circuit. Int. J. Electron. Commun. **51**, 131–138 (1997)
7. Aziz-Alaoui, M.A.: Differential equations with multispiral attractors. Int. J. Bifur. Chaos **9**, 1009–1039 (1999)
8. Aziz-Alaoui, M.A.: Multispiral chaos. In: 2nd International Conference on Control of Oscillations and Chaos, vol. 1, July 5–7, St. Petersberg, Russsia, pp. 88–91 (1999)
9. Arena, P., Baglio, S., Fortuna, L., Manganaro, G.: State controlled CNN: a new strategy for generating high complex dynamics. IEICE Trans. Fund. **E79-A**, 1647–1657 (1996)
10. Zhang, J., Zuo, J., Wang, M., Guo, Y., Xie, Q., Hou, J.: Design and application of multiscroll chaotic attractors based on a novel multi-segmented memristor. Chaos Solit. Fract. **181**, 114676 (2024)
11. Sprott, J.C.: Some simple chaotic flows. Phys. Rev. E **50**, R647 (1994)
12. Suykens, J.A.K., Vandewalle, J.: Between n-double sinks and n-double scrolls ($n = 1, 2, 3, 4, \ldots$). In: International Symposium on Nonlinear Theory and Its Applications, December 5–10, Hawaii, USA, pp. 829–834 (1993)
13. Yu, S., Lü, J., Chen, G., Yu, X.: Design and implementation of grid multiwing butterfly chaotic attractors from a piecewise Lorenz system. IEEE Trans. Circuits Syst.-II **57**, 803–807 (2010)
14. Ozoguz, S., Elwakil, A.S., Kennedy, M.P.: Experimental verification of the butterfly attractor in a modified Lorenz system. Int. J. Bifur. Chaos **12**, 1627–1632 (2002)
15. Arena, P., Baglio, S., Fortuna, L., Manganaro, G.: Generation of n-double scrolls via cellular neural networks. Int. J. Circuit Th. Appl. **24**, 241–252 (1996)
16. Lü, J., Chen, G.: Generating multiscroll chaotic attractors: theories, methods and applications. Int. J. Bifur. Chaos **16**, 775–858 (2006)
17. Yalcin, M.E., Suykens, J.A.K., Vandewalle, J.: Experimental confirmation of 3- and 5-scroll attractors from a generalized Chua's circuit. IEEE Trans. Circuits Syst.-I **47**, 425–429 (2000)
18. Yalcin, M.E., Suykens, J.A.K., Vandewalle, J.: On the realization of n-scroll attractors. In: IEEE International Symposium on Circuits and Systems, vol. 5, May 30–June 2, Orlando, Florida, USA, pp. 483–486 (1999)
19. Yalcin, M.E.: Cellular Neural Networks, Multi-scroll Chaos and Synchronization. Theory Applications and Implementations, World Scientific, Singapore (2005)
20. Zhong, G., Man, K., Chen, G.: A systematic approach to generating n-scroll attractors. Int. J. Bifur. Chaos **12**, 2907–2915 (2002)
21. Yu. S., Tang, W.K.S., Chen. G.: Generation of $n \times m$-scroll attractors under a Chua's circuit framework. Int. J. Bifur. Chaos **17**, 3951–3964 (2007)
22. Yu, S., Qiu, S., Lin, Q.: New results of study on generating multiple-scroll chaotic attractors. Sci. in China, Ser. F: Inf. Sci. **46**, 104–115 (2003)

Chapter 3
Generating n-Scroll Chaotic Attractors with Nonlinear Modulating Functions

Abstract This chapter introduces the nonlinear modulation function method for designing n-scroll chaotic attractors, highlighting its versatility in generating complex chaotic dynamics. Several approaches, using for example trigonometric functions, nonlinear transconductors and piecewise exponential functions, are able to create a wide range of multiscroll attractors with customizable features. The chapter also explores practical implementations, such as memristor-based systems and electronic circuits, demonstrating these methods through numerical simulations and experimental implementations. The findings offer valuable insights into the design and control of chaotic systems, contributing to advancements in analyzing nonlinear dynamics with potential engineering applications.

3.1 Trigonometric Function Methods

This section shows that n-scroll attractors can be efficiently produced using trigonometric functions. This method simplifies the construction procedure as an alternative to the conventional techniques introduced in the last chapter and enabling the creation of complex chaotic behaviors with adjustable scroll numbers and orientations.

3.1.1 Sine Function Method

Tang et al. [1] modified Chua's circuit by substituting the nonlinear characteristic function with a sine function, yielding

$$\begin{cases} \dot{x} = \alpha\left(y - f\left(x\right)\right), \\ \dot{y} = x - y + z, \\ \dot{z} = -\beta y, \end{cases} \tag{3.1}$$

J. Lü et al., *Chaotic Systems with Multiscroll Attractors*,
https://doi.org/10.1007/978-3-032-09854-2_3

where

$$f(x) = \begin{cases} b\pi/2a\,(x - 2ac), & \text{if } x \geq 2ac, \\ -b\sin\left(\frac{\pi x}{2a} + d\right), & \text{if } -2ac < x < 2ac, \\ b\pi/2a\,(x + 2ac,), & \text{if } x \leq 2ac, \end{cases}$$

where

$$d = \begin{cases} \pi, & \text{for } c \text{ is even}, \\ 0, & \text{for } c \text{ is odd}, \end{cases}$$

in which $\alpha, \beta, a, b, c, d$ are real parameters.

System (3.1) can generate chaotic attractors with $(c + 1)$ scrolls. For example, with parameters $\alpha = 10.814$, $\beta = 14.0, a = 1.3$, and $b = 0.11$, the system yields 8-scroll and 9-scroll chaotic attractors when $c = 7$ and $c = 8$, respectively, as illustrated in Figs. 3.1a and b.

Tang et al. [1] further developed an electronic circuit to experimentally verify these n-scroll chaotic attractors. Figure 3.2 presents the circuit diagram, with the associated dynamical equations

$$\begin{cases} \dfrac{dv_1}{dt} = \dfrac{1}{RC_1}(v_2 - v_1) - \dfrac{g(v_1)}{C_1}, \\ \dfrac{dv_2}{dt} = \dfrac{1}{RC_2}(v_1 - v_2) + \dfrac{iL}{C_2}, \\ \dfrac{diL}{dt} = -\dfrac{v_2}{L}, \end{cases} \tag{3.2}$$

where

$$g(v_1) = \begin{bmatrix} ((b\pi/2a) - 1/R)v_1 - b\sin((\pi v_1/2a) + d) \\ -(b\pi/4a) \times (|v_1 + 2ac| - |v_1 - 2ac|) \end{bmatrix}.$$

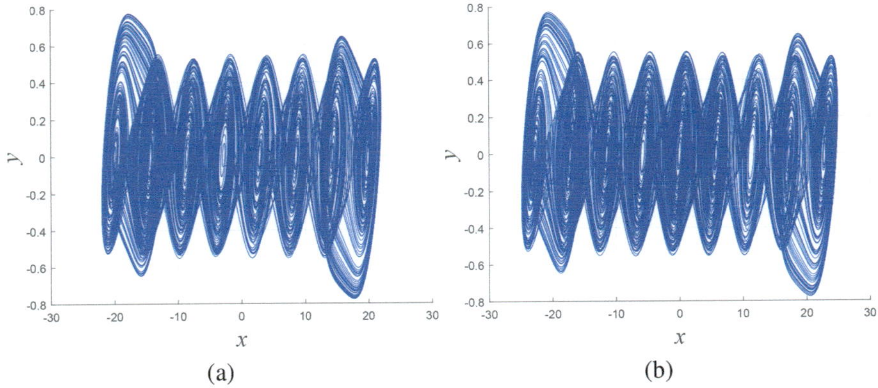

Fig. 3.1 n-Scroll chaotic attractors generated by the sine function method [1]. **a** 8-Scroll chaotic attractor. **b** 9-Scroll chaotic attractor

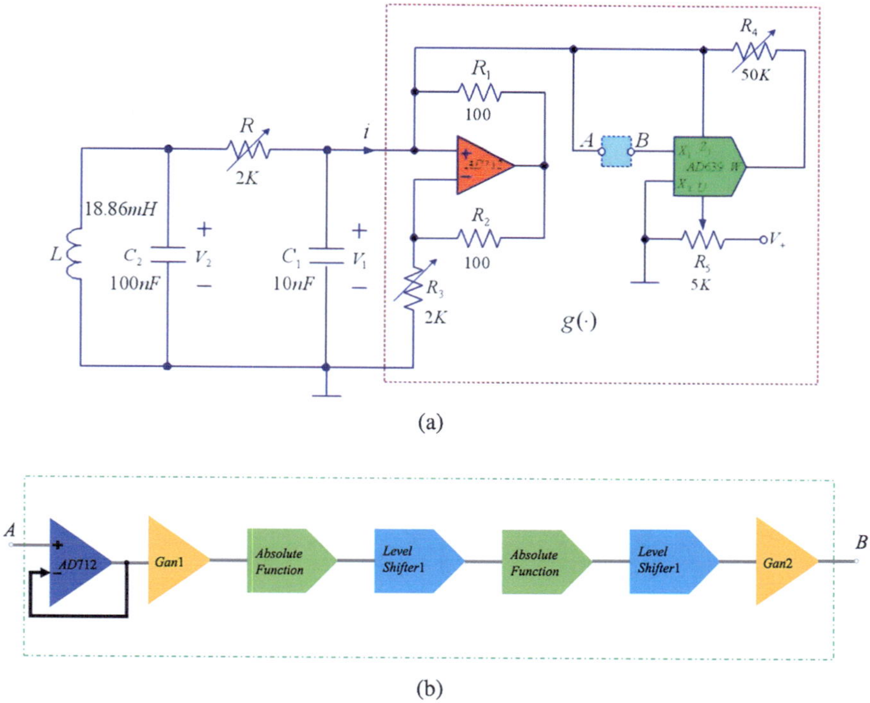

Fig. 3.2 Circuit diagram for generating n-scroll chaotic attractors. **a** Modified Chua's circuit. **b** Sine or cosine function generator inserted into A–B. Reprinted from [1], with permission from the IEEE

As depicted in Fig. 3.2a, the negative resistor $g(\cdot)$ comprises two parallel components: one with a linear negative slope in its v-i characteristic and the other being an R-port with a v-i characteristic governed by a sine function. The circuit utilizes the AD639 trigonometric function chip, which natively supports an angular input range of $\pm 500°$, allowing it to produce an attractor with up to 4-scrolls.

To extend the angular input range, for $x \in [-(2m+1)\pi, (2m+1)\pi]$, the following relationship is applied:

$$\cos(x) = \cos(|x|) = \sin\left(m\pi + \frac{\pi}{2} - |x|\right) = \sin(y).$$

Here, $y = m\pi + \pi/2 - |x|$ and $y \in [-(2+1/2)\pi, (m+1/2)\pi]$. This setup can effectively double the number of scrolls generated by the AD639 using an absolute value operation and voltage adjustment. The enhancement is shown in Fig. 3.2b, which is included in the A-B module shown in Fig. 3.2a. The experimental results shown in Fig. 3.3 displays the chaotic attractors with scroll counts ranging from 6 to 9.

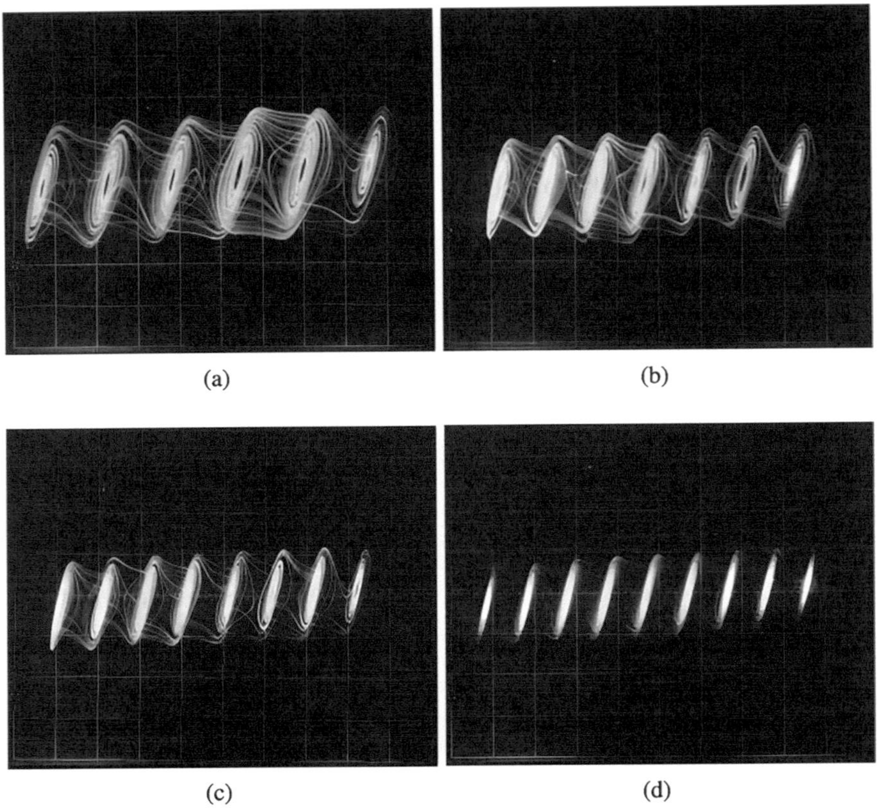

Fig. 3.3 Experimental observations of the sine function method. **a** 6-Scrolls. **b** 7-Scrolls. **c** 8-Scrolls. **d** 9-Scrolls. Reprinted from [1], with permission from the IEEE

3.1.2 Tangent Function Method

This section extends the memristor-based chaotic system, initially developed with a sine function, to include other trigonometric functions such as the tangent function. Sun et al. proposed an approach enables the creation of a more diverse array of memristor-based chaotic systems, offering new opportunities to explore more chaotic dynamics shaped by various trigonometric modulations [2].

The initial value problem is given by

$$\dot{x} = H(x), \quad x(t_0) = x_0, \tag{3.3}$$

where $H(\cdot)$ is the system function, $t_0 \in \mathbf{R}$ represents the initial time and $x_0 = (x_{01}, x_{02}, x_{03}, \ldots, x_{0n})^{\mathrm{T}} \in \mathbf{R}^n$ denotes the initial state. Let function $\varphi(t, t_0, x_0)$ be a solution to the initial value problem (3.3). Assume that $\Omega(t)$ is a compact subset of $\mathbf{R}^n$ for any $t \in \mathbf{R}$, and let $d(x, A) := \inf_{x' \in A} \|x - x'\|$ define the distance between a point x and a set A, where $\|\cdot\|$ is the Euclidean norm in $\mathbf{R}^n$. Assume also that $\Omega_\varepsilon := \{x : d(x, \Omega(t)) < \varepsilon\}$ for a constant $\varepsilon > 0$.

Now, consider the initial value problem (3.3). If the following condition holds:

$$\lim_{t\to\infty} d\big(\varphi(t, t_0, x_0), \Omega(t)\big) = 0, \quad \forall x_0 \in \mathbf{R}^n \setminus \Omega(t),$$

that is, for any $\varepsilon > 0$, there exists $T > t_0$ such that $\varphi(t, t_0, x_0) \in \Omega_\varepsilon$ for all $t \geq T$, then the set $\Omega(t)$ is an ultimate bounded limiting set for the system (3.3). Additionally, if $\varphi(t, t_0, x_0) \in \Omega_\varepsilon$ for all $x_0 \in \Omega(t_0)$, then $\Omega(t)$ is referred to as a positive invariant set for the system (3.3).

If a positively invariant set $\Omega(t)$ of an autonomous system is bounded for all $t \geq t_0$, then the basin of attraction of the corresponding attractor will be contained within this bounded set for all $t \geq t_0$.

An approach for designing autonomous memristor-based chaotic systems with infinitely many chaotic attractors has been explored, in 3-D, 4-D, and 5-D settings for case studies. Consider an autonomous system $\dot{x} = H(x)$, where $H(\cdot)$ will be defined below, and the basin of attraction of the attractor is contained in the following bounded set:

$$\left\{(x_1, x_2, \ldots, x_n) : \max_{1\leq i\leq n} |x_i| \leq C\right\},$$

with $C > 0$ being a constant. The corresponding coordinate transformation is defined by

$$\begin{aligned} x_1 &= K_1 u_1, \\ x_2 &= K_2 u_2, \\ &\vdots \\ x_n &= K_n u_n. \end{aligned}$$

The system $\dot{u} = H(u)$ is thus obtained, where the constant $K_i > 0$ is chosen to be sufficiently large. The basin of attraction is then contained in the set

$$\left\{(u_1, u_2, \ldots, u_n) : |u_i| \leq \frac{C}{K_i}, \; i = 1, 2, \ldots, n\right\}.$$

By appropriately selecting the parameter K_i, it is possible to construct a system such that the continuous functions $\varphi_i(u_i)$ have infinitely many simple zeros, thereby generating infinitely many chaotic attractors,

The proposed method is applied to construct an autonomous memristor-based chaotic system with infinitely many chaotic attractors, demonstrated using a 3-D, 4-D, and 5-D memristor-based chaotic system. Specifically, Sun et al. [2] constructed a 3-D memristor-based chaotic system described by

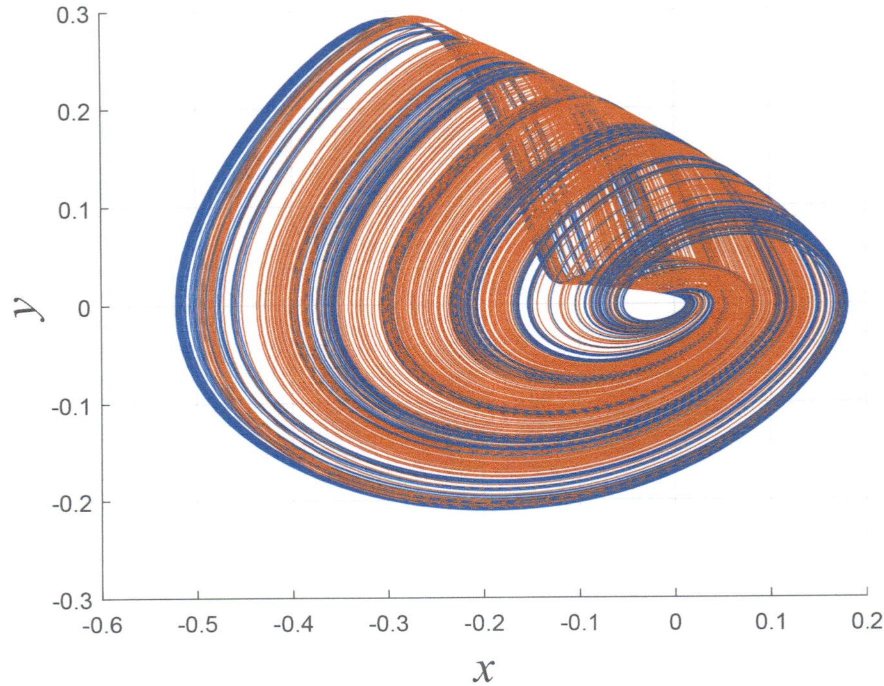

Fig. 3.4 Phase portraits of system (3.4) with $K = 6$ and initial condition $(0.1, 0.1, 0.1)$ [2]

$$\begin{cases} \dot{u}_1 = \tan(u_2), \\ \dot{u}_2 = -\frac{1}{3}\tan(u_1) + \frac{1}{2}\tan(u_2) - \frac{1}{2}K^2\tan(u_2)\tan^2(u_3), \\ \dot{u}_3 = -\tan(u_2) - 0.6\tan(u_2) + K\tan(u_2)\tan(u_3), \end{cases} \tag{3.4}$$

where $K = 6$. The equilibria of system (3.4) are $(m\pi, m\pi, m\pi)$, $m = 0, 1, 2, \ldots, n$, and the system is chaotic as shown in Fig. 3.4.

A 4-D memristor-based chaotic system example is

$$\begin{cases} \dot{u}_1 = 2\tan(u_1) + 10\tan(u_2) - 15K^2\tan(u_1)\tan^2(u_4), \\ \dot{u}_2 = \tan(u_1) - \tan(u_2) + \tan(u_3), \\ \dot{u}_3 = -14\tan(u_2) - 0.1\tan(u_3), \\ \dot{u}_4 = \tan(u_1). \end{cases} \tag{3.5}$$

With $K = 3$, system (3.5) is chaotic, as shown in Fig. 3.5. The equilibrium points of system (3.5) are $(m\pi, m\pi, m\pi, c)$, where c is a real constant, $m = 0, 1, 2, \ldots, n$.

A 5-D memristor-based chaotic system is

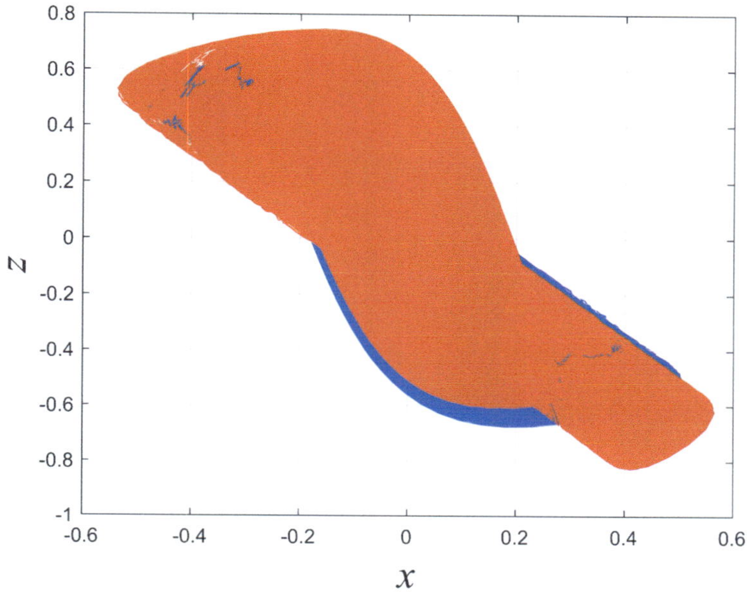

Fig. 3.5 Phase portraits of system (3.5) with $K = 3$ and initial condition $(0.1, -0.1, 0.1, 0.1)$ [2]

$$\begin{cases} \dot{u}_1 = 10.8\tan(u_1) + 9\tan(u_2) - 5.4K^2\tan(u_1)\tan^2(u_5), \\ \dot{u}_2 = -\tan(u_3) + \tan^2(u_4), \\ \dot{u}_3 = 30\left(\tan(u_2) - \tan(u_1) - \tan(u_3)\right), \\ \dot{u}_4 = -15\tan(u_2), \\ \dot{u}_5 = \tan(u_1). \end{cases} \tag{3.6}$$

With $K = 8$, system (3.6) exhibits chaotic behavior, as shown in Fig. 3.6. The equilibria of system (3.6) are $(m\pi, m\pi, m\pi, m\pi, c)$, where c is a real constant, $m = 0, 1, 2, \ldots, n$.

By integrating periodic trigonometric functions into the above 3-D, 4-D, and 5-D memristor-based chaotic systems, an infinite variety of attractors with similar characteristics can be generated.

3.1.3 Nonlinear Combination Method

Lai et al. [3] proposed a simple chaotic system that supports an infinite number of coexisting chaotic attractors. This system, composed of five terms with two nonlinearities, possesses an infinite set of unstable equilibria due to the sinusoidal nonlinearity.

The mathematical model of the system is described by

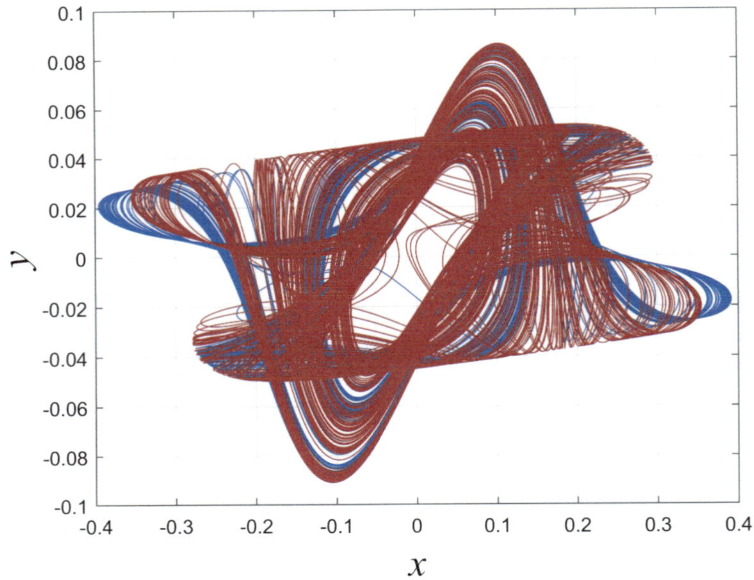

Fig. 3.6 Phase portraits of system (3.6) with $K = 8$ and initial condition $(0, 0, 0, 0.1, 0)$ [2]

$$\begin{cases} \dot{x}_1 = a\,(x_2 - x_1)\,, \\ \dot{x}_2 = bx_1 \sin(x_3), \\ \dot{x}_3 = c - x_1 x_2, \end{cases} \tag{3.7}$$

where x_1, x_2, x_3 are state variables and a, b, c are positive constants.

In fact, system (3.7) can produce infinitely many coexisting chaotic attractors from different initial conditions, even when the parameter values remain fixed. For example, with $a = 1$, $b = 12$, and $c = 0.40$, numerical simulations reveal that the system can generate infinitely many chaotic attractors along the x_3-axis, from the initial conditions

$$x_0 = \{(x_1(0), x_2(0), x_3(0)) \,|\, x_1(0) = x_2(0) = 1,\ x_3(0) = s\pi\},$$

where $s = 0, \pm 2, \pm 4, \pm 6, \ldots$.

Figure 3.7 illustrates seven coexisting chaotic attractors, P_1–P_7, generated from the initial conditions $(1, 1, 0)$, $(1, 1, 2\pi)$, $(1, 1, 4\pi)$, $(1, 1, 6\pi)$, $(1, 1, 8\pi)$, $(1, 1, 10\pi)$, and $(1, 1, 12\pi)$. Furthermore, additional coexisting attractors can be generated by selecting more initial conditions.

An analog electronic emulator for system (3.7) is presented in Fig. 3.8, utilizing standard components such as resistors, capacitors, TL082 operational amplifiers, and AD633 analog multipliers, all powered by symmetric $\pm 18V$ voltages. A voltage-controlled oscillator is used to implement the sine function. In the diagram, the state variables x_1, x_2, and x_3 are represented by the output voltages of the integrators.

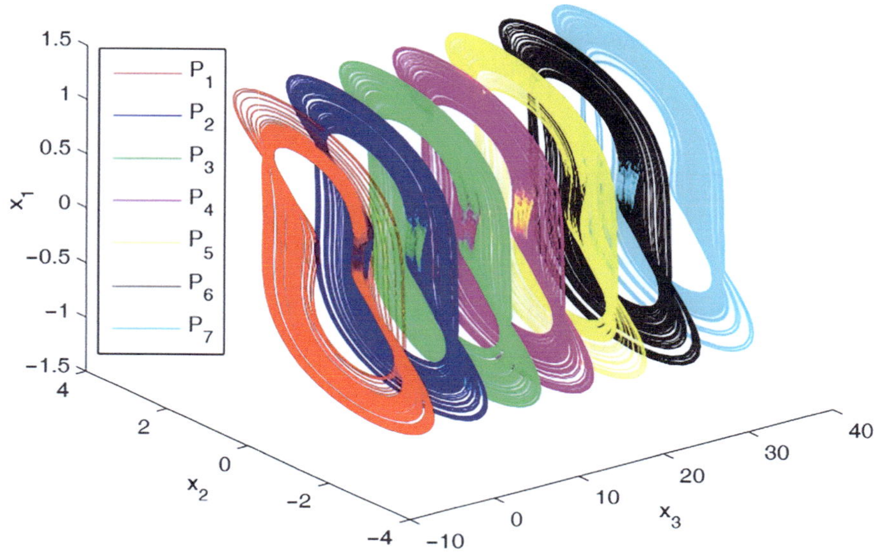

Fig. 3.7 Seven coexisting chaotic attractors P_1–P_7 in system (3.7) generated from initial values $(1, 1, 0)$, $(1, 1, 2\pi)$, $(1, 1, 4\pi)$, $(1, 1, 6\pi)$, $(1, 1, 8\pi)$, $(1, 1, 10\pi)$, and $(1, 1, 12\pi)$. Reprinted from [3], with permission from the IEEE

Corresponding to the circuit diagram, the system equations are derived based on fundamental electrical principles, as follows:

$$\begin{cases} x_1 = \dfrac{R_4}{R_3}\dfrac{1}{RC}\displaystyle\int\left(\dfrac{R}{R_2}x_2 - \dfrac{R}{R_1}x_1\right)dt, \\ x_2 = \dfrac{1}{R_7C_2}\displaystyle\int\left(\dfrac{R_6}{R_5}x_1\sin(x_3)\right)dt, \\ x_3 = \dfrac{R_{11}}{R_{10}}\dfrac{1}{RC_3}\displaystyle\int\left(\dfrac{R}{R_9}V_c - \dfrac{R}{R_8}x_1x_2\right)dt. \end{cases} \tag{3.8}$$

Let $C_1 = C_2 = C_3 = C$, $R_7 = R_8 = R_9 = R$, $R_3 = R_4 = R_{10} = R_{11} = 10\,\text{k}\Omega$, and $R_1 = R_2 = R_a$, with the parameters defined as $a = R/R_a$, $b = R_6/R_5$, $c = V_c$, and the time scaling factor $\tau = t/RC$. Then, it can be verified that the differential equations governing system (3.8) are equivalent to those of system (3.7). With the circuit parameters set to $C = 10\,\text{nF}$, $R = 100\,\text{k}\Omega$, $R_a = 100\,\text{k}\Omega$, $R_5 = 1.47\,\text{k}\Omega$, $R_6 = 10\,\text{k}\Omega$, and $V_c = 0.4\,\text{V}$, a chaotic attractor is observed, as depicted in Fig. 3.9.

By altering the initial conditions, coexisting multiple attractors can be generated. In Fig. 3.10, up to six coexisting chaotic attractors and six coexisting limit cycles are observed for two different sets of parameters: $R_5 = 1.47\,\text{k}\Omega$ with $V_c = 0.4\,\text{V}$ and $R_5 = 1\,\text{k}\Omega$ with $V_c = 0.7\,\text{V}$. The initial conditions for the capacitor voltages C_1, C_2,

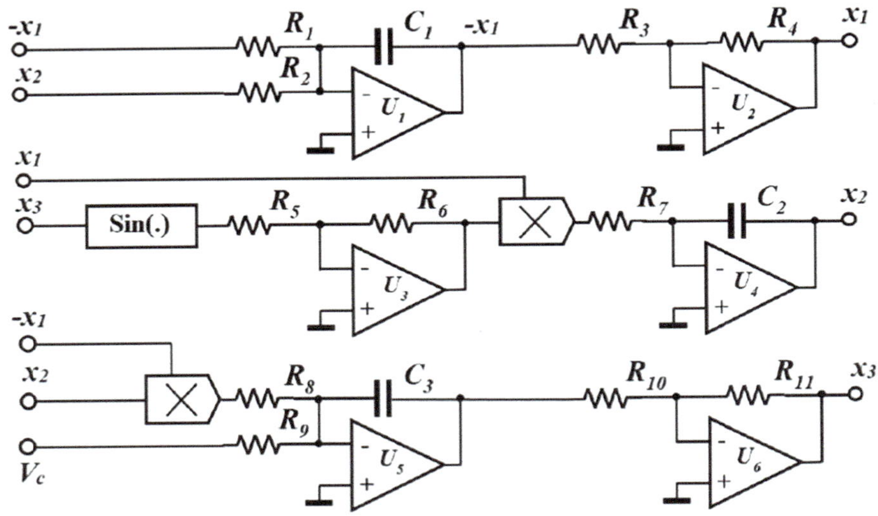

Fig. 3.8 Circuit schematic diagram of system (3.7). Reprinted from [3], with permission from the IEEE

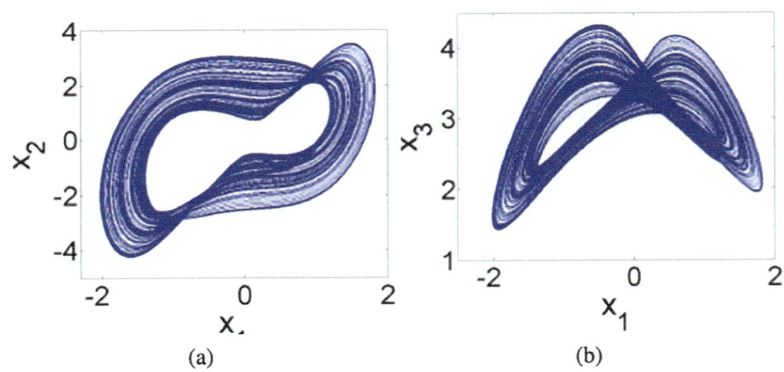

Fig. 3.9 Phase portraits of system (3.7) obtained from the circuit simulation in PSPICE. **a** x_1-x_2 plane. **b** x_1-x_3 plane. Reprinted from [3], with permission from the IEEE

and C_3 are chosen as $(1, 1, -6\pi)$, $(1, 1, -4\pi)$, $(1, 1, -2\pi)$, $(1, 1, 0)$, $(1, 1, 2\pi)$, and $(1, 1, 4)$.

The system (3.7) can be realized using different microcontrollers. A practical implementation with an Arduino Mega 2560 board connected to R-2R DAC is presented in Fig. 3.11.

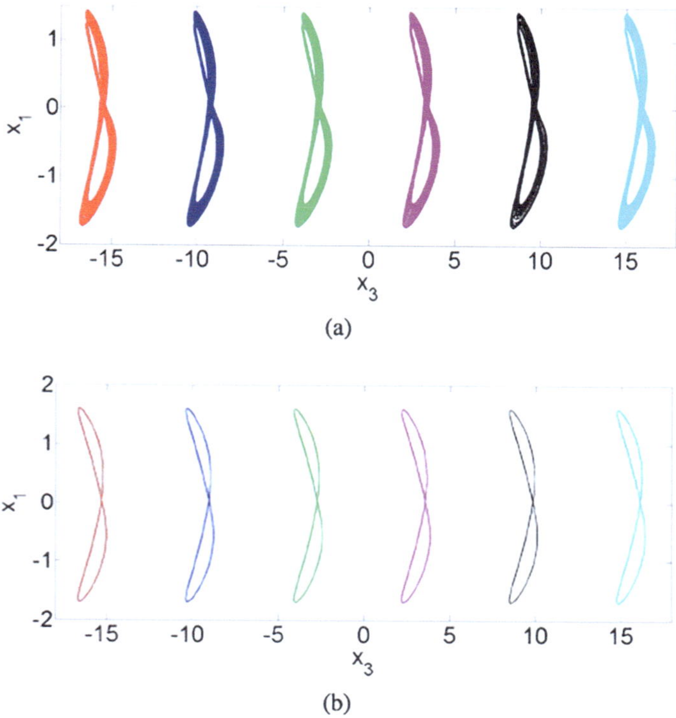

Fig. 3.10 Circuit simulation in PSPICE. **a** Six coexisting chaotic attractors. **b** Six coexisting limit cycles. Reprinted from [3], with permission from the IEEE

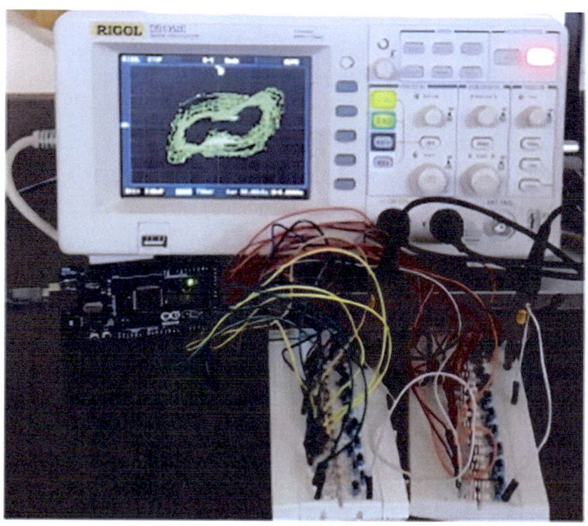

Fig. 3.11 Experimental setup for the digital implementation of system (3.7) with an Arduino Mega 2560 board. Reprinted from [3], with permission from the IEEE

3.1.4 Hyperbolic Tangent Function Method

In this section, the hyperbolic tangent function method is introduced for generating n-scroll chaotic attractors.

Chen et al. [4] developed a chaotic model using some hyperbolic tangent functions to generate multi-directional multi-scroll attractors. Further analysis of the system's dynamics confirmed various types of chaotic attractors, including m-scroll, $m \times n$-grid scroll, and $m \times n \times l$-grid scroll attractors. These findings offer valuable insights into the complex behaviors and intrinsic structures of multi-scroll chaotic systems.

Specifically, the system is described by

$$\begin{cases} \dot{x} = y, \\ \dot{y} = z - y, \\ \dot{z} = -by - \alpha f_M(x), \end{cases} \tag{3.9}$$

where α and b are positive constants, and $f_M(x) = x - g_M(x)$ with

$$g_M(x) = p\left[\sum_{i=0}^{M} \tanh(n(x + 2\pi)) + \sum_{i=0}^{M} \tanh(n(x - 2\pi)) - \tanh(nx)\right], \tag{3.10}$$

for generating $2M + 2$-scroll attractors, or with

$$g_M(x) = p\left[\sum_{i=0}^{M} \tanh(n(x + 1 + 2\pi)) + \sum_{i=0}^{M} \tanh(n(x - 1 - 2\pi))\right], \tag{3.11}$$

for generating $2M + 3$-scroll attractors. Here, the constants $p > 0$ and $n > 0$ are positive, M is a nonnegative integer, and the hyperbolic tangent function is defined as $\tanh(x) = (e^x - e^{-x})/(e^x + e^{-x})$.

The hyperbolic tangent function $\tanh(x)$ and its series expansion given by Eq. (3.10), with $p = 5, n = 30$, and $M = 1$ in Eq. (3.11), are used to generate the chaotic dynamics. By applying system (3.9) along with either Eq. (3.10) or Eq. (3.11), attractors with either an even or an odd number of scrolls can be generated. For example, with parameters $a = 1,\ b = 0.7,\ p = 5,\ n = 30$, and $M = 4$, the system (3.9) produces 10-scroll chaotic attractors, as shown in Fig. 3.12.

For simplicity of discussions, in the following, only the system (3.9) with Eq. (3.10) will be considered.

Chen et al. [4] introduced grid multi-scroll chaotic attractors in both the x- and y-directions. To extend the scrolls along the y-direction, an additional function $f_N(y)$ is incorporated into system (3.9), resulting in the following formulation:

$$\begin{cases} \dot{x} = f_N(y), \\ \dot{y} = z - f_N(y), \\ \dot{z} = -bf_N(y) - \alpha f_M(x), \end{cases} \tag{3.12}$$

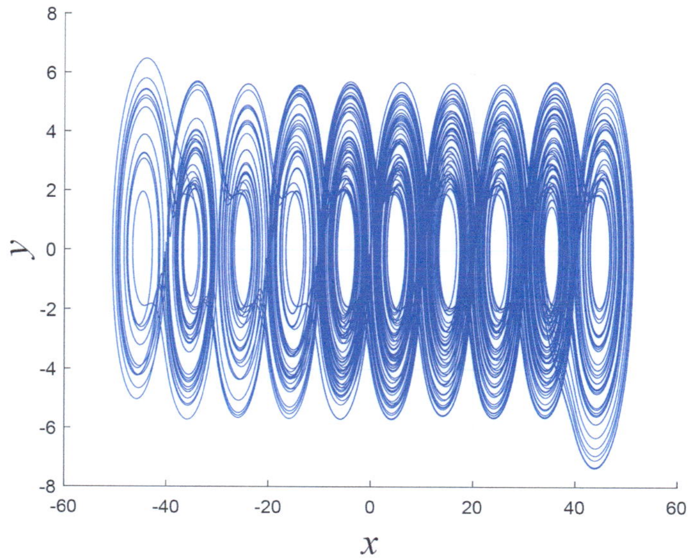

Fig. 3.12 10-scroll chaotic attractors of system (3.9) with Eq. (3.10) on the x-y plane [4]

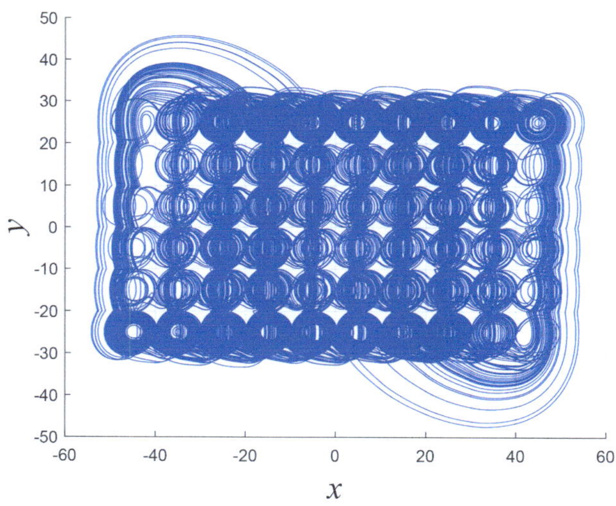

Fig. 3.13 10×5-scroll chaotic attractor on the x-y plane [4]

where $f_M(x)$ and $f_N(y)$ are defined by Eq. (3.10) or Eq. (3.11).

By using system (3.12) along with the functions $f_M(x)$ and $f_N(y)$, one is able to generate 10×5-scroll attractors with the parameters $a = 1$, $b = 0.7$, $p = 5$, and $n = 30$, as shown in Fig. 3.13.

To extend the scrolls along the z-direction, an additional nonlinear function $f_L(z)$ is introduced into the second equation of system (3.9), resulting in the following

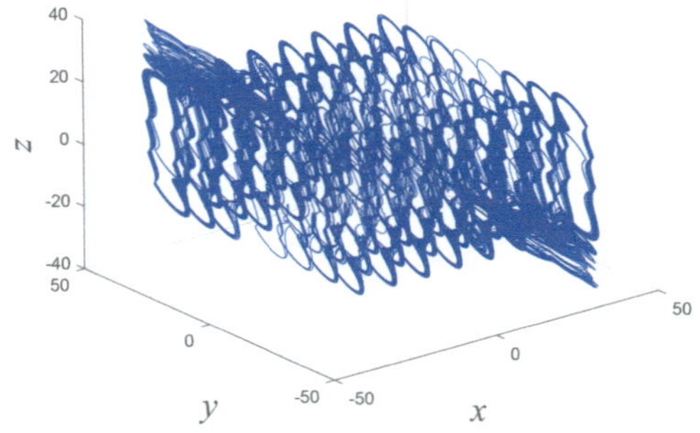

Fig. 3.14 $8 \times 8 \times 5$-scroll chaotic attractor on the x-y plane [4]

modified system:

$$\begin{cases} \dot{x} = f_N(y), \\ \dot{y} = f_L(y) - f_N(y), \\ \dot{z} = -bf_N(y) - \alpha f_M(x), \end{cases} \tag{3.13}$$

where α, b are positive constants, and $f_M(x)$, $f_N(y)$, and $f_L(z)$ are defined by Eq. (3.10) or Eq. (3.11).

With $a = 1, b = 0.7, p = 5, n = 30, M = N = 3$, and $L = 1$, an $8 \times 8 \times 5$-grid scroll chaotic attractor can be generated in the x-y-z space, as shown in Fig. 3.14.

3.2 Nonlinear Transconductor Method

In this section, the nonlinear transconductor approach [5, 6] is introduced for generating n-scroll chaotic attractors.

Ozoguz et al. [5] designed a generator for creating n-scroll chaotic attractors, where the nonlinearity is represented by a smooth hyperbolic tangent function, as described by

$$\begin{pmatrix} \dot{x} \\ \dot{y} \\ \dot{z} \end{pmatrix} = \begin{pmatrix} 0 & 1 & 0 \\ 0 & 0 & 1 \\ 0 & -a & -a \end{pmatrix} \begin{pmatrix} x \\ y \\ z \end{pmatrix} + \begin{pmatrix} 0 \\ 0 \\ -af(x) \end{pmatrix}, \tag{3.14}$$

where

$$f(x) = \sum_{j=-N}^{M} (-1)^{j-1} \tanh k \left(x - o_j\right), \tag{3.15}$$

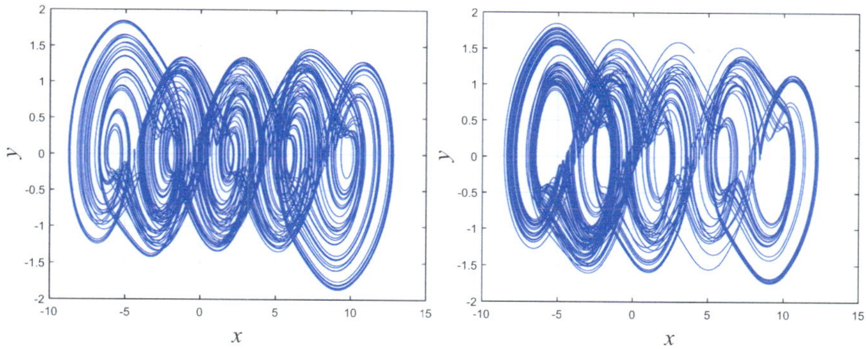

Fig. 3.15 5-scroll chaotic attractors [5]. **a** 5-scroll attractor generated using the hyperbolic tangent function. **b** 5-scroll attractor generated using the sign function

and M, N are odd integers. System (3.14) can generate an $(M + N + 2)/2$-scroll chaotic attractor.

With $a = 0.25$, $M = 5$, $N = 3$, $k = 2$, and $o_j = 2j$, $j \in [-N, M]$, system (3.14) has a 5-scroll chaotic attractor, as shown in Fig. 3.15a.

Figure 3.16a presents the circuit diagram for generating n-scroll chaotic attractors. It consists of a unity-gain voltage buffer, a current feedback operational amplifier (CFOA), and a transconductor $g(V_{C1})$. In Fig. 3.16b, the transconductor is realized by alternating the bipolar D-P cells.

It is noted that the smooth hyperbolic tangent function $f(x)$ in Eq. (3.14) can be replaced by a switching sign function [6], as follows:

$$f(x) = \sum_{j=-N}^{M} \operatorname{sgn} k\left(x - o_j\right). \tag{3.16}$$

With $a = 0.25$, $M = 5$, $N = 3$, $k = 2$, and $o_j = 2j$, $j \in [-N, M]$, system (3.14) with Eq. (3.15) can generate a 5-scroll chaotic attractor, as shown in Fig. 3.15b.

To implement this square-wave-shaped nonlinearity, the circuit diagram in Fig. 3.16b is modified as shown in Fig. 3.17. In this modified design, a high-gain push-pull inverter made of two MOS transistors is used to create the sign function.

To generate $n \times m$-grid scroll chaotic attractors, Salama et al. [6] further extended system (3.16) to the following model:

$$\begin{pmatrix} \dot{x} \\ \dot{y} \\ \dot{z} \end{pmatrix} = \begin{pmatrix} 0 & 1 & 0 \\ 0 & 0 & 1 \\ 0 & -a & -a \end{pmatrix} \begin{pmatrix} x \\ y \\ z \end{pmatrix} + \begin{pmatrix} -af(y) \\ 0 \\ -af(x) \end{pmatrix}, \tag{3.17}$$

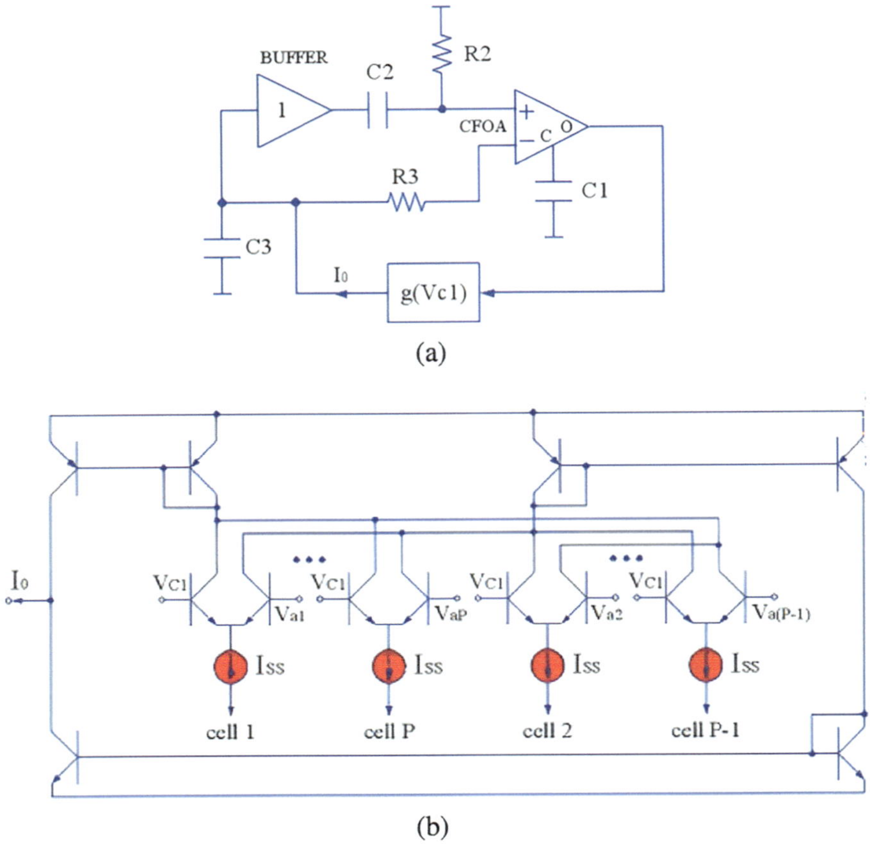

Fig. 3.16 Circuit diagram for generating n-scroll chaotic attractors. **a** Circuit diagram. **b** Realization of $g(V_{C1})$. Reprinted from [7], with permission from World Scientific Pub. Co.

where $f(x)$ and $f(y)$ are defined by Eq. (3.15) or Eq. (3.16). When $a = 0.4$, system (3.17) with the sign function (3.16) can generate a 2×2-grid scroll chaotic attractor, as shown in Fig. 3.18.

In Fig. 3.16, both the CFOA and the buffer are implemented using AD844 operational amplifiers. The NPN transistors are sourced from the LM3046 transistor array, while the PNP transistors are BC557 models. The circuit operates with a $\pm 5V$ power supply. Capacitors are specified to be 3.3 nF, with resistances of $R_2 = 9\,\text{k}\Omega$ and $R_3 = 2.2\,\text{k}\Omega$. The tail current I_{SS} is set to $40\,\mu A$. The control voltages $(V_1, \ldots, V_9)$ are set to $(-500, -380, -200, 0, 200, 380, 500, 640, 800\,\text{mV})$.

Figure 3.19a shows the experimentally observed 5-scroll chaotic attractor generated by the hyperbolic tangent nonlinear transconductor. Similarly, Fig. 3.19b displays the 5-scroll chaotic attractor generated by the sign nonlinear transconductor.

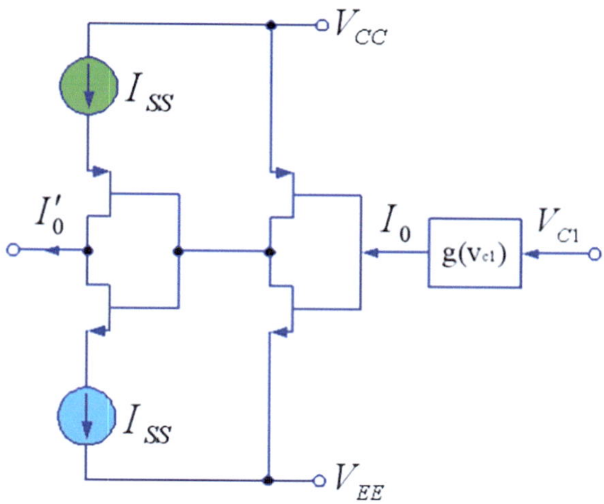

Fig. 3.17 Realization of the square-wave transconductor. Reprinted from [7], with permission from World Scientific Pub. Co.

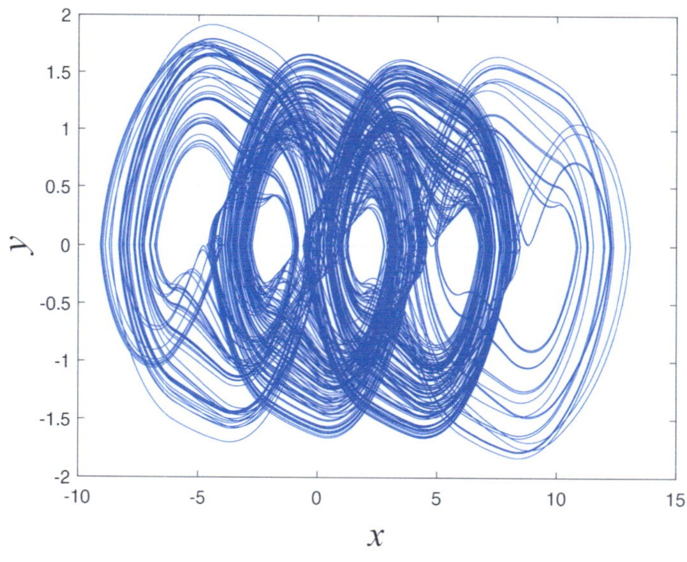

Fig. 3.18 2 × 2-scroll chaotic attractor [6]

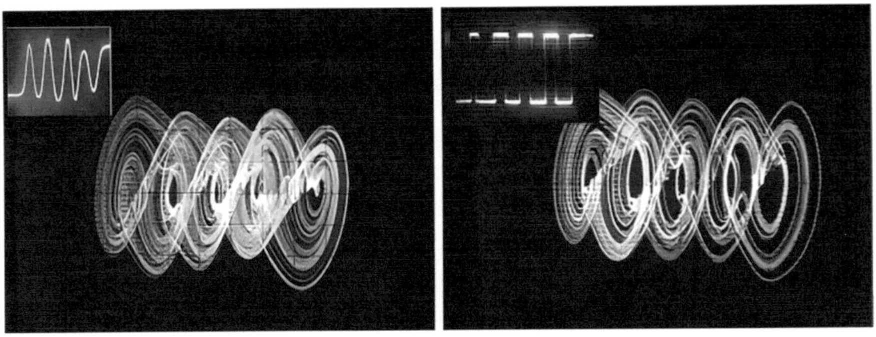

(a) (b)

Fig. 3.19 Experimentally observed n-scroll chaotic attractors (x-axis: 200 mV/div; y-axis: 50 mV/div). **a** $f(x)$ given by Eq. (3.15). **b** $f(x)$ given by Eq. (3.16). Reprinted from [7], with permission from World Scientific Pub. Co.

3.3 General Nonlinear Modulating Function Method

This section presents a generalized nonlinear modulating function approach to generating n-scroll chaotic attractors from a standard jerk circuit. This method provides a flexible framework for designing chaotic systems with any desired number of scrolls, thereby greatly broadening the variety of chaotic behaviors that can be realized by jerk circuits [8–10].

The general form of the jerk circuit is described by

$$\dddot{x} + \beta\ddot{x} + \gamma\dot{x} = f(x), \tag{3.18}$$

where β and γ are real parameters, $f(x)$ represents a nonlinear function, $\dot{x} = dx/d\tau$ is the velocity, $\ddot{x} = d^2x/d\tau^2$ is the acceleration, and $\dddot{x} = d^3x/d\tau^3$ is the jerk.

3.3.1 *Modulating Function Method*

To generate n-scroll chaotic attractors from Eq. (3.18), Yu et al. [9, 10] introduced a swing modulating function based on a double sawtooth wave, represented by

$$f(x) = |F(x)|\,\mathrm{sgn}(x) - x, \tag{3.19}$$

where $F(\cdot)$ can be an n-dimensional autonomous function or a nonautonomous function produced by outer signals, or even a constant in the simplest case. Here, assume that $|F(\cdot)| = |A\sin(ax)|$, where $A, a > 0$ are parameters.

This technique offers an effective way to design multi-scroll chaotic systems with increased complexity.

3.3.2 Adjustable Sawtooth Wave Function Method

Yu et al. [9, 10] also utilized an adjustable sawtooth wave to generate n-scroll chaotic attractors from Eq. (3.18). These attractors can be categorized into two distinct cases, depending on whether the number of scrolls is even or odd, each displaying a unique chaotic attractor with a distinct structure. (1) Even number of scrolls:

$$\begin{aligned} f_1(x) = & A_0 \operatorname{sgn}(x) + \sum_{i=1}^{M} \left[\frac{A_{i-1} + A_i}{2} \operatorname{sgn}\left(x - \frac{2}{B} \sum_{j=0}^{i-1} A_j \right) \right] \\ & + \sum_{i=1}^{M} \left[\frac{A_{i-1} + A_i}{2} \operatorname{sgn}\left(x + \frac{2}{B} \sum_{j=0}^{i-1} A_j \right) \right] - Bx, \end{aligned} \tag{3.20}$$

where all parameters $A_i > 0, i = 0, 1, 2, \ldots$, and $B \in [0.7, 1.2]$, which can generate $(2M + 2)$-scrolls in the attractor, $M = 1, 2, 3, \ldots$.

(2) Odd number of scrolls:

$$\begin{aligned} f_2(x) = & \sum_{i=1}^{M} \left\{ \frac{A_{i-1} + A_i}{2} \times \operatorname{sgn}\left[x - \frac{1}{B} \left(2 \sum_{j=0}^{i-1} A_j - A_0 \right) \right] \right\} \\ & + \sum_{i=1}^{M} \left\{ \frac{A_{i-1} + A_i}{2} \times \operatorname{sgn}\left[x + \frac{1}{B} \left(2 \sum_{j=0}^{i-1} A_j - A_0 \right) \right] \right\} - Bx, \end{aligned} \tag{3.21}$$

where all parameters $A_i > 0, i = 0, 1, 2, \ldots$, and $B \in [0.7, 1.2]$, which can generate $(2M + 1)$-scrolls in the attractor, $M = 1, 2, 3, \ldots$.

3.3.3 Adjustable Triangular Wave Function Method

Yu et al. [9, 10] furthermore introduced a PWL function with tunable breakpoints and slopes to generate n-scroll chaotic attractors from Eq. (3.18). The approach differentiates between two cases: one for even scroll numbers and the other for odd numbers, with each case producing a chaotic attractor with a distinct structure.
(1) Even number of scrolls:

$$f_1(x) = \sum_{n=-M}^{M} \frac{A}{2\alpha_n} \left[\left| \left(x - \frac{2A_n}{B} \right) + \alpha_n \right| - \left| \left(x - \frac{2A_n}{B} \right) - \alpha_n \right| \right] - Bx, \tag{3.22}$$

where parameters $A > 0$, $0.8 \leq B \leq 1.2$, and $\alpha_n \in (0, 3A/10B]$, $n = 0, \pm 1, \ldots, \pm M$, with $M = 1, 2, \ldots$, which can create a $(2M + 1)$-scroll chaotic attractor.
(2) Odd number of scrolls:

$$f_2(x) = \sum_{n=-M, n\neq 0}^{M} \frac{A}{2\alpha_n} \left[\left| \left(x - \frac{A}{B}\left(2n - \frac{|n|}{n} \right) \right) + \alpha_n \right| \right. \\ \left. - \left| \left(x - \frac{A}{B}\left(2n - \frac{|n|}{n} \right) \right) - \alpha_n \right| \right] - Bx, \tag{3.23}$$

where parameters $A > 0$, $0.8 \leq B \leq 1.2$, and $\alpha_n \in (0, 3A/10B]$, $n = 0, \pm 1, \ldots, \pm M$, with $M = 1, 2, \ldots$, which can create a $(2M + 2)$-scroll chaotic attractor.

3.3.4 Adjustable Transconductor Wave Function Method

Yu et al. [9, 10] later developed a technique using an adjustable transconductor wave function for generating n-scroll chaotic attractors from Eq. (3.18). Similarly to the previously discussed methods, the approach considers two distinct cases: one for even numbers of scrolls and the other for odd numbers, with each case leading to a unique chaotic attractor in a distinct structure.
(1) Even number of scrolls:

$$f_1(x) = \sum_{n=-M}^{M} A \tanh \left[C_n \left(x - \frac{2nA}{B} \right) \right] - Bx, \tag{3.24}$$

where A, B, C_n are adjustable parameters and $M \in N$.
(2) Odd number of scrolls:

$$f_2(x) = \sum_{n=-M, n\neq 0}^{M} A \tanh \left[C_n \left(x - \left(2n - \frac{|n|}{n} \right) \frac{A}{B} \right) \right] - Bx, \tag{3.25}$$

where A, B, C_n are adjustable parameters and $M \in N$.

For clarity, define several types of multiscroll attractors. Type I: multiscroll attractors, where the scroll sizes gradually increase from the center to both sides; Type II: multiscroll attractors, where the scroll sizes gradually decrease from the center to both sides; Type III: multiscroll attractors with alternating small and large scrolls; Type IV: multiscroll attractors, where all scrolls are of the same size.

Yu et al. [9, 10] designed a circuit diagram to physically realize various multiscroll chaotic attractors. Figure 3.20 illustrates the circuitry, which consists of five functional components. Part I: integrator N_0; Part II: sawtooth or triangular wave generator N_1; Part III: buffer N_2; Part IV: switch linkages, including K, K_{10}, K_{11}, K_{12}, K_{13}, K_{14}, and K_{15}; Part V: voltage-current conversion resistors $R_{10}, \ldots, R_{15}$.

Let $R_{3j} = 1k\Omega$ and $R_{2j} = 200k\Omega$, $0 \leq j \leq 5$. When K is switched on, N_1 generates a triangular wave; when K is switched off, N_1 creates a sawtooth wave. All experimental parameters are given in [10]. Figure 3.21 shows the experimental observations of 12-scroll chaotic attractors with various sizes.

Furthermore, the systematic nonlinear modulation function approach presented here allows for customizing various characteristics of n-scroll chaotic attractors, such

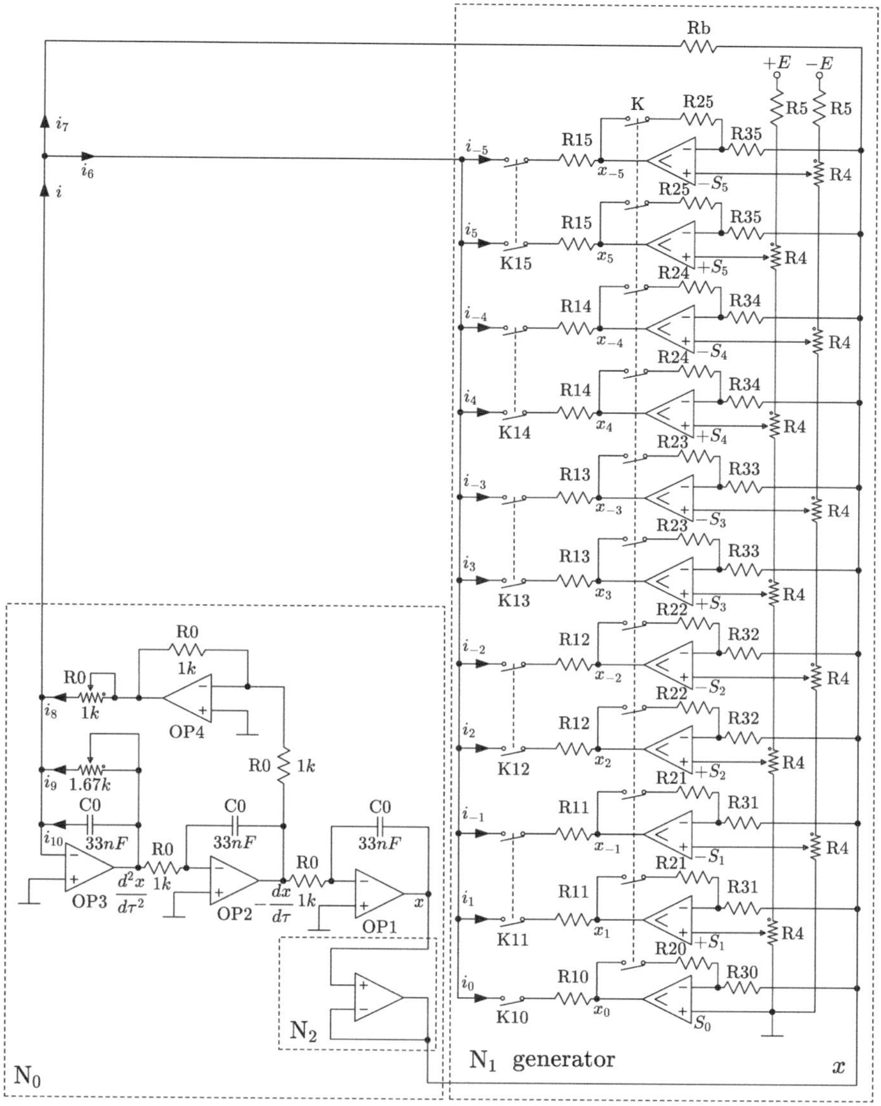

Fig. 3.20 Circuit diagram for generating n-scroll attractors. Reprinted from [10], with permission from the IEEE

as the swings, widths, slopes, breakpoints, equilibria, and shapes. This flexibility is made possible by utilizing adjustable sawtooth waves, triangular waveforms, and transconductor wave functions, which enables precise design and control over chaotic dynamics.

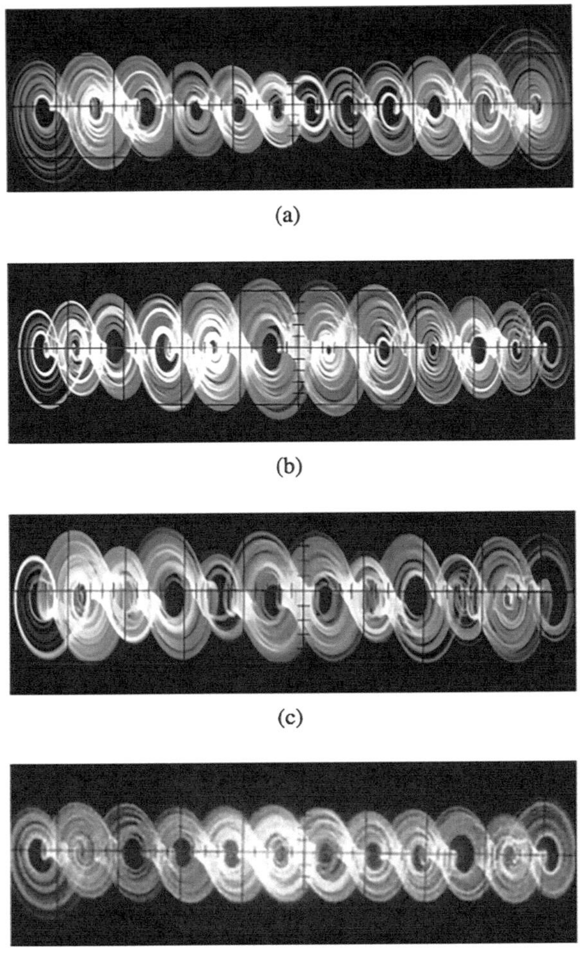

Fig. 3.21 Experimental observations of 12-scroll chaotic attractors. **a** Type I, $x = 1.1\,\text{V/div}$, $y = 0.4\,\text{V/div}$. **b** Type II, $x = 1.05\,\text{V/div}$, $y = 0.5\,\text{V/div}$. **c** Type III, $x = 0.8\,\text{V/div}$, $y = 0.4\,\text{V/div}$. **d** Type IV, $x = 0.66\,\text{V/div}$, $y = 0.33\,\text{V/div}$. Reprinted from [10], with permission from the IEEE

3.4 Generating Multi-scroll Chaotic Attractors with Exponential Functions

Nonlinear functions play a crucial role in the generation of multi-scroll chaotic attractors. Building on this idea, Gao et al. [11] introduced another method for creating multi-scroll chaotic attractors, by incorporating a piecewise exponential function as the nonlinear component into a jerk system and extending it to a multi-segment exponential function series. This approach enables the generation of chaotic attractors with a wide range of scroll numbers.

The piecewise exponential function is defined as

$$f(x) = \begin{cases} 1 - e^{-x}, & \text{if } x \geq 0, \\ -1 + e^{x}, & \text{if } x < 0. \end{cases} \tag{3.26}$$

In this approach, the base e of the natural exponential function can be substituted with any other positive constant greater than 1. Also, Eq. (3.26) can be extended into two different multi-segment exponential function series, defined respectively by

$$F(x) = f(x) + \sum_{n=1}^{r} f(x \pm 2n\tau), \tag{3.27}$$

or

$$F(x) = \sum_{n=1}^{r} f(x \pm (2n-1)\tau), \tag{3.28}$$

where the parameter r is a positive integer and τ is a positive real number. Moreover, the function $f(x)$ is a piecewise exponential function defined in Eq. (3.26). By varying the values of r, Eqs. (3.27) and (3.28) can be used to construct different segmented function series.

When applied to a generalized chaotic system, this approach increases the number of index-2 equilibrium points, thereby allowing control over the number of scrolls. Specifically, the number of scrolls is determined by the above two equations: Using Eq. (3.27) results in an even number of scrolls, while using Eq. (3.28) generates an odd number.

Building upon the jerk system

$$\begin{cases} \dot{x} = y, \\ \dot{y} = z, \\ \dot{z} = -a(x + y + z) + bf(x), \end{cases} \tag{3.29}$$

where a and b are adjustable parameters, and $f(x)$ can either be the exponential function in Eq. (3.26), or the exponential functions in Eqs. (3.27) and (3.28). By varying the values of a, b and $f(x)$, Eq. (3.29) can generate attractors with different numbers of scrolls.

(1) Double-scroll attractors:
When $f(x)$ is defined by Eq. (3.26), and the parameters are set as $a = 0.5$ and $b = 20$, Eq. (3.29) generates a double-scroll chaotic attractor, as shown in Fig. 3.22a.

(2) Even-scroll attractors:
When $f(x)$ is defined by Eq. (3.27) and suitable parameters are chosen, Eq. (3.29) can generate a $(2r+2)$-scroll attractor. For example, let $b = a\tau$, $a = 0.5$, $\tau = 100$, and $r = 2$. Then, Eq. (3.29) can generate a 6-scroll attractor, as shown in Fig. 3.22b.

(3) Odd-scroll attractors:

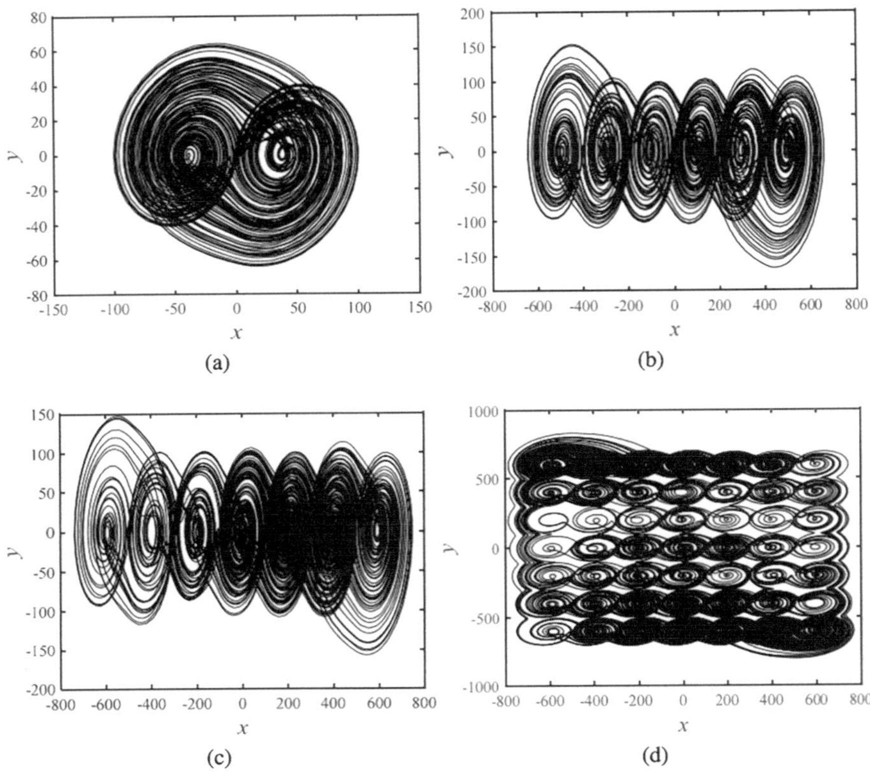

Fig. 3.22 Multi-scroll chaotic attractors. **a** Double-scroll attractor. **b** 6-scroll attractor. **c** 7-scroll attractor. **d** 2-D grid scroll attractor. Reprinted from [11], with permission from the IEEE

Similarly, when $f(x)$ is defined by Eq. (3.28) and suitable parameters are chosen, Eq. (3.29) can generate a $(2r+1)$-scroll attractor. Let $b = a\tau$, $a = 0.5$, $\tau = 100$, and $r = 3$. Then, a 7-scroll chaotic attractor can be generated, as shown in Fig. 3.22c.

(4) 2-D Grid Scroll Attractors:
When $f(x)$ is defined by Eq. (3.27) or Eq.(3.28), and two of the directions x, y or z are selected to extend the phase space of Eq. (3.29), 2-D grid scroll attractors can be generated. For instance, along the x- and y-directions, and setting the parameters to match those of a 7-scroll attractor, a grid 7×7-scroll attractor can be generated, as shown in Fig. 3.22d.

3.5 Conclusion

This chapter studies the nonlinear modulation function methods as an effective approach to generating n-scroll chaotic attractors. The flexibility and scalability of these techniques are demonstrated via producing diverse chaotic attractors by utilizing various nonlinearities, including trigonometric functions, nonlinear transconductors, and piecewise exponential functions.

Practical implementations, such as memristor-based circuits and electronic prototypes, verify the feasibility and effectiveness of the nonlinear modulation function methods. Numerical simulations and experimental demonstrations confirm the ability of the techniques to customize attractor properties, such as the scroll count, symmetry and topology. These methods significantly advance the design and implementation of multiscroll chaotic systems, with promising potentials for applications in secure communications, signal processing and nonlinear control technologies.

References

1. Tang, W.K.S., Zhong, G.Q., Chen, G., Man, K.F.: Generation of n-scroll attractors via sine Function. IEEE Trans. Circuits Syst.-I **48**, 1369–1372 (2001)
2. Sun, J., Zhao, X., Fang, J., Wang, Y.: Autonomous memristor chaotic systems of infinite chaotic attractors and circuitry realization. Nonlinear Dyn. **94**, 2879–2887 (2018)
3. Lai, Q., Kuate, P.D.K., Liu, F., Iu, H.H.C.: An extremely simple chaotic system with infinitely many coexisting attractors. IEEE Trans. Circuits Syst.-II **67**, 1129–1133 (2019)
4. Chen, Z., Wen, G., Zhou, H., Chen, J.: Generation of grid multi-scroll chaotic attractors via hyperbolic tangent function series. Optik **130**, 594–600 (2017)
5. Ozoguz, S., Elwakil, A.S., Salama, K.N.: n-scroll chaos generator using nonlinear transconductor. Electron. Lett. **38**, 685–686 (2002)
6. Salama, K.N., Ozoguz, S., Elwakil, A.S.: Generation of n-scroll chaos using nonlinear Transconductors. In: IEEE International Symposium on Circuits and Systems, vol. 3, May 25–28, Bangkok, Thailand, pp. 176–179 (2003)
7. Lü, J., Chen, G.: Generating multiscroll chaotic attractors: theories, methods and applications. Int. J. Bifur. Chaos **16**, 775–858 (2006)
8. Linz, S.J., Sprott, J.C.: Elementary chaotic flow. Phys. Lett. A **259**, 240–245 (1999)
9. Yu, S., Lü, J., Leung, H., Chen, G.: N-scroll chaotic attractors from a general jerk circuit. In: IEEE International Symposium on Circuits and Systems, May 23–26, Kobe, Japan, pp. 1473–1476 (2005)
10. Yu, S., Lü, J., Leung, H., Chen, G.: Design and implementation of n-scroll chaotic attractors from a general jerk circuit. IEEE Trans. Circuits Syst.-I **52**, 1459–1476 (2005)
11. Gao, Y., Li, Q., Li, X., Qian, G.: Construction of multi-scroll chaotic attractors with exponential function. In: IEEE International Conference on Digital Signal Processing, pp. 542–544 (2016)

Chapter 4
Generating Multidirectional Multiscroll Chaotic Attractors

Abstract This chapter introduces some methods for designing chaotic systems with multidirectional multiscroll chaotic attractors using basic circuits such as step circuits, hysteresis circuits and saturated circuits. These circuits are used in appropriate combinations to create 1-D, 2-D, and 3-D chaotic attractors with scalable complexity. Designs are verified through numerical simulations and experimental circuit implementation. Applications include the generation of n-scroll chaotic attractors and grid-scroll chaotic attractors in multidimensional systems. These findings provide significant advancements in chaos theory and offer practical tools for applications in secure communications, nonlinear control and complex system modeling.

4.1 Step Function Method

This section explores different techniques for generating multidirectional multiscroll chaotic attractors through the use of a step circuit.

4.1.1 A Step Function Circuit

Yalcin et al. [1] developed a family of scroll and grid-scroll attractors utilizing the step circuit. This family includes 1-D n-scroll, 2-D $n \times m$-grid scroll, and 3-D $n \times m \times l$-grid scroll chaotic attractors. The state equations governing this family of systems are given as follows:

$$\dot{X} = AX + B\sigma\left(CX\right). \tag{4.1}$$

Here, $X = (x, y, z)^T$, with

$$A = \begin{pmatrix} 0 & 1 & 0 \\ 0 & 0 & 1 \\ -a & -a & -a \end{pmatrix}, \quad B = \begin{pmatrix} b_y & 0 & 0 \\ 0 & b_z & 0 \\ 0 & 0 & a \end{pmatrix}, \quad C = \begin{pmatrix} 0 & 1 & 0 \\ 0 & 0 & 1 \\ 1 & 0 & 0 \end{pmatrix},$$

J. Lü et al., *Chaotic Systems with Multiscroll Attractors*,
https://doi.org/10.1007/978-3-032-09854-2_4

in which a, b_x, b_y, b_z are real parameters, and $\sigma(\cdot)$ will be specified below.

There are three different cases.

(1) 1-D n-scroll chaotic attractors:

$$\begin{cases} b_y = b_z = 0, \\ \sigma(\cdot) = \begin{pmatrix} 0 \\ 0 \\ f_1(\cdot) \end{pmatrix}, \end{cases}$$

in which

$$f_1(x) = \sum_{i=1}^{M_x} g_{\frac{(-2i+1)}{2}}(x) + \sum_{i=1}^{N_x} g_{\frac{(2i-1)}{2}}(x), \quad g_\theta(\xi) = \begin{cases} 1, & \text{if } \xi \ge 0, \theta > 0, \\ 0, & \text{if } \xi < 0, \theta > 0, \\ 0, & \text{if } \xi \ge 0, \theta < 0, \\ -1, & \text{if } \xi < 0, \theta < 0, \end{cases}$$

belonging to the sector [0, 2].

(2) 2-D $n \times m$-grid scroll chaotic attractors:

$$\begin{cases} b_y = -1, b_z = 0, \\ \sigma(\cdot) = \begin{pmatrix} f_1(\cdot) \\ 0 \\ f_2(\cdot) \end{pmatrix}, \end{cases}$$

where $f_1(\cdot)$ is defined as above, $f_2(x) = \sum_{i=1}^{m-1} \beta g_{pi}(x)$, with $\beta = M_y + N_y + 1$ and $p_i = M_y + 0.5(i-1)(M_y + N_y + 1)$, which belongs to the sector $\left[0, \left(M_y + N_y + 1\right) / \left(M_y + 0.5\right)\right]$ for some constants M_y, N_y.

(3) 3-D $n \times m \times l$-grid scroll chaotic attractors:

$$\begin{cases} b_y = -1, b_z = 0, \\ \sigma(\cdot) = \begin{pmatrix} f_1(\cdot) \\ f_1(\cdot) \\ f_3(\cdot) \end{pmatrix}, \end{cases}$$

where $f_1(\cdot)$ is defined as above, $f_3(x) = \sum_{i=1}^{k-1} \gamma g_{n_l}(x)$, with $\gamma = \rho + \xi + 1$ and $n_l = \rho + 0.5 + (l-1)(\rho + \xi + 1)$, in which

$$\rho = \left| \min_{i,j} \left\{ u_i^{eq,y} + u_j^{eq,z} \right\} \right|,$$

$$\xi = \left| \max_{i,j} \left\{ u_i^{eq,y} + u_j^{eq,z} \right\} \right|,$$

and $u_i^{eq,y}$, $u_j^{eq,z}$ are the vectors for the y and z variables, which belong to the sector $[0, ((\xi + \rho + 1) / (\rho + 0.5))]$.

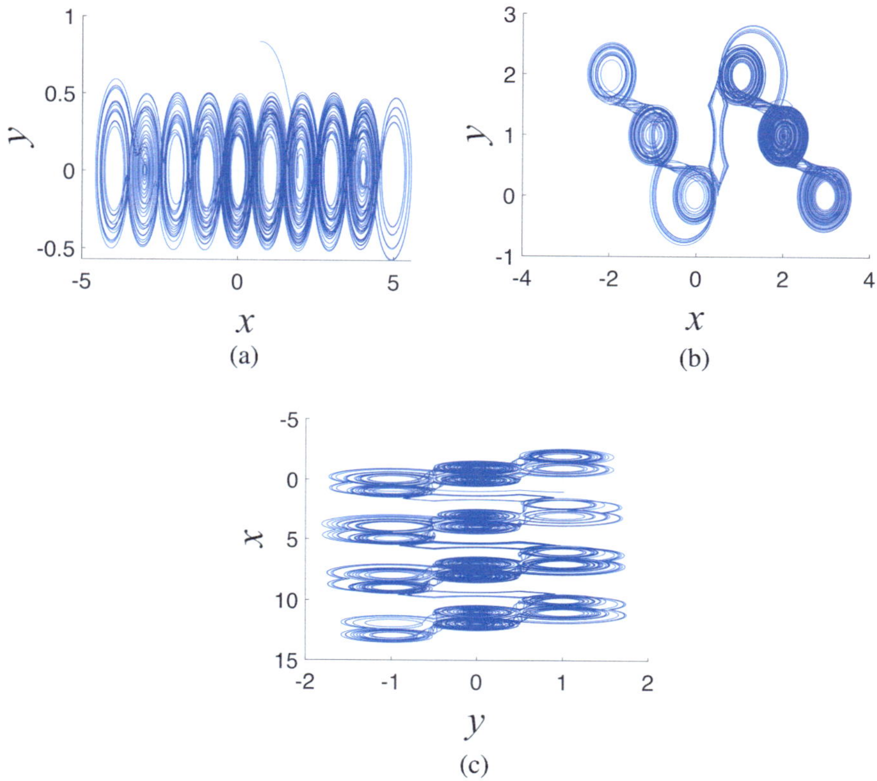

Fig. 4.1 Phase portraits of multiscroll chaotic attractors generated by a step circuit. **a** 1-D 10-scroll attractor. **b** 2-D 2 × 3-grid scroll attractor. **c** 3-D 4 × 3 × 2-grid scroll attractor

Figure 4.1a presents a 10-scroll chaotic attractor with the parameters $M_x = 0$, $N_x = 1$, $a = 0.8$, $M = 4$, $N = 5$, and initial state $X_0 = [0.6721, 0.8381, 0.0196]$. Figure 4.1b [1] displays a 2×3 grid-scroll chaotic attractor, where $M_y = 0$, $N_y = 2$ and $m = 2$. Finally, Fig. 4.1c depicts a $4 \times 3 \times 2$ grid-scroll chaotic attractor, where $M_y = 1$, $N_y = 1$, $M_x = 0$, $N_x = 1$, and $k = 4$.

Yalcin et al. [1] also developed a circuit diagram to experimentally demonstrate the multi-scroll chaotic attractors. Figure 4.2 is the layout of the circuit. The subcircuits enclosed by dashed lines in the upper-left, upper-right, and lower sections generate multi-scroll attractors in the x, y, and z directions, respectively. CFOA are implemented using AD844, whereas LM311 comparators are employed. The voltage sources are set to $\pm 15V$, with $C_1 = C_2 = C_3 = 1\text{nF}$ and $R_1 = 5.1\,\text{k}\Omega$. When $R_2 = R_4 = 8\,\text{k}\Omega$ and $R_{x1} = R_{x2} = R_{x3} = R_{x4} = 70\,\text{k}\Omega$, the circuit can be modified by removing the subcircuits within the dashed lines at the upper-right and lower sections, meanwhile adding two additional comparators in the upper-left subcircuit, which can generate a 5-scroll chaotic attractor, as shown in Fig. 4.3a.

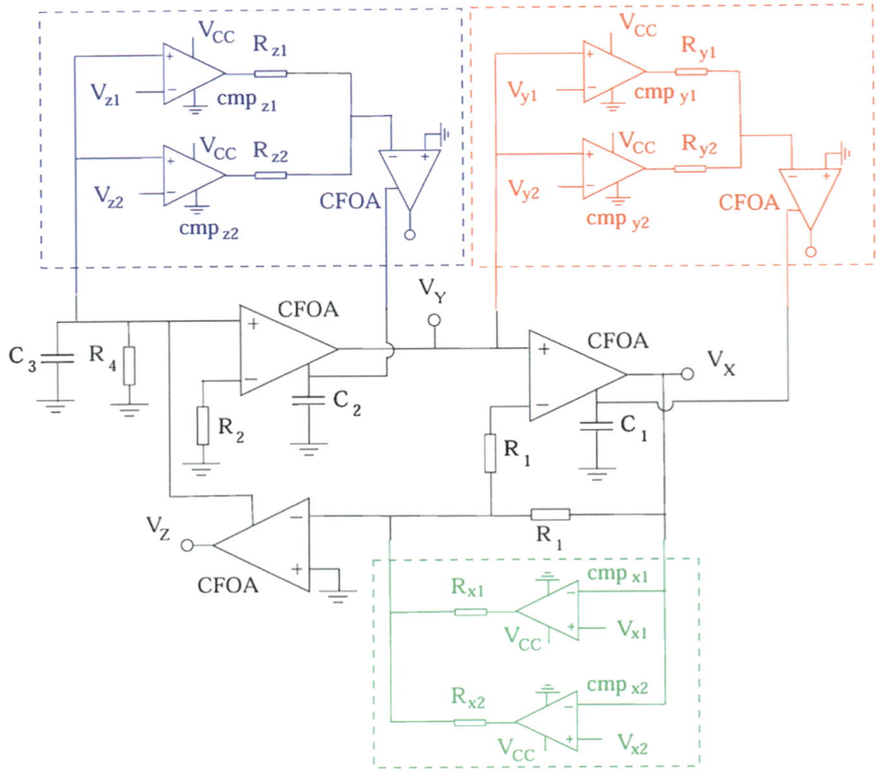

Fig. 4.2 Circuit diagram for realizing multiscroll attractors. Reprinted from [1], with permission from World Scientific Pub. Co.

With $R_2 = R_4 = 12\,\text{k}\Omega$, $R_{x1} = 28\,\text{k}\Omega$, $R_{x2} = 30\,\text{k}\Omega$, $R_{y1} = 90\,\text{k}\Omega$, and $R_{y2} = 80\,\text{k}\Omega$, the circuit is modified by removing the subcircuit at the lower section and adding one comparator in the upper-right subcircuit, which generates a 3×3-scroll chaotic attractors, as shown in Fig. 4.3b.

Lastly, with $R_2 = R_4 = 8.3\,\text{k}\Omega$, $R_{x1} = 19\,\text{k}\Omega$, $R_{y1} = 47\,\text{k}\Omega$, and $R_{z1} = 50\,\text{k}\Omega$, the circuit is modified by removing the comparators comp_{x2}, comp_{y2}, and comp_{z2}, resulting in a $2 \times 2 \times 2$-scroll chaotic attractors, as shown in Fig. 4.3c.

4.1.2 A 3-D Step Function System

Ma et al. [2] introduced a chaotic system to generate multiscroll chaotic attractors, using the step function as the nonlinear element in the system.

This is a 3-D autonomous system, with dimensionless normalized state equations

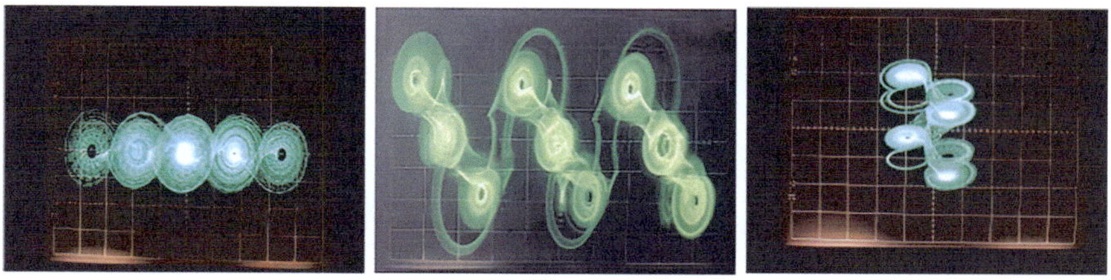

(a) (b) (c)

Fig. 4.3 Experimental observations of multiscroll chaotic attractors generated by a step circuit. **a** 1-D 5-scroll ($a = 0.64$, $x = 0.5\,\text{V/div}$, $y = 0.5\,\text{V/div}$). **b** 2-D 3×3-grid scroll ($x = 1\,\text{V/div}$, $y = 0.5\,\text{V/div}$). **c** 3-D $2 \times 2 \times 2$-grid scroll ($x = 1\,\text{V/div}$, $y = 2\,\text{V/div}$). Reprinted from [1], with permission from World Scientific Pub. Co.

$$\begin{cases} \dot{x} = A\left[y - bx - 0.5\,(a - b)\,f\,(x)\right], \\ \dot{y} = x - y + z, \\ \dot{z} = -By, \end{cases} \tag{4.2}$$

where the step function is

$$f\,(x) = \sum_{i=-N}^{N} \operatorname{sgn}\,(x - 2i). \tag{4.3}$$

With $a = -3/7, b = 3/7, A = 9$, and $B = 15$, the system is capable of producing multiscroll chaotic attractors. The step function features a simple structure, allowing for the generation of multiscroll chaotic attractors by adjusting the parameter N to control the number of scrolls after selecting other appropriate system parameters. Notably, the system can generate 8-scroll chaotic attractors when $N = 3$.

For the system described by Eq. (4.2), the Runge–Kutta integration method is employed over the time interval $T = (0, 800)$, with initial conditions $\{x, y, z\} = \{0.02, 0.01, 0.03\}$. The simulation results, shown in Fig. 4.4a–d, are the phase diagrams on the X-Y plane, for 2-, 4-, 6-, and 8-scrolls, respectively. These results confirm the system's ability of generating multiscroll chaotic attractors, spanning from 2-scrolls to 8-scrolls.

To generate an 8-scroll attractor as an example, the modular circuit design is applied to Eq. (4.2) through differential-integral transformations. The resulting expression is given by

$$\begin{cases} x = \int A\,(y - bx - 0.5\,(a - b)\,f\,(x))\,d\tau, \\ y = \int (x - y + z)\,d\tau, \\ z = \int (-By)\,d\tau. \end{cases} \tag{4.4}$$

Based on Eq. (4.2), the circuit design is shown in Fig. 4.5. The circuit is divided into two parts: Fig. 4.5a presents the basic Chua's circuit, which includes a reversed-

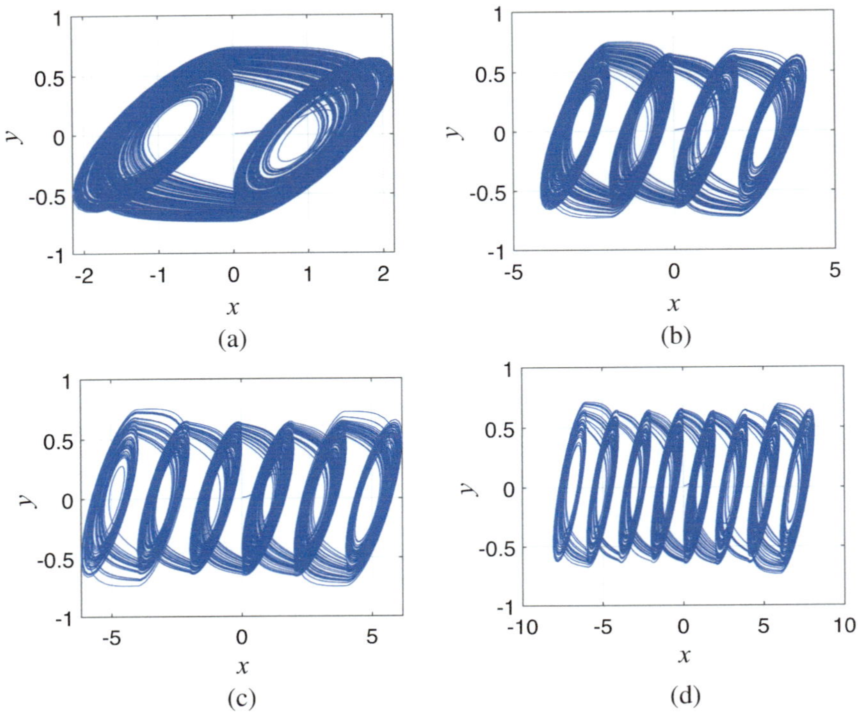

Fig. 4.4 Simulation results of 2- to 8-scroll chaotic attractors [2]. **a** 2-scroll. **b** 4-scroll. **c** 6-scroll. **d** 8-scroll

phase order, an integrator and an inverter, while Fig. 4.5b depicts the step function generating circuit, responsible for generating multiscroll chaotic attractors. The control switches, labeled S_1 to S_6, are used to regulate the system dynamics. Additionally, the number of scrolls can be adjusted by toggling these switches on or off.

The hardware circuits shown in Fig. 4.5 produce the experimental outcomes displayed in Fig. 4.6. These results were captured using a digital camera connected to an analog oscilloscope. By adjusting the switch-off states, the number of scrolls can be modified.

Figure 4.6 illustrates that the hardware circuit design is capable of generating up to 8-scroll attractors. The consistency between theoretical analysis, simulations and hardware experiments demonstrates the system's feasibility. Moreover, its physical implementation highlights its potential for future applications in engineering and technologies.

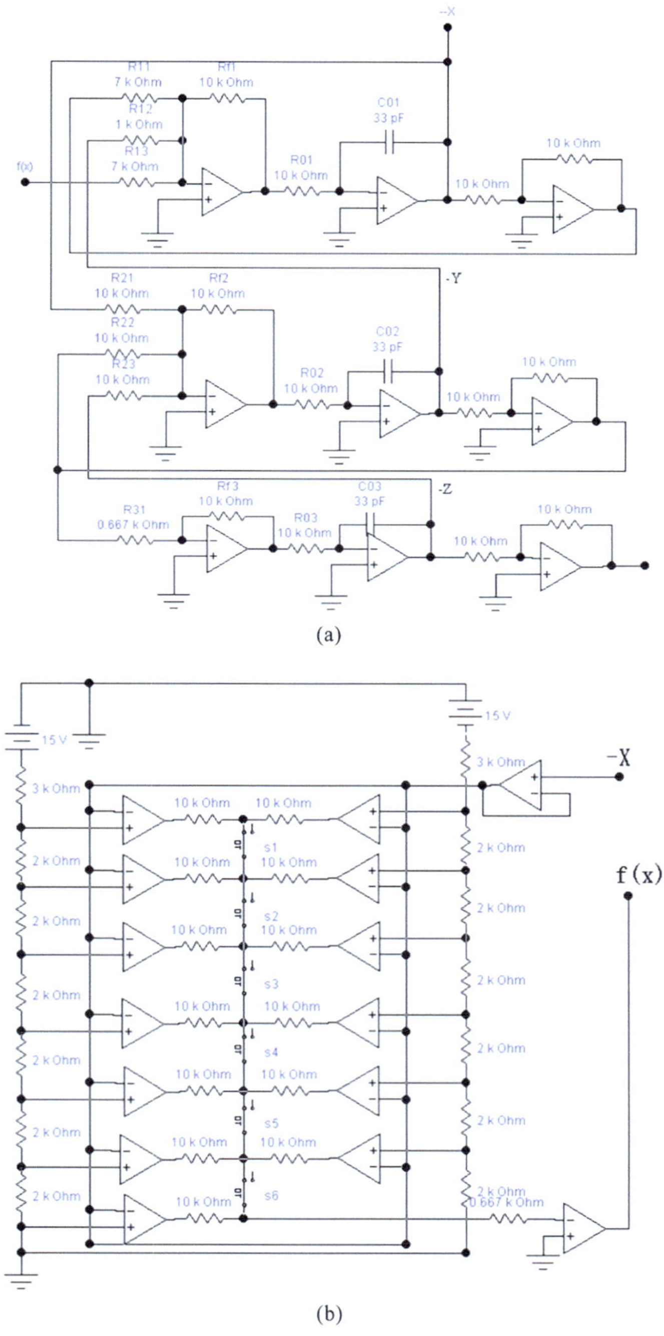

Fig. 4.5 Hardware circuit diagram. **a** The Chua's circuit. **b** The step function generating circuit. Reprinted from [2], with permission from World Scientific Pub. Co.

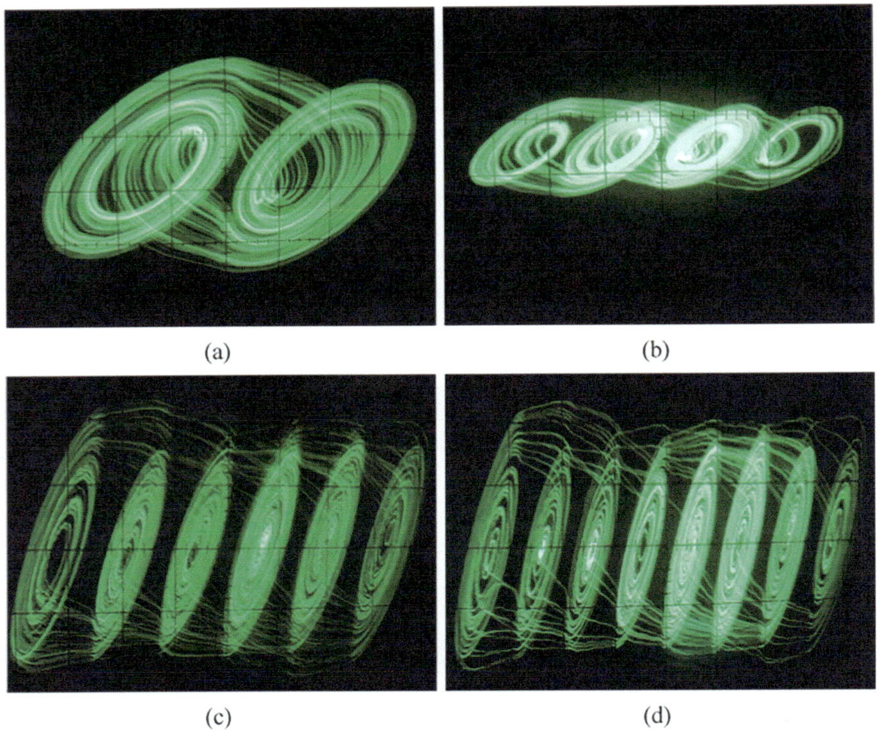

Fig. 4.6 Hardware circuit experimental results. **a** 2-scroll. **b** 4-scroll. **c** 6-scroll. **d** 8-scroll. Reprinted from [2], with permission from World Scientific Pub. Co.

4.2 Hysteresis Series Method

This section presents a structured method for generating multidirectional multiscroll chaotic attractors using hysteresis series. The approach covers two scenarios: controlling a 2-D linear autonomous system and a 3-D linear autonomous system. Furthermore, a technique for creating $n \times m$-scroll attractors based on a two-port network is introduced.

4.2.1 A 2-D Hysteresis System

This subsection outlines a design for generating multiscroll chaotic attractors from a 2-D linear autonomous system using hysteresis series. This technique can produce 1-D n-scroll and 2-D $n \times m$-grid scroll chaotic attractors.

Han et al. [3] introduced a 2-D hysteresis-based multiscroll chaotic system, described by

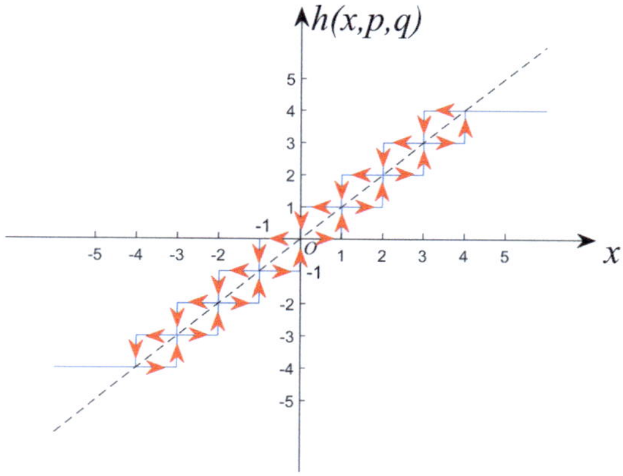

Fig. 4.7 Phase portrait of the hysteresis series. Reprinted from [4], with permission from World Scientific Pub. Co.

$$\dot{X} = AX + B\theta\left(CX\right), \tag{4.5}$$

where $X = (x, y)^T$ and

$$A = \begin{pmatrix} 0 & 1 \\ -a & b \end{pmatrix}, \quad B = \begin{pmatrix} -1 & 0 \\ -b & a \end{pmatrix}, \quad C = \begin{pmatrix} 0 & 1 \\ 1 & 0 \end{pmatrix},$$

with $\theta(\cdot)$ to be specified below.

There are three cases to consider.

(1) 1-D horizontal n-scroll chaotic attractors:

$$\theta\left(\cdot\right) = \begin{pmatrix} 0 \\ h\left(x, p_1, q_1\right) \end{pmatrix}. \tag{4.6}$$

The hysteresis series function $h(x, p_1, q_1)$ is defined as follows:

$$h\left(x, p_1, q_1\right) = \begin{cases} -p_1, & \text{if } x < -p_1 + 1, \\ i, & \text{if } i - 1 < x < i + 1, \\ q_1, & \text{if } x > -q_1 - 1, \end{cases} \tag{4.7}$$

where $i = -p_1 + 1, \ldots, x > -q_1 - 1$.

Figure 4.7 shows the phase portrait of the hysteresis series, where all equilibrium points are positioned along the x-axis.

(2) 1-D vertical n-scroll chaotic attractors:

$$\theta\left(\cdot\right) = \begin{pmatrix} h\left(x, p_2, q_2\right) \\ 0 \end{pmatrix}, \tag{4.8}$$

where $h(y, p_2, q_2)$ is similarly described by Eq. (4.6), with equilibrium points located in the y-axis.

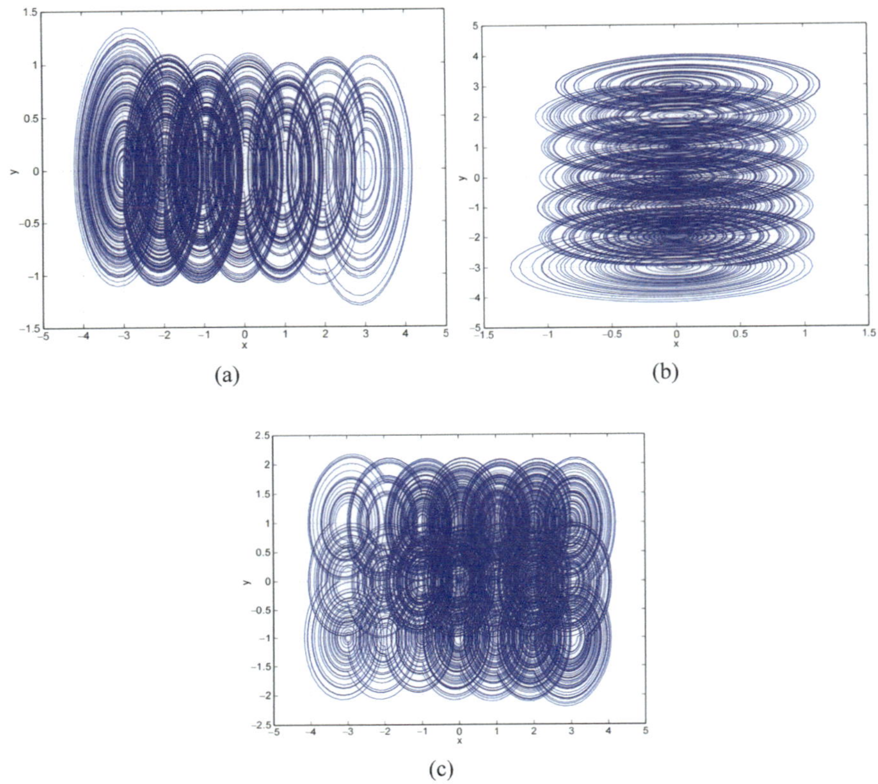

Fig. 4.8 Hysteresis chaotic attractors generated from a 2-D linear autonomous system. **a** 1-D horizontal 7-scroll attractor. **b** 1-D vertical 7-scroll attractor. **c** Hysteresis chaotic attractors generated from a 2-D linear autonomous system. Reprinted from [4], with permission from World Scientific Pub. Co.

(3) 2-D $n \times m$-grid scroll chaotic attractors:

$$\theta(\cdot) = \begin{pmatrix} h(x, p_2, q_2) \\ h(x, p_1, q_1) \end{pmatrix}. \tag{4.9}$$

The functions $h(x, p_1, q_1)$ and $h(y, p_2, q_2)$ are similarly defined by Eq. (4.5), with equilibrium points being at integer positions on the x-y plane.

For instance, with $a = 1$, $b = 0.125$, and $p = q = 3$, system (4.5) can generate a 1-D horizontal 7-scroll chaotic attractor, as shown in Fig. 4.8a. Similarly, with $a = 1$, $b = 0.125$, and $p = q = 3$, the system generates a 1-D vertical 7-scroll chaotic attractor, as shown in Fig. 4.8b. Additionally, with $a = 1$, $b = 0.125$, $p_1 = q_1 = 3$, and $p_2 = q_2 = 1$, system (4.5) generates a 2-D 7×3-grid scroll chaotic attractors, as shown in Fig. 4.8c. This hysteresis chaotic system (4.5) offers flexible control over the number, position and orientation of the scrolls.

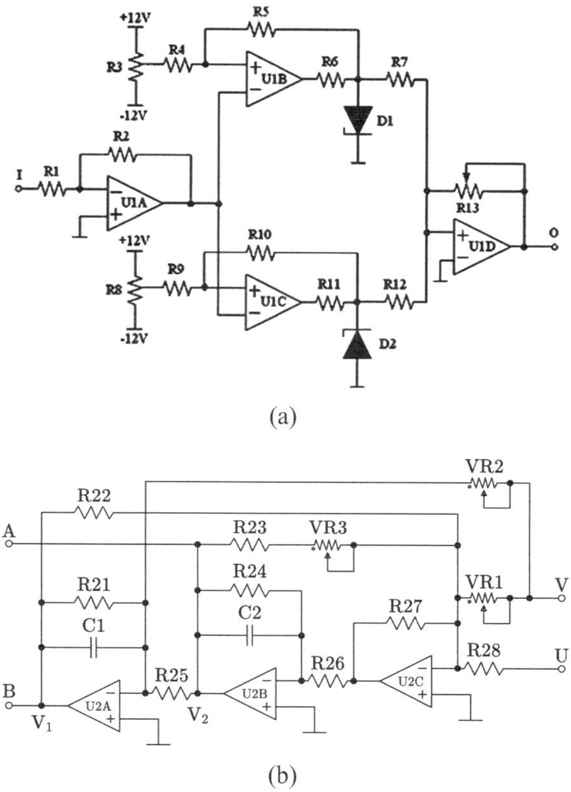

Fig. 4.9 Circuitry for realizing $n \times m$-grid scroll chaotic attractors. **a** Hysteresis series building block. **b** Second-order system. Reprinted from [4], with permission from World Scientific Pub. Co.

Figure 4.9 shows the circuit used to implement $n \times m$-grid scroll chaotic attractors. The circuit consists of two main parts: (a) a hysteresis series building block and (b) a second-order system as described by Han et al. [3]. The circuit parameters are as follows: $R_1 = R_2 = R_4 = R_7 = R_9 = R_{12} = 10\,\mathrm{k}\Omega$, $R_5 = R_{10} = R_{25} = R_{26} = 100\,k\Omega$, $R_{22} = R_{23} = R_{28} = 220\,k\Omega$, and $R_{21} = R_{24} = 3.3\,\mathrm{M}\Omega$. Potentiometers R_3, R_8, and R_{11} are rated at $50, 50$, and 20kΩ, respectively, with VR_1, VR_2, and VR_3 set at $2\,\mathrm{M}\,\Omega$, 300, and 300kΩ, respectively. Capacitors $C_1 = 0.01\,\mu F$ and $C_2 = 0.001\,\mu F$, along with diodes D_1 and D_2, are included. All operational amplifiers are LM324. Figure 4.10 demonstrates a 2-D 5×3-grid scroll chaotic attractor generated using this circuit.

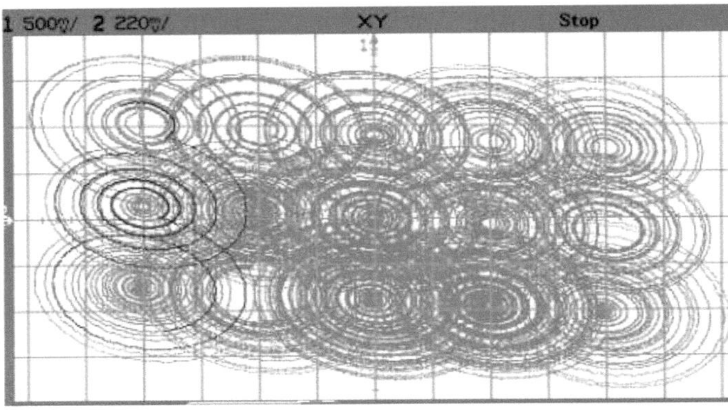

Fig. 4.10 Observed 5×3-grid scroll chaotic attractor. Reprinted from [4], with permission from World Scientific Pub. Co.

4.2.2 A 3-D Hysteresis System

This subsection introduces a systematic approach for generating multidirectional multiscroll chaotic attractors from a 3-D linear autonomous system using hysteresis series [5–9]. This method facilitates the generation of 1-D n-scroll, 2-D $n \times m$-grid scroll, and 3-D $n \times m \times l$-grid scroll chaotic attractors. The system is described by [6, 7]

$$\dot{X} = AX + B\theta\left(X\right). \tag{4.10}$$

Here, $X = (A, B, C)^T$ is the state vector, $B = -A$, with

$$A = \begin{pmatrix} 0 & 1 & 0 \\ 0 & 0 & 1 \\ -a & -b & -c \end{pmatrix},$$

and $\theta(\cdot)$ will be specified below.

There are three cases to consider.

(1) 1-D hysteresis n-scroll chaotic attractors:

$$\theta\left(X\right) = \begin{pmatrix} h\left(x, p_1, q_1\right) \\ 0 \\ 0 \end{pmatrix},$$

where the hysteresis series function $h(x, p_1, q_1)$ is described by Eq. (4.6).

(2) 2-D hysteresis $n \times m$-grid scroll chaotic attractors:

$$\theta(X) = \begin{pmatrix} h(x, p_1, q_1) \\ h(x, p_2, q_2) \\ 0 \end{pmatrix},$$

where the hysteresis series functions $h(x, p_1, q_1)$ and $h(y, p_2, q_2)$ are similarly described by Eq. (4.6).

(3) 3-D hysteresis $n \times m \times l$-grid scroll chaotic attractors:

$$\theta(X) = \begin{pmatrix} h(x, p_1, q_1) \\ h(x, p_2, q_2) \\ h(x, p_3, q_3) \end{pmatrix},$$

where the hysteresis series functions $h(x, p_1, q_1)$, $h(y, p_2, q_2)$, and $h(z, p_3, q_3)$ are similarly described by Eq. (4.6).

Figure 4.11 shows the multidirectional multi-scroll chaotic attractors generated by the hysteresis-controlled system (4.10): (a) a 1-D 7-scroll chaotic attractor with parameters $a = 0.8$, $b = 0.72$, $c = 0.6$, and $p = q = 3$; (b) a 2-D 5×7-grid scroll chaotic attractor with $a = 0.8$, $b = 0.7$, $c = 0.6$, $p_1 = q_1 = 2$, and $p_2 = q_2 = 3$; (c) and (d) a 3-D $5 \times 8 \times 3$-grid scroll chaotic attractor with $a = 0.8$, $b = 0.72$, $c = 0.66$, $p_1 = q_1 = 2$, $p_2 = 3$, $q_2 = 4$, and $p_3 = q_3 = 1$.

Figure 4.12 illustrates the dynamic behaviors of a 3-scroll hysteresis chaotic attractor. In particular, Lü et al. [6, 7] developed a 2-D Poincaré return map to precisely capture the chaotic behavior of this 3-scroll attractors. The map is expressed as follows:

$$f : \Omega_1 \to \Omega_1, \quad (y_0, x_0) \to \begin{cases} (y_2, x_2), & \text{for } (y_0, x_0) \in H_0, \\ (-y_3, -z_3), & \text{for } (y_0, x_0) \in \bar{H}_0, \end{cases} \tag{4.11}$$

where $\bar{H}_0 = \Omega_1 - H_0$ and H_0, Ω_1 are defined in [6, 7].

It is important to note that a precise mathematical formulation for this map can be derived using an exact solution, as detailed in [6, 7]. However, in practical applications, numerical techniques like the Newton–Raphson method are commonly used to solve the equations.

The Jacobi matrix of this function f is given by

$$Df = \begin{cases} \begin{pmatrix} \dfrac{\partial y_2}{\partial y_0} & \dfrac{\partial y_2}{\partial z_0} \\ \dfrac{\partial z_2}{\partial y_0} & \dfrac{\partial z_2}{\partial z_0} \end{pmatrix}, & \text{for } (y_0, z_0) \in H_0, \\ -\begin{pmatrix} \dfrac{\partial y_3}{\partial y_0} & \dfrac{\partial y_3}{\partial y_0} \\ \dfrac{\partial z_3}{\partial y_0} & \dfrac{\partial z_3}{\partial z_0} \end{pmatrix}, & \text{for } (y_0, z_0) \in \bar{H}_0. \end{cases} \tag{4.12}$$

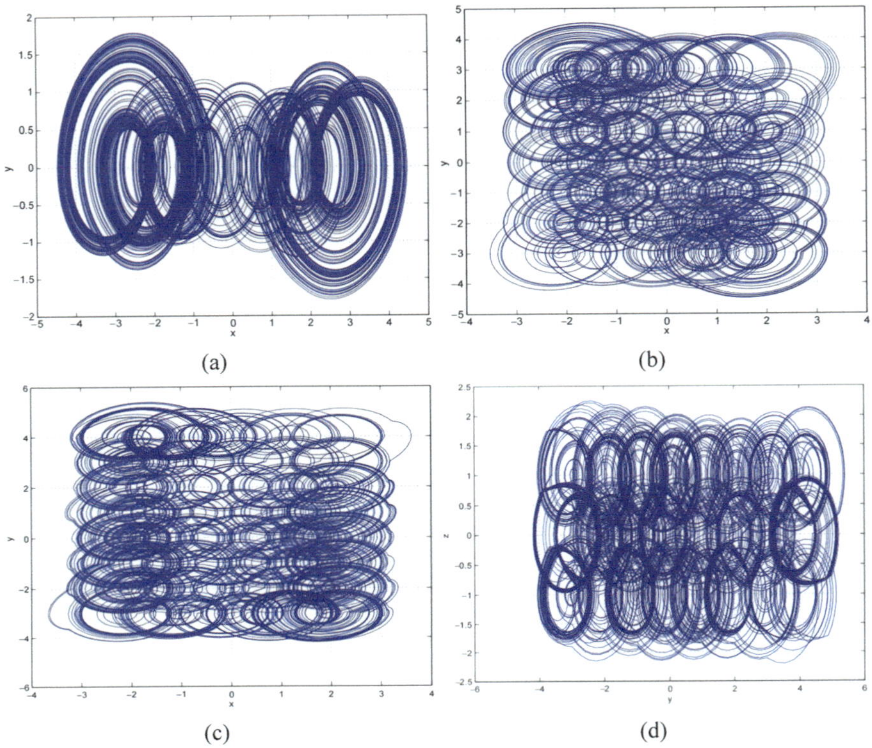

Fig. 4.11 Multiscroll chaotic attractors generated by hysteresis series function from a 3-D linear autonomous system. **a** 1-D 7-scroll attractors. **b** 2-D 5 × 7-grid scroll attractors. **c** 3-D 5 × 8 × 3-grid scroll attractors in x-y plane. **d** 3-D 5 × 8 × 3-grid scroll attractors on the y-z plane. Reprinted from [4], with permission from World Scientific Pub. Co.

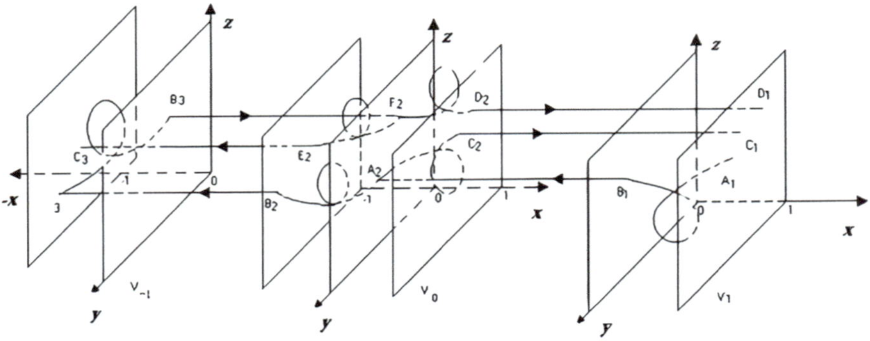

Fig. 4.12 Hysteresis phase space. Reprinted from [4], with permission from World Scientific Pub. Co.

Using the Jacobi matrix, precise mathematical formulations for the Lyapunov exponents λ_1 and λ_2 (with $\lambda_1 \geq \lambda_2$) of the Poincaré map f can be derived. However, these formulations are quite intricate. In practice, numerical techniques are generally used to compute λ_1 and λ_2. When $0 < \lambda_1 < +\infty$, the 3-scroll hysteresis system (4.10) exhibits chaotic behavior. Similarly, the conditions for generating 2-D $n \times m$-grid scroll and 3-D $n \times m \times l$-grid scroll chaotic attractors can be rigorously established by constructing a 2-D Poincaré return map.

Lü et al. [6, 10] also developed a block circuit diagram to experimentally verify these multidirectional hysteresis multiscroll chaotic attractors. Figure 4.13 shows this circuit diagram, which can physically generate 1-D 5 to 11-scroll, 2-D 3×5 to 3×11-grid scroll, and 3-D $3 \times 3 \times 5$ to $3 \times 3 \times 11$-grid scroll chaotic attractors, by adjusting the switches. This was the first experimental demonstration of 2-D 3×11-grid scroll and 3-D $3 \times 3 \times 11$-grid scroll chaotic attractors. Furthermore, this design method provides a theoretical basis for the hardware implementation of chaotic attractors with multiple directions and a large number of scrolls.

When the switches K_1, K_2, K_3, K_4 are turned off, the circuit in Fig. 4.14 can generate 1-D 5 to 11-scroll chaotic attractors. When K_1 and K_3 are on, K_2 and K_4 are off, the circuit generates 2-D 3×5 to 3×11-grid scroll chaotic attractors. Finally, when all the switches K_1, K_2, K_3, K_4 are on, the circuit generates 3-D $3 \times 3 \times 5$ to $3 \times 3 \times 11$-grid scroll chaotic attractors.

Figure 4.14 displays the experimental observations of a 3-D $3 \times 3 \times 11$-grid scroll chaotic attractors.

4.3 Saturated Function Series Method

This section offers a concise overview of a multidirectional saturated multiscroll chaotic system [5–8, 11, 12].

Lü et al. [5] constructed a 3-D saturated multiscroll chaotic system, described by

$$\dot{X} = AX + B\varphi(CX). \tag{4.13}$$

Here, $X = (x, y, z)^T$ is the state vector, with

$$A = \begin{pmatrix} 0 & 1 & 0 \\ 0 & 0 & 1 \\ -a & -b & -c \end{pmatrix}, \quad B = \begin{pmatrix} 0 & \frac{d_2}{b} & 0 \\ 0 & 0 & -\frac{d_3}{c} \\ d_1 & d_2 & d_3 \end{pmatrix}, \quad C = \begin{pmatrix} 1 & 0 & 0 \\ 0 & 1 & 0 \\ 0 & 0 & 1 \end{pmatrix},$$

and $\varphi(\cdot)$ will be specified in the following.

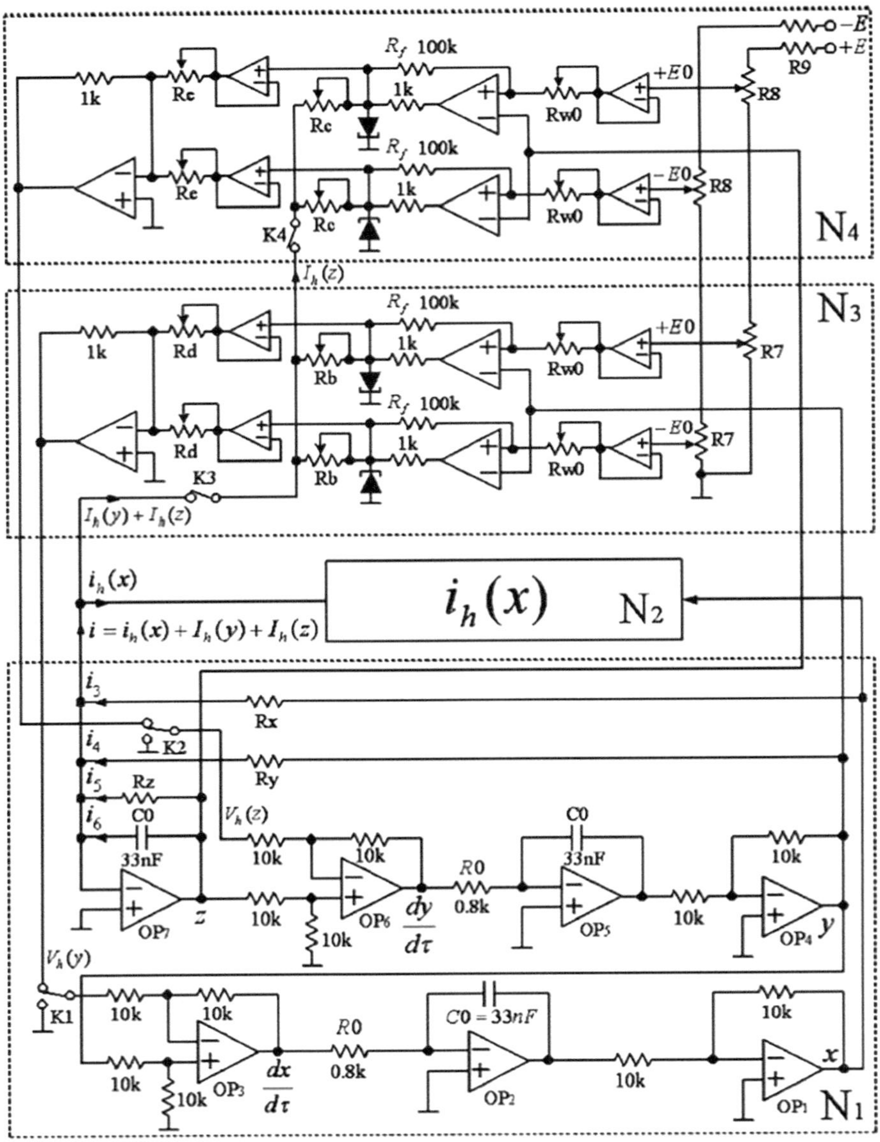

Fig. 4.13 Observed 5 × 3-grid scroll chaotic attractor. Reprinted from [7], with permission from the IEEE

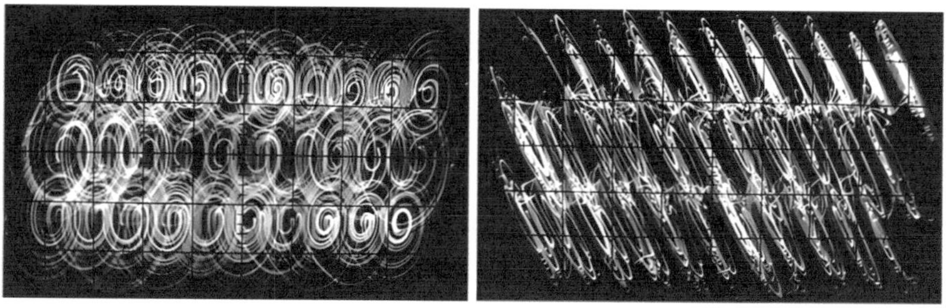

(a) (b)

Fig. 4.14 Experimental observations of a 3-D $3 \times 3 \times 11$-grid scroll chaotic attractor. **a** x-y plane, where $x = 1.44$ V/div, $y = 0.8$ V/div. **b** x-z plane, where $x = 1.44$ V/div, $z = 0.8$V/div. Reprinted from [7], with permission from the IEEE

This system can generate three types of attractors, as discussed below.

(1) 1-D saturated n-scroll chaotic attractors:

$$\varphi(CX) = \begin{pmatrix} f(x; k_1, h_1, p_1, q_1) \\ 0 \\ 0 \end{pmatrix},$$

where the saturated function series $f(x; k_1, h_1, p_1, q_1)$ is defined by

$$f(x; k_1, h_1, p_1, q_1) = \begin{cases} (2q_1 + 1) k_1, & \text{if } x > q_1 h_1 + 1, \\ k_1 (x - i h_1) + 2 i k_1, & \text{if } \begin{cases} |x - i h_1| \le 1, \\ -p_1 \le i \le q_1, \end{cases} \\ (2i + 1) k_1, & \text{if } \begin{cases} 1 < x - i h_1 < h_1 - 1, \\ -p_1 \le i \le q_1 - 1, \end{cases} \\ -(2p_1 + 1) k_1, & \text{if } x < -p_1 h_1 - 1. \end{cases} \tag{4.14}$$

Figure 4.15 displays a saturated function series with $k = 1,\ h = 4$.

(2) 2-D saturated $n \times m$-grid scroll chaotic attractors:

$$\varphi(CX) = \begin{pmatrix} f(x; k_1, h_1, p_1, q_1) \\ f(x; k_2, h_2, p_2, q_2) \\ 0 \end{pmatrix},$$

where the saturated function series $f(x; k_1, h_1, p_1, q_1)$ and $f(x; k_2, h_2, p_2, q_2)$ are similarly described by Eq. (4.14).

(3) 3-D saturated $n \times m \times l$-grid scroll chaotic attractors:

$$\varphi(CX) = \begin{pmatrix} f(x; k_1, h_1, p_1, q_1) \\ f(x; k_2, h_2, p_2, q_2) \\ f(x; k_3, h_3, p_3, q_3) \end{pmatrix},$$

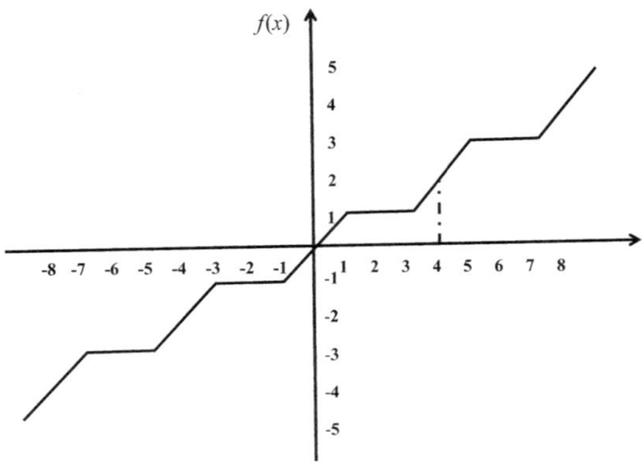

Fig. 4.15 Saturated function series with $k = 1,\ h = 4$. Reprinted from [4], with permission from World Scientific Pub. Co.

where the saturated function series $f\ (x; k_1, h_1, p_1, q_1)$, $f\ (x; k_2, h_2, p_2, q_2)$, and $f\ (x; k_3, h_3, p_3, q_3)$ are similarly described by Eq. (4.14).

With $a = b = c = d_1 = 0.7,\ k_1 = 9,\ h_1 = 18$, and $p_1 = q_1 = 2$, system (4.13) can generate a 1-D 6-scroll chaotic attractors, as shown in Fig. 4.16a. In another case, with $a = b = c = d_1 = d_2 = 0.7,\ k_1 = k_2 = 50,\ h_1 = h_2 = 100$, and $p_1 = q_1 = p_2 = q_2 = 2$, the system generates a 2-D 6×6 grid-scroll chaotic attractors, as shown in Fig. 4.16b. Moreover, with $a = d_1 = 0.7, b = c = d_2 = d_3 = 0.8, k_1 = 100, h_1 = 200, k_2 = k_3 = 40, h_2 = h_3 = 80$, and $p_1 = p_2 = p_3 = q_1 = q_2 = q_3 = 2$, the system generates a 3-D $6 \times 6 \times 6$ grid-scroll chaotic attractors, as shown in Fig. 4.16b and c.

Additionally, Lü et al. [5] developed a 2-D Poincaré return map to rigorously demonstrate the chaotic behavior of these saturated double-scroll attractors. The map is defined by

$$F : M_1 \to M_1,\quad (y_0, x_0) \to \begin{cases} (y_1, z_1), & \text{for } (y_0, x_0) \in M_1 - M_1^-, \\ (y_2, z_2), & \text{for } (y_0, x_0) \in \Gamma, \\ -(y_3, z_3), & \text{for } (y_0, x_0) \in M_1 - \Gamma, \end{cases} \tag{4.15}$$

where M_1, M_1^-, Γ are defined in [5].

The Jacobi matrix of this map F is expressed as follows:

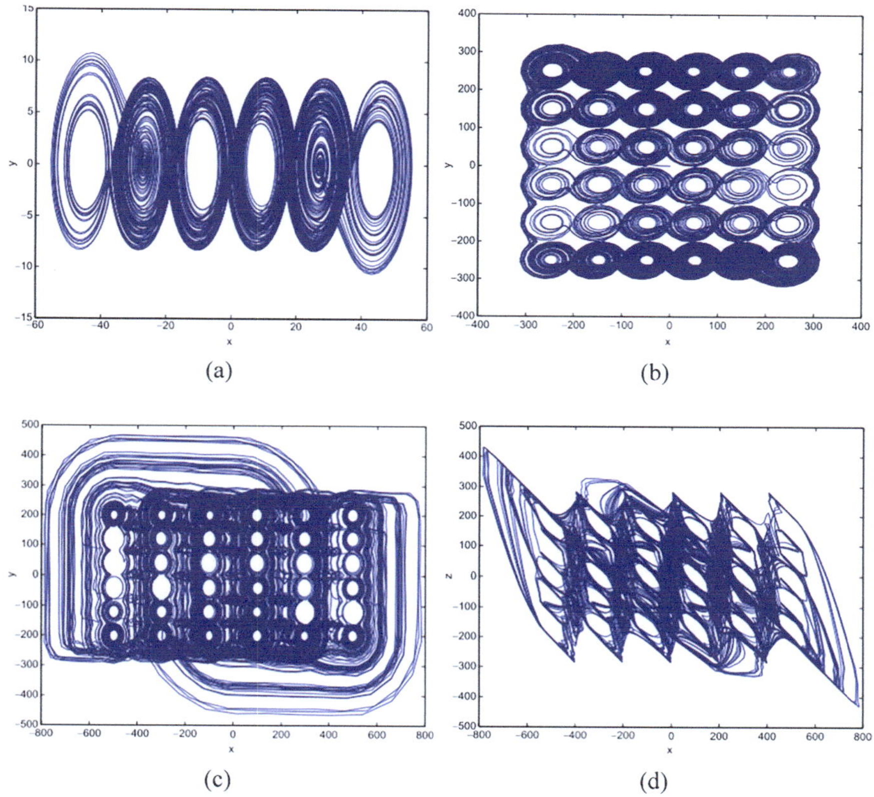

Fig. 4.16 Phase portraits of saturated multiscroll chaotic attractors. **a** 1-D 6-scroll attractors. **b** 2-D 6×6-scroll attractors. **c** 3-D $6 \times 6 \times 6$-grid scroll attractors in the x-y plane. **d** 3-D $6 \times 6 \times 6$-grid scroll attractors in the x-z plane. Reprinted from [4], with permission from World Scientific Pub. Co.

$$DF = \begin{cases} \begin{pmatrix} \dfrac{\partial y_1}{\partial y_0} & \dfrac{\partial y_1}{\partial z_0} \\ \dfrac{\partial z_1}{\partial y_0} & \dfrac{\partial z_1}{\partial z_0} \end{pmatrix}, & \text{for } (y_0, z_0) \in M_1 - M_1^-, \\ \begin{pmatrix} \dfrac{\partial y_2}{\partial y_0} & \dfrac{\partial y_2}{\partial z_0} \\ \dfrac{\partial z_2}{\partial y_0} & \dfrac{\partial z_2}{\partial z_0} \end{pmatrix}, & \text{for } (y_0, z_0) \in \Gamma, \\ -\begin{pmatrix} \dfrac{\partial y_3}{\partial y_0} & \dfrac{\partial y_3}{\partial z_0} \\ \dfrac{\partial z_3}{\partial y_0} & \dfrac{\partial z_3}{\partial z_0} \end{pmatrix}, & \text{for } (y_0, z_0) \in M_1^- - \Gamma. \end{cases} \tag{4.16}$$

The exact mathematical expressions for the Lyapunov exponents LE_1 and LE_2 of the mapping F can be derived from the Jacobian matrix of F. However, these formulations are quite complex. In practice, numerical methods are commonly used to compute the maximum Lyapunov exponent LE_1. If $0 < LE_1 < +\infty$, the system described by Eq. (4.13) is chaotic. Furthermore, the chaotic nature of the attractor can be rigorously confirmed using the Homoclinic type of Shilnikov Theorem [13, 14].

Furthermore, Lü et al. [7] introduced a block circuit diagram to generate multidirectional saturated multiscroll chaotic attractors, as shown in Fig. 4.17. The circuit is composed of five main components: N_1, N_2, N_3, N_4, and a set of switches. Specifically, N_1 contains three integrators, two inverter-amplifiers and two subtractor amplifiers; N_2 is responsible for generating the x-directional current saturated function series $i_x(x)$ and the x-directional voltage saturated function series $f_1(x)$; N_3 includes the generator for the y-directional current saturated function series $i_y(y)$ and the y-directional voltage saturated function series $f_2(y)$; finally, N_4 generates the z-directional current saturated function series $i_z(z)$ and the z-directional voltage saturated function series $f_3(z)$.

All the required circuit parameters for generating various multiscroll chaotic attractors are provided in [7]. Figure 4.18 presents experimental results, showcasing some of these saturated multiscroll chaotic attractors.

The designed block circuit can physically realize 1-D attractors with 10, 12 and 14 scroll attractors, 2-D attractors with 10, 12, and 14 × 10-grid scroll attractors, and 3-D attractors with 10 × 10 × 10-grid scroll attractors, by adjusting the switches. Notably, this was the first experimental demonstration of 1-D 14-scroll attractors, 2-D 14 × 10-grid scroll attractors, and 3-D 10 × 10 × 10-grid (totalling 1000) scroll chaotic attractors. It is important to recognize that hardware realization of 1-D n-scroll attractors with $n \geq 10$, 2-D $n \times m$-grid scroll attractors with $n, m \geq 10$, and 3-D $n \times m \times l$-grid scroll attractors with $n, m, l \geq 10$, presents significant technical challenges. The circuit design method outlined above provides a theoretical basis for hardware implementation of such multidirectional chaotic attractors with a large number of scroll attractors.

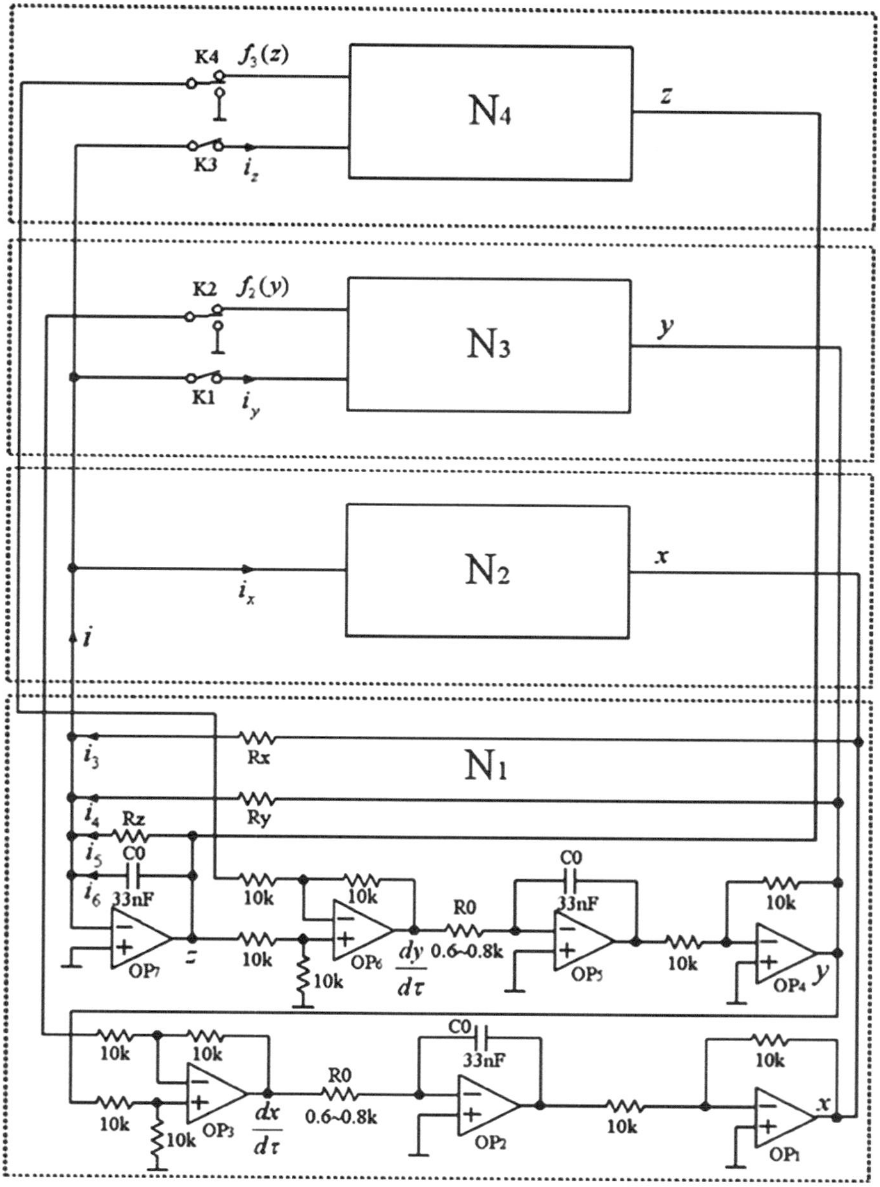

Fig. 4.17 Block circuitry for realizing multidirectional saturated multiscroll chaotic attractors. Reprinted from [7], with permission from the IEEE

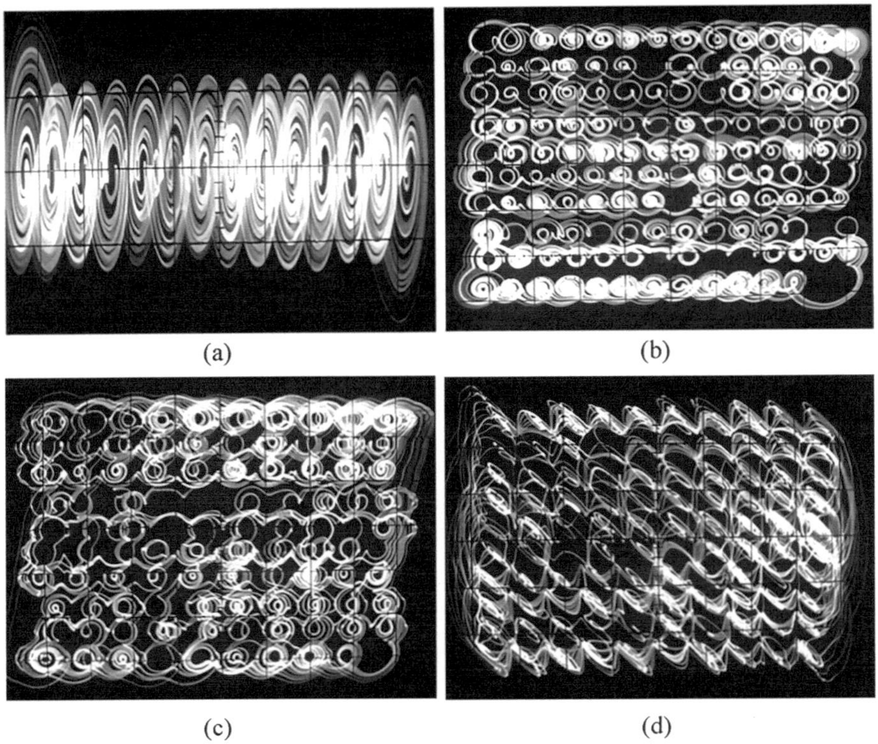

Fig. 4.18 Experimental observations of saturated multiscroll chaotic attractors. **a** 1-D 14-scroll attractors in the x-y plane, where $x = 1.4\,\text{V/div}$, $y = 0.4\,\text{V/div}$. **b** 2-D 14×10-grid scroll attractors on the x-y plane, where $x = 1.4\,\text{V/div}$, $y = 1.6\,\text{V/div}$. **c** 3-D $10 \times 10 \times 10$-grid scroll attractors on the x-y plane, where $x = 1.0\,\text{V/div}$, $y = 0.8\,\text{V/div}$. **d** 3-D $10 \times 10 \times 10$-grid scroll attractors on the x-z plane, where $x = 1.0\,\text{V/div}$, $y = 0.8\,\text{V/div}$. Reprinted from [7], with permission from the IEEE

4.4 Multilevel Pulse Control Method

This section introduces a control method for generating multidirectional multiscroll chaotic attractors.

Anam et al. [15] generated simplified mirror-symmetric, multidirectional multiscroll chaotic attractors using multilevel pulse excitation sources, based on the multilevel-logic pulse control technique.

The approach is illustrated using a chaotic satellite attitude system, described by the Euler equations, as follows:

$$\begin{cases} I_x\dot{\omega}_x = \omega_y\omega_z\left(I_y - I_z\right) + H_x + U_x, \\ I_y\dot{\omega}_y = \omega_x\omega_z\left(I_z - I_x\right) + H_y + U_y, \\ I_z\dot{\omega}_z = \omega_x\omega_y\left(I_x - I_y\right) + H_z + U_z, \end{cases} \tag{4.17}$$

where I_x, I_y, I_z denote the principal moments of inertia, H_x, H_y, H_z represent the perturbing torques, ω_x, ω_y, ω_z indicate the angular velocities, and U_x, U_y, U_z are the three control torques.

The principal moments of inertia and the perturbing disturbance torques for the chaotic satellite attitude system are defined as follows:

$$\begin{bmatrix} H_x \\ H_y \\ H_z \end{bmatrix} = \begin{bmatrix} -1.2 & 0 & \sqrt{6}/2 \\ 0 & 0.35 & 0 \\ -\sqrt{6} & 0 & -0.4 \end{bmatrix} \begin{bmatrix} \omega_x \\ \omega_y \\ \omega_z \end{bmatrix}, \begin{bmatrix} I_x \\ I_y \\ I_z \end{bmatrix} = \begin{bmatrix} 3kg \cdot m^2 \\ 2kg \cdot m^2 \\ 1kg \cdot m^2 \end{bmatrix}. \tag{4.18}$$

Using the defined perturbing torques and moments of inertia in Eq. (4.18), the attitude equations of the satellite system in Eq. (4.17) are rewritten in the following form:

$$\begin{cases} \dot{x} = 1/3yz - ax + 1/\sqrt{6}z, \\ \dot{y} = -xz + by, \\ \dot{z} = xy - \sqrt{6}x - cz, \end{cases} \tag{4.19}$$

where a, b, and c are positive real parameters.

The satellite system in Eq. (4.19) exhibits rotational symmetry under the coordinate transformation $(x, y, z) \rightarrow (-x, y, -z)$, which holds true for all values of the system parameters. This symmetry is crucial for understanding the formation of multiscroll attractors in the satellite system.

The system described by Eq. (4.18) produces a double-wing chaotic attractor when $a = 0.40$, $b = 0.175$, and $c = 0.40$, with its phase trajectories shown in Fig. 4.19.

Anam et al. [15] implemented a non-autonomous pulse control to generate multi-directional mirror symmetric multiscroll chaotic attractors (MSMCA). This method leverages the pulse excitation source technology, typically used in satellite systems, to produce multidirectional MSMCA in system (4.18).

Initially, the multilevel logic pulse signal is defined, as follows:

$$M(t) = \sum_{k=1}^{L} D\text{sgn}\left(\sin\left(\psi t - \omega_k\right)\right), \quad \omega_k = \frac{(k-1)\pi}{L}, \tag{4.20}$$

where $L \geq 2$ represents the number of pulse signals and $M(t)$ denotes the sum of signals with different initial phases ω. The parameters $D \in \mathbf{R}$ and $\psi \in \mathbf{R}^+$ control the amplitude and frequency of each single pulse, respectively. For different values of L, the multilevel pulse signal $M(t)$ is the sum of signals with varying initial phases, where $\omega_k = (k-1)\pi/L$, and is given by

$$M(t) = \begin{cases} \pm D, \pm 3D, \ldots, \pm(L-2)D, \pm LD, & \text{when } D = odd, \\ 0, \pm 2D, \ldots, \pm(L-2)D, \pm LD, & \text{when } D = even. \end{cases} \tag{4.21}$$

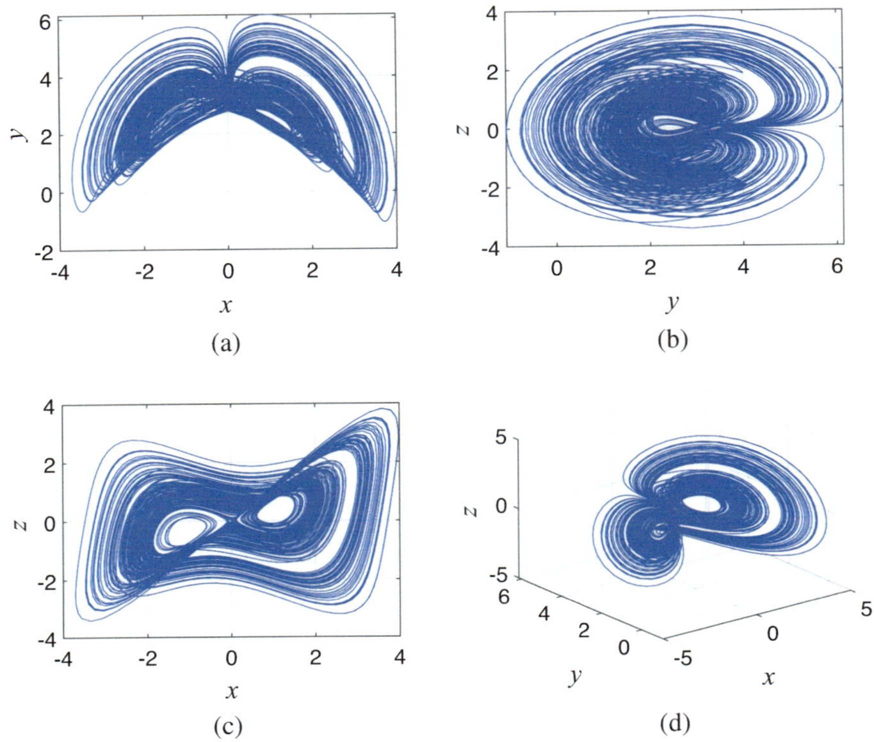

Fig. 4.19 Double-wing chaotic attractors of the satellite system, with $a = 0.40$, $b = 0.175$, $c = 0.40$ [15]. **a** x-y plane. **b** y-z plane. **c** x-z plane. **d** x-y-z space

From Eq. (4.21), it is evident that $M(t)$ depends on three parameters: D, ψ, and L. The parameters D and L together determine the logical values, while the width of each logic level is given by $\tau = T/2L = \pi/L\psi$.

For the mirror-symmetric satellite attractor, introduce the multilevel pulse signal in Eq. (4.20), with the coefficient of each nonlinear term in the satellite system (4.19) defined as follows:

$$\begin{cases} \dot{x} = M(t)\dfrac{yz}{3} - ax + \dfrac{z}{\sqrt{6}}, \\ \dot{y} = -M(t)xz + by, \\ \dot{z} = M(t)xy - \sqrt{6}x - cz. \end{cases} \tag{4.22}$$

By setting $L = D = 1$ and $\psi = 0.02$ in Eq. (4.21), the pulse signal $M(t)$ oscillates between -1 and $+1$. As a result, the system (4.22), starting from the initial condition (3, 4, 2), generates a mirror-symmetric chaotic attractor composed of two symmetric attractors. The projection of this mirror-symmetric satellite system attractor, with

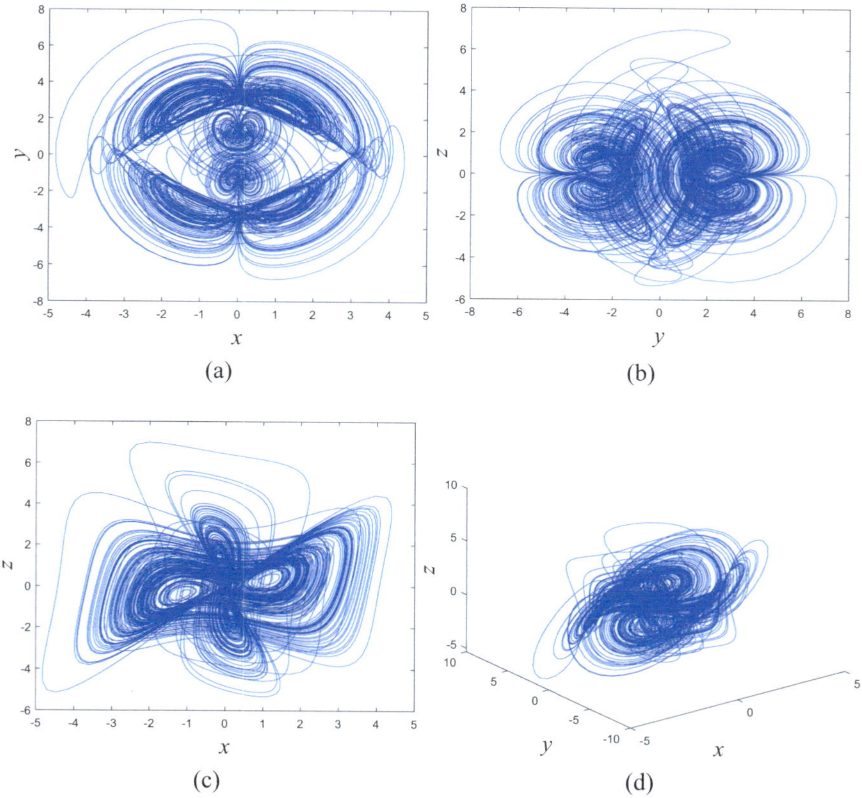

Fig. 4.20 Mirror symmetric satellite system attractor [15]. **a** x-y plane. **b** y-z plane. **c** x-z plane. **d** x-y-z space, with initial condition (3, 4, 2)

parameters $a = 0.40, b = 0.175$, and $c = 0.40$, across various phase planes, is shown in Fig. 4.20.

Now, the system (4.19) takes the following form:

$$\begin{cases} \dot{x} = \dfrac{M(t)}{3} yz - a\delta(x) + \dfrac{1}{\sqrt{6}} z, \\ \dot{y} = M(t)\,\delta(x)\, z + by, \\ \dot{z} = M(t)\,\delta(x)\, y - \sqrt{6}\delta(x) - cz, \end{cases} \tag{4.23}$$

where $\delta(x) = x(t) - M_x(t)$. The state variable $x(t)$ is now controlled by the pulse signal.

The system (4.23) can generate an arbitrary number of 1-D MSMCA by adjusting the control parameters D_x, L_x, ψ_x, L, and D. By choosing the control parameters as $L = D = 1$, $\psi = 0.02 = \psi_x$, $D_x = 2$, and $L_x = 4, 5, 6, 7$, the system (4.23) pro-

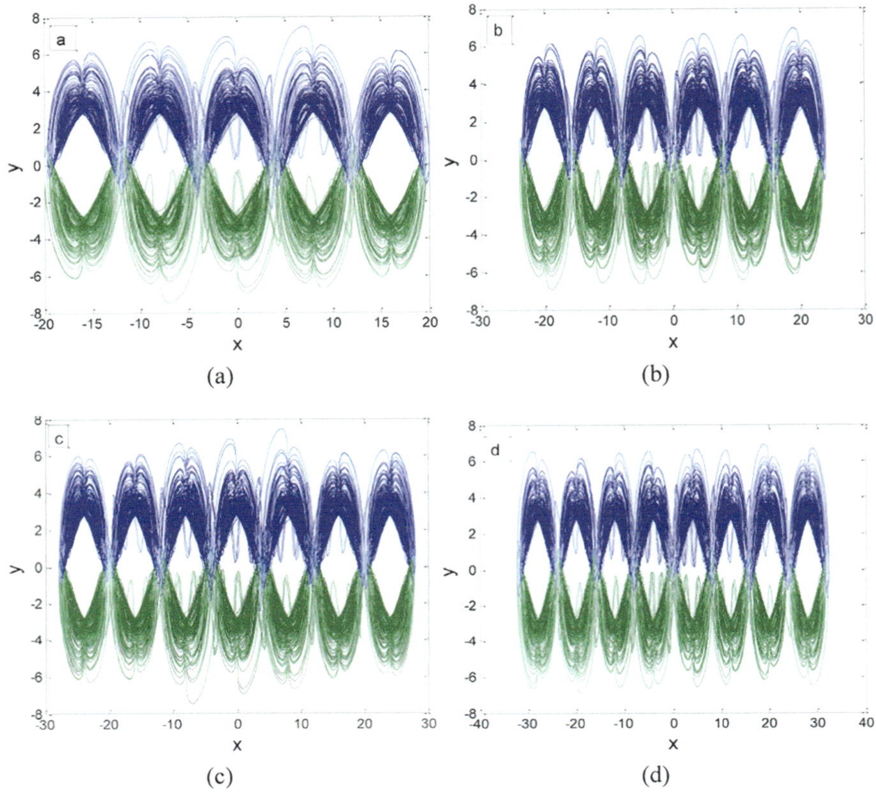

Fig. 4.21 1-D MSMCA in the x-direction. **a** 5 MSMCA. **b** 6 MSMCA. **c** 7 MSMCA. **d** 8 MSMCA. Reprinted from [15], with permission from Elsevier

duces 5–8 mirror-symmetric 1-D multiscroll chaotic attractors in the x-direction, as shown in Fig. 4.21a–d.

The generated 1-D MSMCA in the y- and z-directions, along with their associated dynamical systems, is displayed in Fig. 4.22.

Next, by applying the two pulse control functions and the multilevel logic pulse signal $M(t)$ into system (4.19), one obtains

$$\begin{cases} \dot{x} = \dfrac{M(t)}{3}\delta(y)z - a\delta(x) + \dfrac{1}{\sqrt{6}}z, \\ \dot{y} = M(t)\delta(x)z + b\delta(y), \\ \dot{z} = M(t)\delta(x)\delta(y) - \sqrt{6}\delta(x) - cz. \end{cases} \tag{4.24}$$

Here, multilevel pulse control signal $M_x(t) = (D_x = 4, \psi_x = 0.02, L_x = (2, 2, 3, 3))$, $M_y(t) = (Dy = 4, \psi_y = 0.02, L_y = (1, 2, 1, 2))$, $M(t) = (D = 1 = L, \psi = 0.02)$.

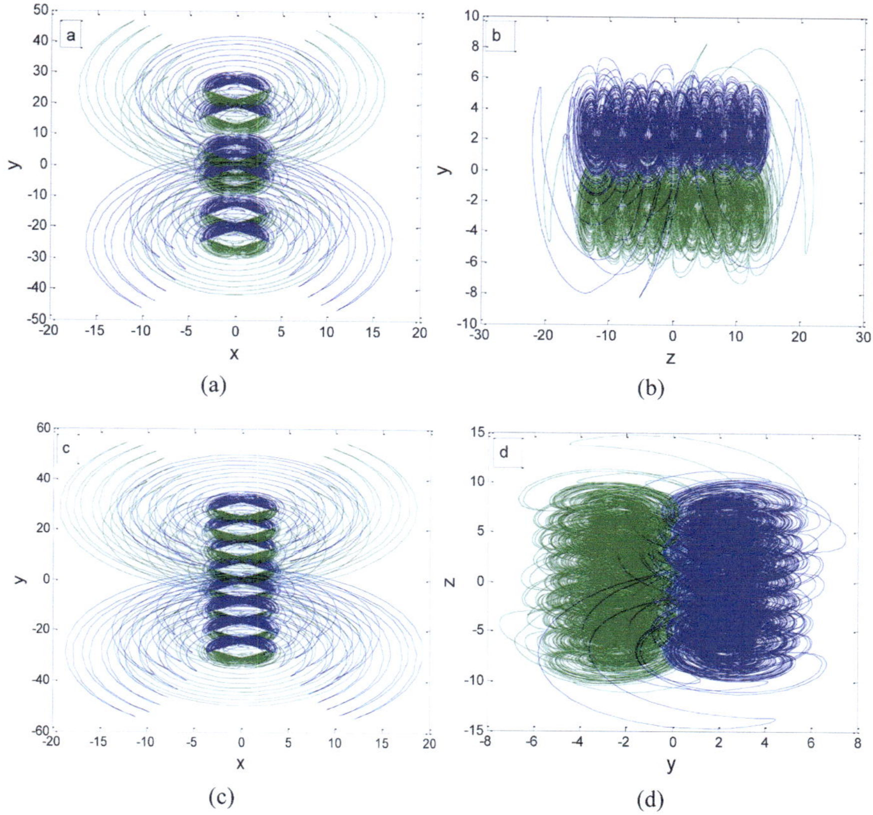

Fig. 4.22 Phase portraits of 1-D MSMCA in y- and z-directions. **a** 7 MSMCA in y-direction. **b** 7 MSMCA in z-direction. **c** 8 MSMCA in y-direction. **d** 8 MSMCA in z-direction. Reprinted from [15], with permission from Elsevier

The 2-D MSMCA with varying control parameters are presented in Fig. 4.23a–d.

For the 3-D MSMCA, fix all three state variables $x(t)$, $y(t)$, and $z(t)$, and replace them with their corresponding control functions $\delta(x)$, $\delta(y)$, and $\delta(z)$, respectively. The system (4.19) then takes the following form:

$$\begin{cases} \dot{x} = \dfrac{M(t)}{3}\delta(y)\delta(z) - a\delta(x) + \dfrac{1}{\sqrt{6}}\delta(z), \\ \dot{y} = M(t)\delta(x)\delta(z) + b\delta(y), \\ \dot{z} = M(t)\delta(x)\delta(y) - \sqrt{6}\delta(x) - c\delta(z). \end{cases} \tag{4.25}$$

Here, $M_x(t) = (D_x = 8,\ \psi x = 0.02,\ L_x = (1, 1))$, $M_y(t) = (D_y = 8, \psi_y = 0.02, L_y = (1, 2))$, $M_z(t) = (D_z = 4, \psi_z = 0.02, L_z = (1, 1))$, $M = L = 1$, and $\psi = 0.02$.

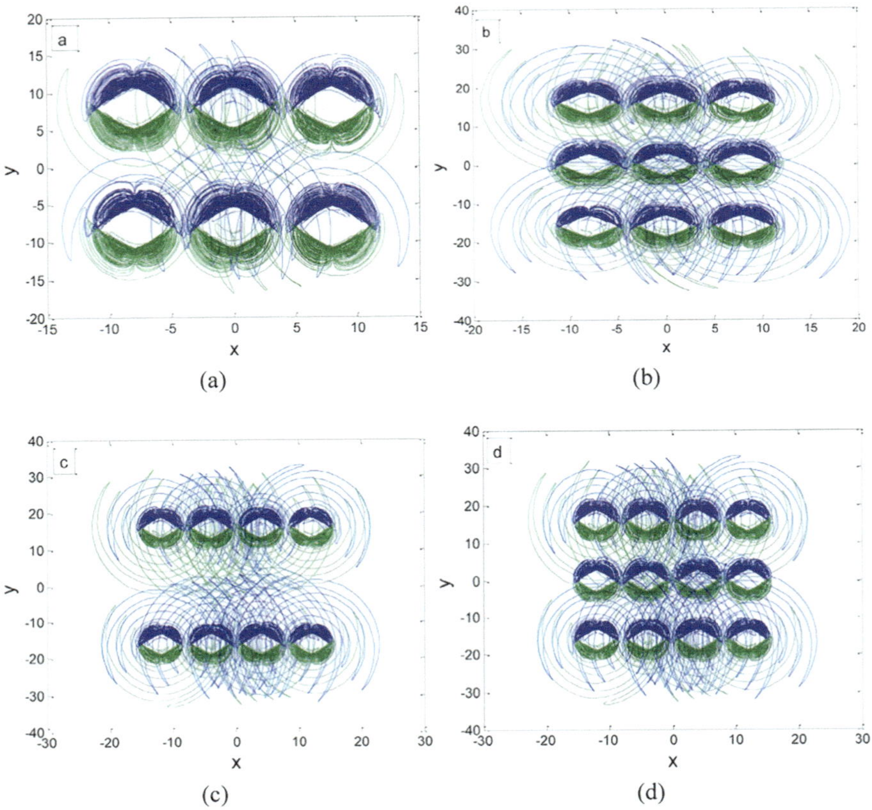

Fig. 4.23 2-D MSMCA on the x-y plane. **a** 3×2 MSMCA. **b** 3×3 MSMCA. **c** 4×2 MSMCA. **d** 4×3 MSMCA. Reprinted from [15], with permission from Elsevier

System (4.25), along with the specified control parameter values, generates arbitrary 3-D MSMCA, as shown in Fig. 4.24.

4.5 Multipulse Control Method

This section introduces a multipulse control method for generating multi-directional multi-double-scroll (MDMDS) chaotic attractors.

Hong et al. [16] developed a multipulse control method for generating MDMDS chaotic attractors without modifying the original nonlinear functions. By applying this technique to a 3-D double-scroll chaotic system, a new class of double-scroll grid attractors can be generated, including 1-D, 2-D, and 3-D double-scroll chaotic attractors.

Fig. 4.24 3-D MSMCA. **a** 2 × 2 × 2 MSMCA. **b** 2 × 3 × 2 MSMCA. Reprinted from [15], with permission from Elsevier

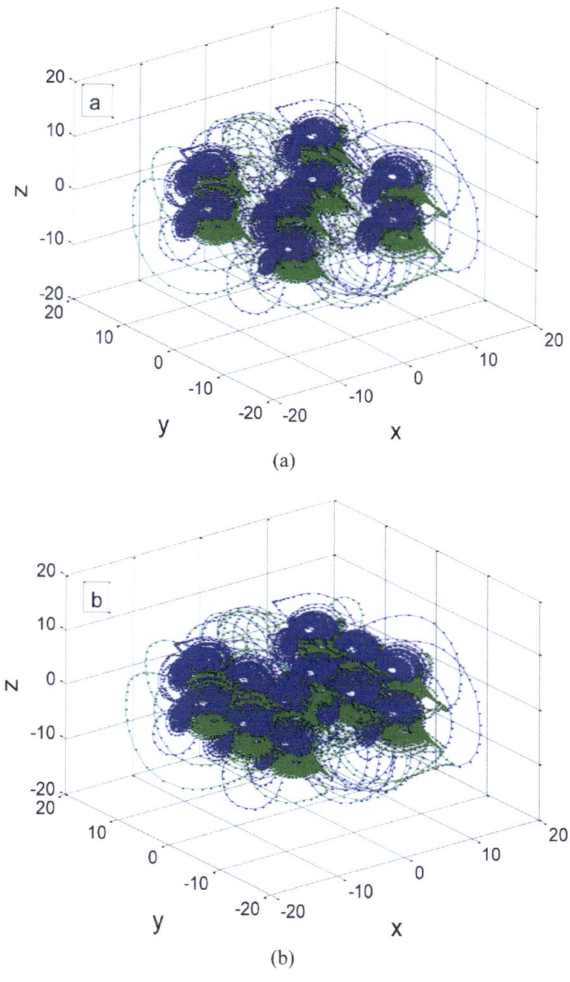

To start, the multi-level logic pulse signals are discussed. This type of signals is formed by combining a series of pulses with different phases, expressed as

$$P\,(t) = \sum_{i=1}^{N} A\text{sgn}\,(\sin\,(\omega t - \phi_i)), \tag{4.26}$$

where $\phi_i = (i-1)\pi/N$, and A and w control the amplitude and frequency of a single pulse, respectively. By choosing different values for N, the corresponding $(N+1)$-level logic pulse signal can be generated, given by

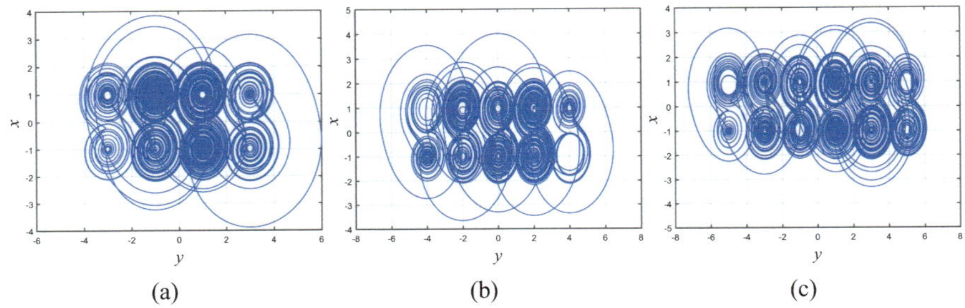

Fig. 4.25 1-D multi-double-scroll attractors on the x-y plane [16]. **a** 4-double-scroll attractors. **b** 5-double-scroll attractors. **c** 6-double-scroll attractors

$$P = \begin{cases} \pm 1A, \pm 3A, \ldots, \pm (n-2)\,A, \pm nA, \text{ if } N = odd, \\ 0A, \pm 2A, \ldots, \pm (n-2)\,A, \pm nA, \quad \text{if } N = even. \end{cases} \tag{4.27}$$

To construct n-double-scroll attractors along the y-axis, Hong et al. [16] suggested to couple the Sprott system, thereby obtaining

$$\begin{bmatrix} \dot{x} \\ \dot{y} \\ \dot{z} \end{bmatrix} = \begin{bmatrix} 0 & 1 & 0 \\ 0 & 0 & 1 \\ a & a & a \end{bmatrix} \begin{bmatrix} x - f(x) \\ y - p(t) \\ z \end{bmatrix}, \tag{4.28}$$

where $f(x)$ is the sgn function.

For instance, with $N = 3$ and $A = 1$, the corresponding attractor is generated, with the phase diagram shown in Fig. 4.25a. Figure 4.25b and c depict the 5-double-scroll and 6-double-scroll attractors with $N = 4$ and $N = 5$, respectively.

Similarly, it is possible to increase the number of scrolls along the x-axis. The multi-double-scroll in the x-direction can be generated by

$$\begin{bmatrix} \dot{x} \\ \dot{y} \\ \dot{z} \end{bmatrix} = \begin{bmatrix} 0 & 1 & 0 \\ 0 & 0 & 1 \\ a & a & a \end{bmatrix} \begin{bmatrix} x - P(x) - f(x - P(x)) \\ y \\ z \end{bmatrix}. \tag{4.29}$$

The corresponding 4-double-scroll attractors are shown in Fig. 4.26a, with parameters $N = 3$, $A = 2$, and $\omega = 0.005$. Additional results are presented in Fig. 4.26b and c, with $N = 4$ and $N = 5$, respectively.

From the same approach, 1-D double-scroll attractors on the x-z plane can be generated, as shown in Fig. 4.27a–c, with the parameters $A = 1.2$, $\omega = 0.005$, and $N = 4$, $N = 5$, and $N = 6$, respectively.

Hong et al. [16] demonstrated that it is also possible to generate 2-D multi-double-scroll attractors by introducing two pulse excitation sources, $P_1(t)$ and $P_2(t)$. The equation for the system is described by

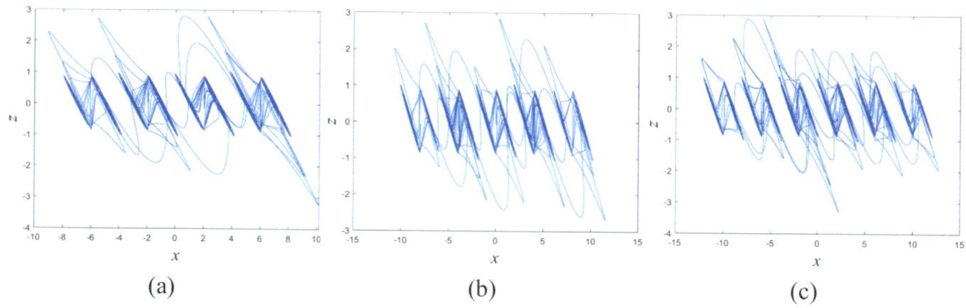

Fig. 4.26 1-D Multi-double-scroll attractors on the z-x plane [16]. **a** 4-double-scroll attractors. **b** 5-double-scroll attractors. **c** 6-double-scroll attractors

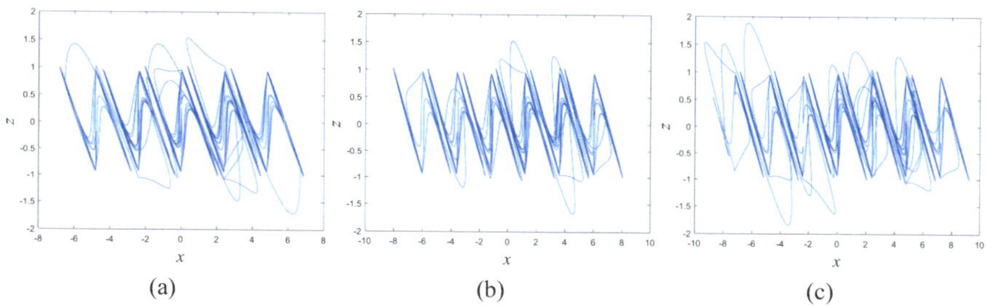

Fig. 4.27 1-D multi-double-scroll attractors on z-x plane [16]. **a** 5-double-scroll attractors. **b** 6-double-scroll attractors. **c** 7-double-scroll attractors

$$\begin{bmatrix} \dot{x} \\ \dot{y} \\ \dot{z} \end{bmatrix} = \begin{bmatrix} 0 & 1 & 0 \\ 0 & 0 & 1 \\ a & a & a \end{bmatrix} \begin{bmatrix} x - P_1(x) - f(x - P_1(x)) \\ y - P_2(x) \\ z \end{bmatrix}, \tag{4.30}$$

where

$$P_1(t) = \sum_{i=1}^{N_1} A_1 \text{sgn}(\sin(\omega_1 t - \phi_{i1})),$$

and

$$P_2(t) = \sum_{i=1}^{N_2} A_2 \text{sgn}(\sin(\omega_2 t - \phi_{i2})).$$

Here, $\phi_{i1} = (i-1)\pi/N_1$ and $\phi_{i2} = (i-1)\pi/N_2$.

Figure 4.28a shows 2×3-grid double-scroll attractors generated with the parameters $N_1 = 1$, $A_1 = 1.2$, $\omega_1 = 0.005$, $N_2 = 2$, $A_2 = 2$, and $\omega_2 = 0.003$. By introducing a pulse along the y-axis, a 2×4 double-scroll can be created, as shown in Fig. 4.28b. In contrast, adding a pulse along the x-axis results in 3×3 double-scroll attractors, as shown in Fig. 4.28c. This demonstrates that the number of scrolls can be easily modified by adding or removing the pulse sources in an appropriate direction.

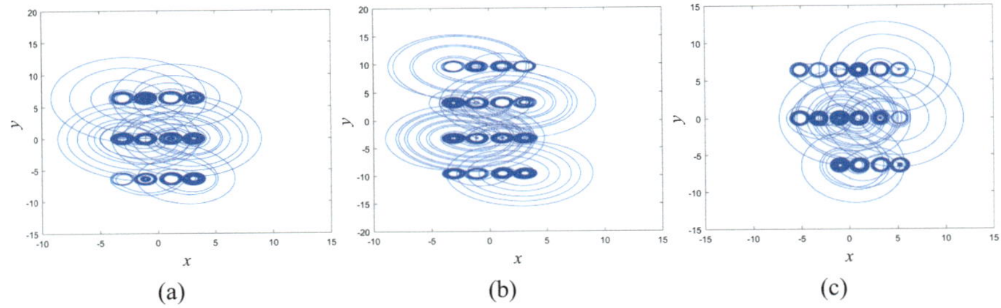

Fig. 4.28 2-D multi-double-scroll attractors on the x-y plane [16]. **a** 2 × 3-double-scroll attractors. **b** 2 × 4-double-scroll attractors. **c** 3 × 3-double-scroll attractors

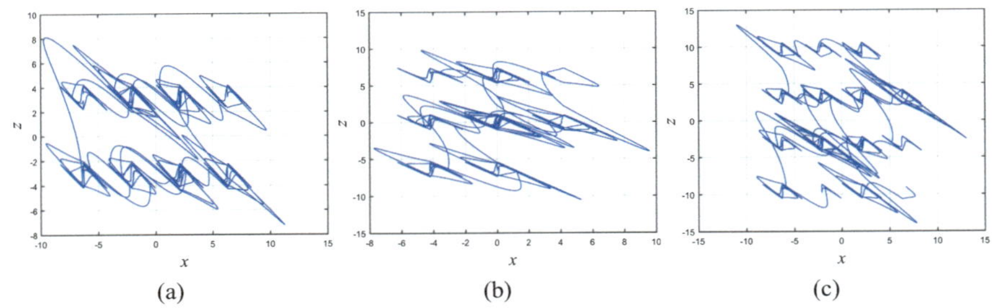

Fig. 4.29 2-D multi-double-scroll attractors on the x-z plane [16]. **a** 4 × 2 double-scroll attractors. **b** 3 × 3 double-scroll attractors. **c** 4 × 4 double-scroll attractors

2-D multi-double-scroll attractors situated on the x-z plane can also be generated, by

$$\begin{bmatrix} \dot{x} \\ \dot{y} \\ \dot{z} \end{bmatrix} = \begin{bmatrix} 0 & 1 & 0 \\ 0 & 0 & 1 \\ a & a & a \end{bmatrix} \begin{bmatrix} x - P_1(x) - f\,(x - P_1(x)) \\ y \\ z - P_2(x) \end{bmatrix}. \tag{4.31}$$

Figure 4.29 displays the 4 × 2, 3 × 3, and 4 × 4 double-scroll attractors on the x-z plane.

The system can be further modified by introducing an additional pulse-exciting source, $P_3(t)$, leading to

$$\begin{bmatrix} \dot{x} \\ \dot{y} \\ \dot{z} \end{bmatrix} = \begin{bmatrix} 0 & 1 & 0 \\ 0 & 0 & 1 \\ a & a & a \end{bmatrix} \begin{bmatrix} x - P_1(x) - f\,(x - P_1(x)) \\ y - P_2(x) \\ z - P_3(x) \end{bmatrix}, \tag{4.32}$$

where

$$P_3\,(t) = \sum_{i=1}^{N_3} A_3 \mathrm{sgn}\,(\sin\,(\omega_3 t - \phi_{i3})),$$

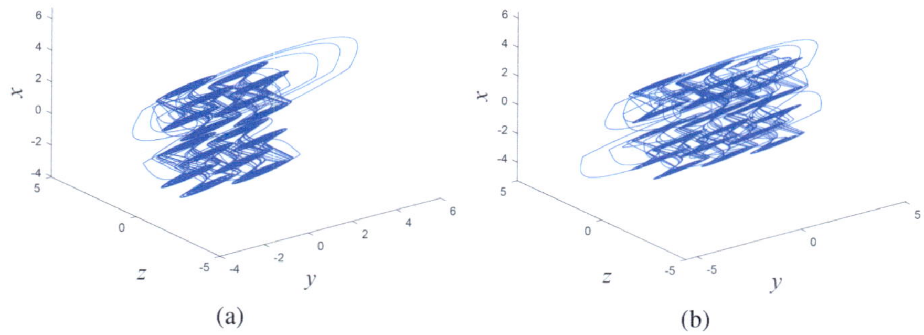

Fig. 4.30 3-D multi-double-scroll attractors [16]. **a** $2 \times 2 \times 2$ double-scroll attractors. **b** $3 \times 2 \times 2$ double-scroll attractors

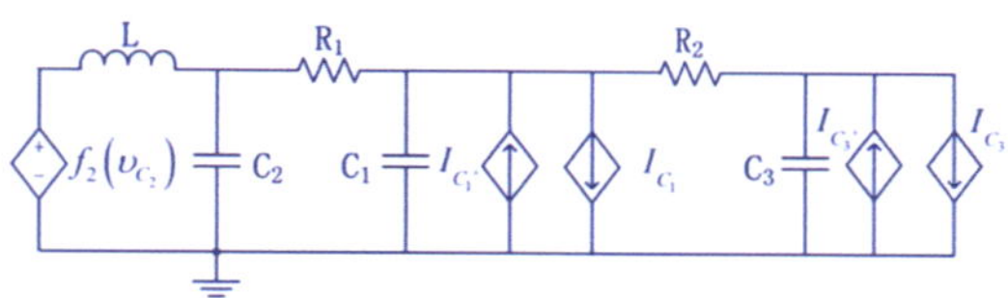

Fig. 4.31 Fourth-order Chua's circuit. Reprinted from [17], with permission from World Scientific Pub. Co.

with $\phi_{i3} = (i-1)\pi/N_3$.

Some examples of 3-D double-scroll grid attractors are shown in Fig. 4.30: Fig. 4.30a displays a $2 \times 2 \times 2$ double-scroll grid attractors with parameters $N_1 = N_2 = N_3 = 1$, $W_1 = 0.002$, $W_2 = 0.003$, $W_3 = 0.005$, $A_1 = 2$, $A_2 = 1.2$, and $A_3 = 1$; Fig. 4.30b shows a $3 \times 2 \times 2$ double-scroll grid attractors, where $N_1 = N_3 = 1$, $N_2 = 2$, $W_1 = 0.002$, $W_2 = 0.003$, $W_3 = 0.005$, $A_1 = 2$, $A_2 = 1.2$, and $A_3 = 1$.

4.6 Modified Chua's Circuit Framework

This section introduces the design framework using the modified Chua's circuit to generate multi-directional multiscroll (MDMS) chaotic attractors.

Wang et al. [17] constructing a higher-order Chua's circuit by using a dual-port RCL network. This higher-order circuit can generate (MDMS) chaotic attractors through the inclusion of an RC structure and some appropriate nonlinear functions.

A simple approach to constructing a higher-order Chua's circuit, suitable for an RCL structure and capable of generating MDMS chaotic attractors, is to add an energy storage device to the existing lower-order Chua's circuit. A fourth-order Chua's circuit is designed by incorporating a nonlinear controlled source at the output, as illustrated in Fig. 4.31.

The complete set of state equations for the fourth-order Chua's circuit are as follows:

$$\begin{cases} C_1 \dfrac{dv_{C1}}{d\tau} = \dfrac{1}{R_1}(v_{C2} - v_{C1}) + \dfrac{1}{R_2}(v_{C3} - v_{C1}) + I_{C_1^+} + I_{C_1^-}, \\ C_2 \dfrac{dv_{C2}}{d\tau} = \dfrac{1}{R_1}(v_{C1} - v_{C2}) + i_L, \\ L \dfrac{di_L}{d\tau} = -[v_{C2} - f_2(v_{C2})], \\ C_3 \dfrac{dv_{C3}}{d\tau} = \dfrac{1}{R_2}(v_{C1} - v_{C3}) + I_{C_3^+} + I_{C_3^-}, \end{cases} \tag{4.33}$$

where $I_{C_1^+} = [v_{C1} + \xi f_1(v_{C1})]/R_1$, $I_{C_1^-} = [f_2(v_{C2}) + \sigma f_3(v_{C3})]/R_1$, $I_{C_3^+} = f_3(v_{C3})/R_2$, $I_{C_3^-} = \lambda v_{C1} + \varsigma f_1(v_{C1})/R_2$, and

$$f_m(x) = A_m \left\{ \begin{array}{l} \sum\limits_{i=1}^{M} \mathrm{sgn}[x + (2i-1)A_m] \\ + \sum\limits_{i=1}^{M} \mathrm{sgn}[x - (2i-1)A_m] \end{array} \right\}, \quad \text{for} \quad M \geq 1.$$

Let $v_{C1} = x$, $v_{C2} = y$, $i_L = z/R_1$, $v_{C3} = w$, $\tau = R_1 C_2 t$, $a = C_2/C_1$, $\xi = \sigma = R_1/R_2$, $b = \xi C_2/C_1$, $c = C_2 R_1^2/L$, $\varsigma = 1 - \lambda$, and $d = \varsigma e = \varsigma C_2 R_1/(C_3 R_2)$. Then, Eq. (4.23) can be described in the dimensionless form, as follows:

$$\begin{cases} \dfrac{dx}{dt} = a(y - f_2(y)) - b(x - f_1(x) - (w - f_3(w))), \\ \dfrac{dy}{dt} = x - y + z, \\ \dfrac{dz}{dt} = -c(y - f_2(y)), \\ \dfrac{dw}{dt} = d(x - f_1(x)) - e(w - f_3(w)). \end{cases} \tag{4.34}$$

Here, $f_3(w)$ is one of the functions in $f_m(\chi)$. The coefficients for the state variables are chosen as follows: $a = 15, b = 2.5, c = 16, d = 2.5$, and $e = 4$. For the nonlinear function $f_1(x)$, the parameters are set to $N = 2$, $A_1 = 0.5$, $M = 2$, and $A_2 = 0.25$ in $f_2(y)$, while $N = 1$ and $A_3 = 0.25$ are used in $f_3(w)$.

The system generates a 6×5 multi-scroll attractors on the x-y plane, as shown in Fig. 4.32a, and a 6×4 multi-scroll attractors on the x-w plane, as shown in Fig. 4.32b. Figure 4.32c displays the phase diagram in the 3-D space.

The system allows for modification of the number of multi-scrolls by adjusting the values of M or N. For example, $M = 4$ in $f_1(x)$, $M = 3$ in $f_2(y)$, and $M = 3$ in $f_3(w)$. Figure 4.33 demonstrates the chaotic attractors after modifying the nonlinear functions. Figure 4.33a and c display the phase diagrams on the x-y plane, x-w plane, and in the x-y-w space, respectively.

Furthermore, this method can be extended to fifth-order and sixth-order circuits, as shown in Fig. 4.34.

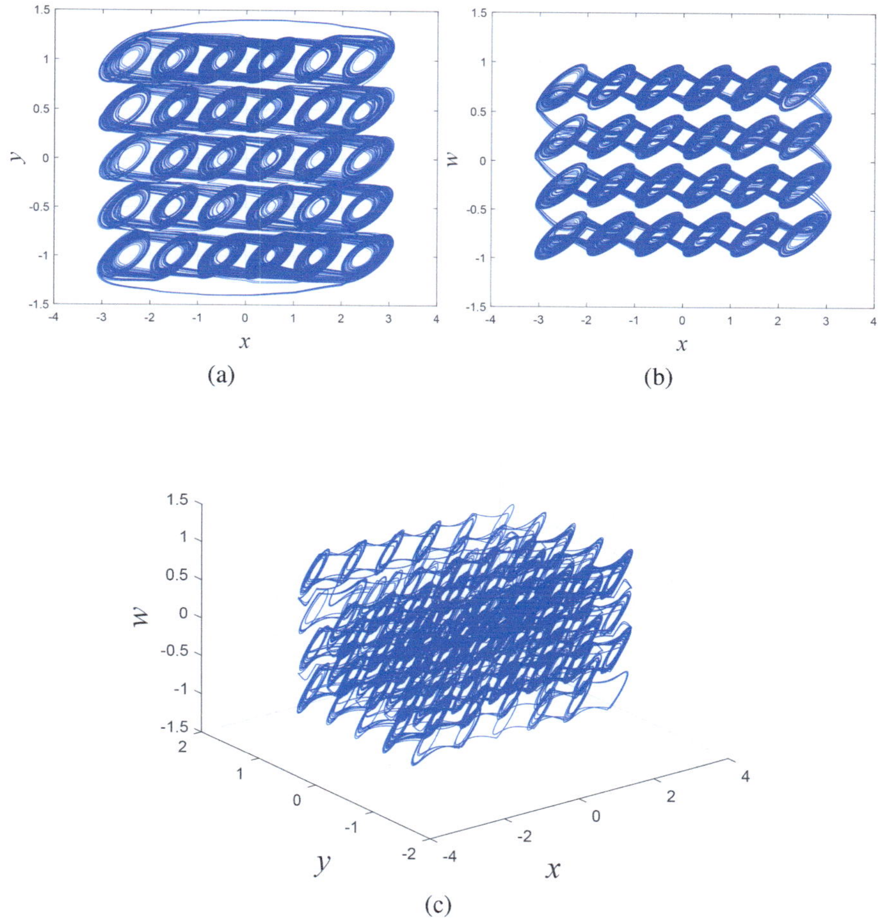

Fig. 4.32 $6 \times 5 \times 4$ multi-scrolls [17]. **a** x-y plane. **b** x-w plane. **c** x-y-w space

After the normalization process for the fourth-order circuit, the equations for normalizing the fifth- and sixth-order Chua's circuits are obtained, as follows:

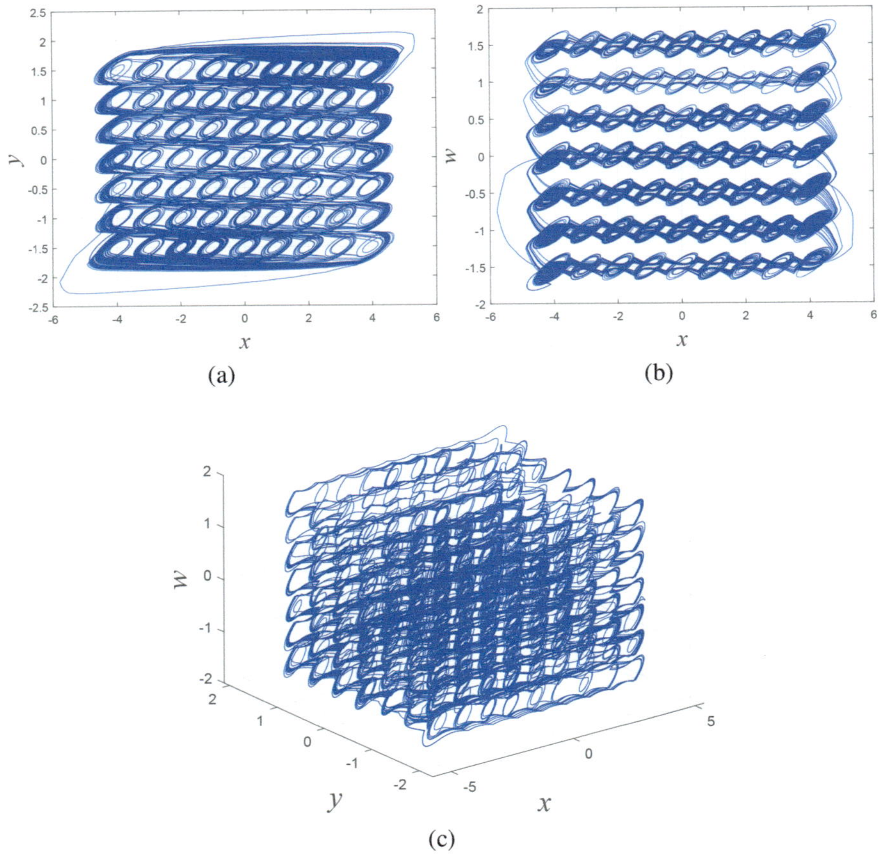

Fig. 4.33 $9 \times 7 \times 7$ multi-scrolls [17]. **a** x-y plane. **b** x-w plane. **c** x-y-w space

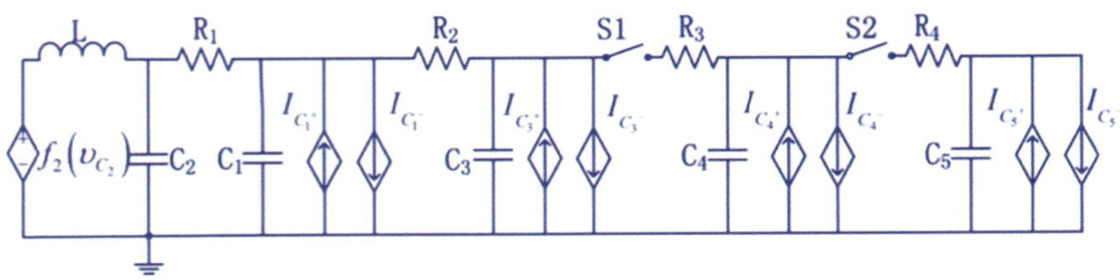

Fig. 4.34 Fourth to sixth-order Chua's circuit. Reprinted from [17], with permission from World Scientific Pub. Co.

$$\begin{cases} \dfrac{dx}{dt} = a\left(y - f_2\left(y\right)\right) - b\left(x - f_1\left(x\right) - \left(\omega - f_3\left(\omega\right)\right)\right), \\ \dfrac{dy}{dt} = x - y + z, \\ \dfrac{dz}{dt} = -c\left(y - f_2\left(y\right)\right), \\ \dfrac{d\omega}{dt} = d\left(x - f_1\left(x\right)\right) - e\left(\omega - f_3\left(\omega\right)\right) + g\left(u - f_4\left(u\right)\right), \\ \dfrac{du}{dt} = h\left(\omega - f_3\left(\omega\right)\right) - k\left(u - f_4\left(u\right)\right), \end{cases} \tag{4.35}$$

and

$$\begin{cases} \dfrac{dx}{dt} = a\left(y - f_2\left(y\right)\right) - b\left(x - f_1\left(x\right) - \left(\omega - f_3\left(\omega\right)\right)\right), \\ \dfrac{dy}{dt} = x - y + z, \\ \dfrac{dz}{dt} = -c\left(y - f_2\left(y\right)\right), \\ \dfrac{d\omega}{dt} = d\left(x - f_1\left(x\right)\right) - e\left(\omega - f_3\left(\omega\right)\right) + g\left(u - f_4\left(u\right)\right), \\ \dfrac{du}{dt} = h\left(\omega - f_3\left(\omega\right)\right) - k\left(u - f_4\left(u\right)\right) + l\left(v - f_5\left(v\right)\right), \\ \dfrac{dv}{dt} = m\left(u - f_4\left(u\right)\right) - n\left(v - f_5\left(v\right)\right), \end{cases} \tag{4.36}$$

where $f_4(u)$ and $f_5(v)$ are one of the $f_m(x)$, respectively.

The first step in determining the parameter values for each component of the sixth-order Chua's circuit is to express the state equation of the circuit as follows:

$$\begin{aligned} C_1\frac{dv_{C1}}{d\tau} &= \frac{1}{R_1}\left(v_{C2} - v_{C1}\right) + \frac{1}{R_2}\left(v_{C3} - v_{C1}\right) + \frac{v_{C1}+\xi f_1(v_{C1})}{R_1} - \frac{f_2(v_{C2})+\sigma f_3(v_{C3})}{R_1}, \\ C_2\frac{dv_{C2}}{d\tau} &= \frac{1}{R_1}\left(v_{C1} - v_{C2}\right) + i_L, \\ L\frac{di_L}{d\tau} &= -\left[v_{C2} - f_2\left(v_{C2}\right)\right], \\ C_3\frac{dv_{C3}}{d\tau} &= \frac{1}{R_2}\left(v_{C1} - v_{C3}\right) + \frac{1}{R_3}\left(v_{C4} - v_{C3}\right) + \frac{\phi v_{C3}+\varepsilon f_3(v_{C3})}{R_2} - \frac{f_1(v_{C1})+\varsigma f_4(v_{C4})}{R_2}, \\ C_4\frac{dv_{C4}}{d\tau} &= \frac{1}{R_3}\left(v_{C3} - v_{C4}\right) + \frac{1}{R_4}\left(v_{C5} - v_{C4}\right) + \frac{\alpha v_{C4}+\beta f_4(v_{C4})}{R_3} - \frac{f_3(v_{C3})+\lambda f_5(v_{C5})}{R_3}, \\ C_5\frac{dv_{C5}}{d\tau} &= \frac{1}{R_4}\left(v_{C4} - v_{C5}\right) + \frac{\rho v_{C5}+\theta f_5(v_{C5})}{R_4} - \frac{f_4(v_{C4})}{R_5}. \end{aligned}$$

Set $a = C_2/C_1 = 15$ with $C_1 = 16\,\text{nF}$ and $C_2 = 240\,\text{nF}$. Let $c = C_2R_1^2/L = 16$, $L = 15\,\text{mH}$ after choosing $R_1 = 1\,\text{k}\Omega$. The parameters ξ and σ are both determined as $1/6$, based on $b = \xi C_2/C_1 = 2.5$, $\xi = \sigma = R_1/R_2$, and $R_2 = 6\,\text{k}\Omega$.

The remaining parameters are obtained as follows: $C_3 = 16\,\text{nF}$, $C_4 = 800\,\text{pF}$, $C_5 = 64\,\text{pF}$, $R_3 = 30\,\text{k}\Omega$, and $R_4 = 1500\,\text{k}\Omega$. Through similar calculations, the parameters are determined as: $\alpha = 0.42$, $\beta = 0.6$, $\lambda = 0.02$, $\phi = -0.4$, $\varepsilon = 1.6$, $\varsigma = 0.2$, $\theta = 0.3$, and $\rho = 0.7$.

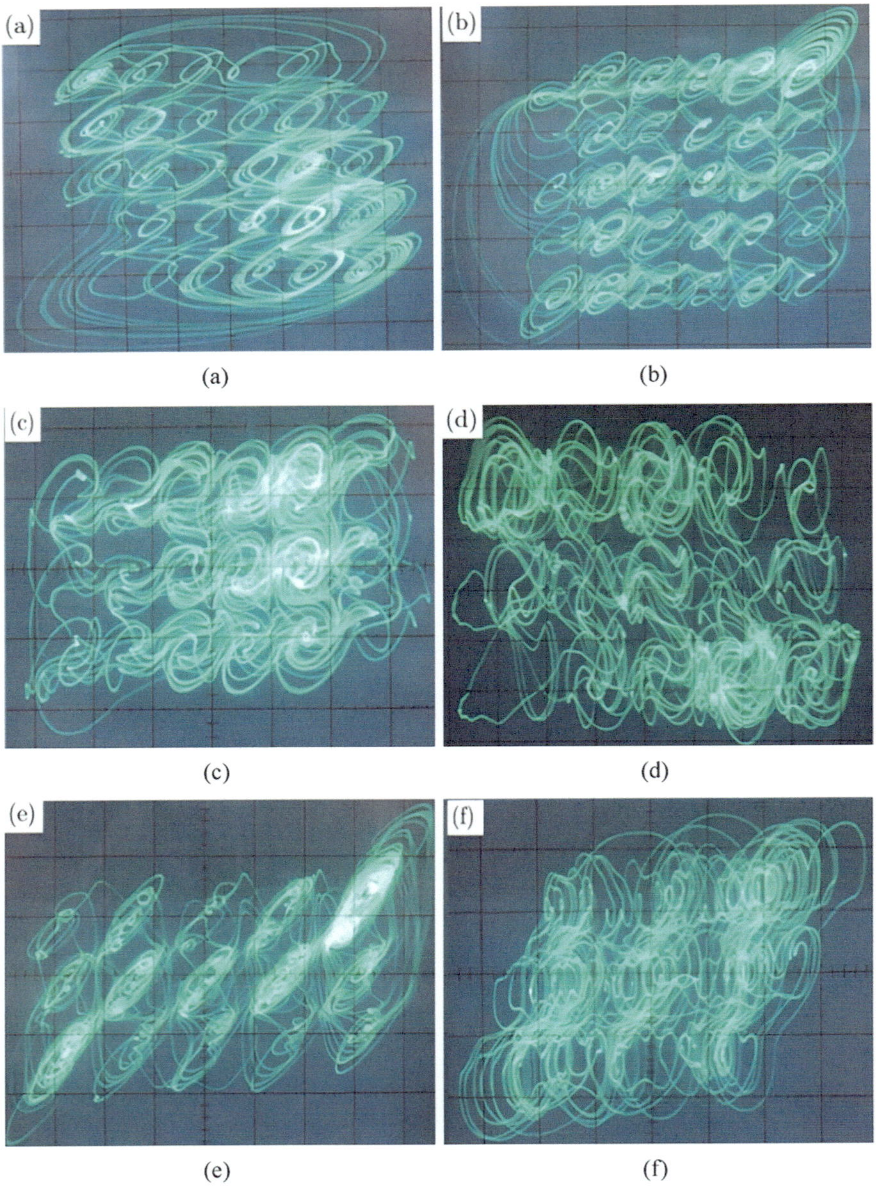

Fig. 4.35 $6 \times 5 \times 5 \times 3 \times 3$ scroll attractors. **a** v_{C1}-v_{C2}. **b** v_{C1}-v_{C3}. **c** v_{C1}-v_{C4}. **d** v_{C2}-v_{C4}. **e** v_{C3}-v_{C4}. **f** v_{C4}-v_{C5}. Reprinted from [17], with permission from World Scientific Pub. Co.

As an example, for generating $6 \times 5 \times 5 \times 3 \times 3$ scrolls, the circuit with stair functions requires appropriate settings. The experimental results are shown in Fig. 4.35.

4.7 Conclusion

This chapter systematically explores the generation of multidirectional multiscroll chaotic attractors using fundamental circuits such as step circuits, hysteresis circuits, and saturated circuits. By leveraging advanced techniques, including step functions, hysteresis series and saturated nonlinear functions, the chapter demonstrates the ability to design chaotic attractors in 1-D, 2-D, and 3-D settings with scalable complexity. Theoretical analyses are supported by numerical simulations and verified through experimental circuit implementations, showcasing the practical feasibility of these methods. The findings contribute significantly to the field of chaos theory, offering robust tools for applications in secure communications, nonlinear system control and the modeling of complex dynamical systems.

References

1. Yalcin, M.E., Suykens, J.A.K., Vandewalle, J., Ozoguz, S.: Families of scroll grid attractors. Int. J. Bifur. Chaos **12**, 23–41 (2002)
2. Ma, Y., Jiang, H., Ju, L.: Circuit implementation of multiscroll chaotic attractors generated using step function. Int. J. Bifur. Chaos **25**, 1550160 (2015)
3. Han, F., Lü, J., Yu, X., et al: Generating multi-scroll chaotic attractors via a linear second-order hysteresis system. In: Dynamics of Continuous, Discrete and Impulsive Systems Series B: Applications and Algorithms, vol. 12, pp. 95–110 (2005)
4. Lü, J., Chen, G.: Generating multiscroll chaotic attractors: theories, methods and applications. Int. J. Bifur. Chaos **16**, 775–858 (2006)
5. Lü, J., Chen, G., Yu, X., Leung, H.: Design and analysis of multiscroll chaotic attractors from saturated function series. IEEE Trans. Circuits Syst.-I **51**, 2476–2490 (2004)
6. Lü, J., Yu, S., Leung, H., Chen, G.: Experimental verification of 3-D hysteresis multi-scroll chaotic attractors. In: IEEE International Symposium on Circuits and Systems, vol. 4, May 23–26, Kobe, Japan, pp. 3391–3394 (2005)
7. Lü, J., Yu, S., Leung, H., Chen, G.: Experimental verification of multidirectional multiscroll chaotic attractors. IEEE Trans. Circuits Syst.-I **53**, 149–165 (2006)
8. Lü, J., Yu, S., Leung, H.: Design of 3-D multi-scroll chaotic attractors via basic circuits. In: 4th International Conference on Engineering Applications and Computational Algorithms, vol. 8, pp. 324–328 (2005)
9. Lü, J., Yu, X., Chen, G.: Switching control for multi-scroll chaos generation: an overview. In: International Conference on Physics and Control, vol. 2, August 20–22, pp. 420–428 (2003)
10. Lai, Q., Kuate, P.D.K., Liu, F., Iu, H.H.C.: An extremely simple chaotic system with infinitely many coexisting attractors. IEEE Trans. Circuits Syst.-II **67**, 1129–1133 (2019)
11. Lü, J., Chen, G., Yu, X., Leung, H.: Generating multi-scroll chaotic attractors via switching control. In: 5th Asian Control Conference, vol. 3, July 20–23, Melbourne, Australia, pp. 1753–1761 (2004)

12. Lü, J., Yu, X., Chen, G.: Generating chaotic attractors with multiple merged basins of attraction: a switching piecewise-linear control approach. IEEE Trans. Circuits Syst.-I **50**, 198–207 (2003)
13. Chen, G., Yu, X.: Chaos Control: Theory and Applications. Springer, Berlin (2003)
14. Silva, C.P.: Shil'nikov's theorem–A tutorial. IEEE Trans. Circuits Syst.-I **40**, 675–682 (1993)
15. Azam, A., Aqeel, M., Sunny, D.A.: Generation of multidirectional mirror symmetric multiscroll chaotic attractors (MSMCA) in double wing satellite chaotic system. Chaos Solit. Fract. **155**, 111715 (2022)
16. Hong, Q., Xie, Q., Xiao, P.: A novel approach for generating multidirection multi-double-scroll attractors. Nonlinear Dyn. **87**, 1015–1030 (2017)
17. Wang, C.H., Xu, H., Yu, F.: A novel approach for constructing high-order Chua's circuit with multi-directional multi-scroll chaotic attractors. Int. J. Bifur. Chaos **23**, 1350022 (2013)

Chapter 5
Generating Multiscroll Chaotic Attractors with Switching Controls

Abstract This chapter explores advanced methods for generating multiscroll chaotic attractors using switching controls. Control strategies, including piecewise linear switching and step-series switching, are introduced, enabling the generation of n-scroll attractors from linear autonomous systems. These approaches are then extended to creating compound chaotic attractors by combining multiple Lorenz-like systems. Practical implementations, such as circuit-based realizations, verify the theoretical frameworks and demonstrate their versatility in generating multiscroll chaotic attractors and various complex dynamical behaviors.

5.1 Generating Chaotic Attractors with PWL Switching Control

This section outlines a control strategy for effectively generating n-scroll chaotic attractors from a 3-D linear autonomous system.

Lü et al. [1, 2] developed a systematic method using switching piecewise linear (PWL) control functions to generate chaotic attractors with multiple combined basins of attraction, based on the following 3-D linear controlled system:

$$\dot{X} = AX + U(X). \tag{5.1}$$

Here, $X = (x, y, z)^T$, $A = \begin{pmatrix} a & b & 0 \\ -b & a & 0 \\ 0 & 0 & c \end{pmatrix}$, and

$$U(X) = f_1(X) = \begin{cases} k \begin{pmatrix} -x \\ -y \\ d \end{pmatrix}, & \text{if} \quad z + \sqrt{x^2 + y^2} > k, \\ 0, & \text{otherwise}, \end{cases}$$

where a, b, c, d, k are real parameters. System (5.1) with controller $U(X) = f_1(X)$ can generate chaotic attractors within an appropriate parameter value range.

J. Lü et al., *Chaotic Systems with Multiscroll Attractors*,
https://doi.org/10.1007/978-3-032-09854-2_5

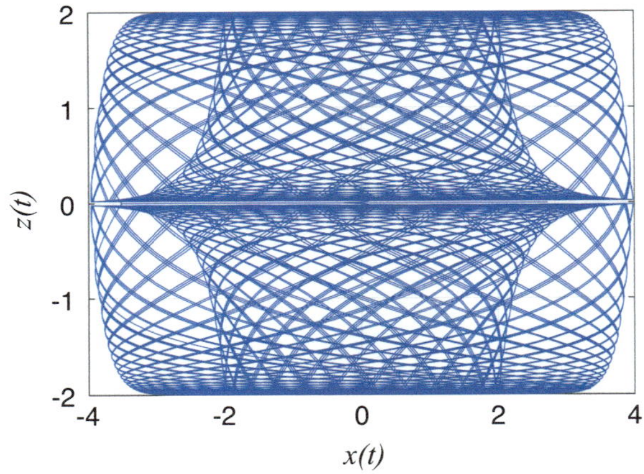

Fig. 5.1 Upper-attractor and lower-attractor [1, 2]

To simultaneously generate two chaotic attractors from system (5.1), the controller $U(X)$ is modified as follows:

$$f_2(X) = \begin{cases} k\begin{pmatrix} -x \\ -y \\ d \end{pmatrix}, & \text{if } z > 0, z + \sqrt{x^2 + y^2} > k, \\ m\begin{pmatrix} -x \\ -y \\ e \end{pmatrix}, & \text{if } z < 0, z - \sqrt{x^2 + y^2} < -m, \\ 0, & \text{otherwise,} \end{cases}$$

where a, b, c, d, k, m, e are real parameters.

System (5.1) utilizes the controller $U(X) = f_2(X)$ to generate two simultaneous chaotic attractors, namely, an upper attractor and a lower attractor. This occurs across a wide range of parameter values. Notably, the line $z = 0$ serves as an invariant manifold for the system when using this switching controller. For instance, when the parameters are set as $a = 3$, $b = 20$, $c = -20$, $d = 10$, $e = -10$, $k = 4$, and $m = 4$, the system successfully produces both the upper and lower chaotic attractors concurrently, as illustrated in Fig. 5.1.

Remark It is possible to generate n distinct chaotic attractors simultaneously, each with different initial conditions, in the switching system (5.1) by applying parallel displacement and rotational transformations. Since all chaotic attractors are confined within a finite sphere, this space can be divided into n distinct subspaces, where the original attractor–whether it is the upper or lower attractor–can be replicated in each subspace.

Note that these n attractors are completely independent, with no trajectories connecting any two distinct attractors. Therefore, to obtain a single one-piece orbit connecting all attractors, some design work is needed.

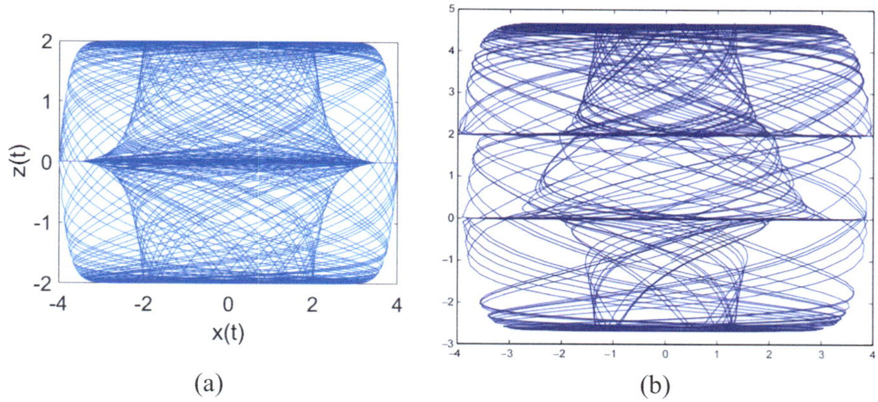

Fig. 5.2 Chaotic attractors with multiple merged basins of attraction in the x-z plane. **a** Attractor with two merged basins of attraction. **b** Attractor with three merged basins of attraction. Reprinted from [3], with permission from World Scientific Pub. Co.

For $n = 2$, to connect the orbits of the upper and lower attractors to form a single chaotic attractor, the controller needs to be designed to guide the orbit of the upper attractor to cross the invariant manifold $z = 0$ from above, entering the subspace below, namely $\{(x, y, z)|z < 0\}$. At the same time, the controller should ensure that the orbit of the lower attractor is able to cross the invariant manifold $z = 0$ from below, returning to the upper subspace $\{(x, y, z)|z > 0\}$ later. Following this idea, $\delta(0, 0, -\text{sgn}(z))$ is used to replace 0 in the controller $f_2(X)$, resulting in the following modified controller:

$$f_2(X) = \begin{cases} k\begin{pmatrix} -x \\ -y \\ d \end{pmatrix}, & \text{if} \quad z > 0, z + \sqrt{x^2 + y^2} > k, \\ m\begin{pmatrix} -x \\ -y \\ e \end{pmatrix}, & \text{if} \quad z < 0, z - \sqrt{x^2 + y^2} < -m, \\ \delta\begin{pmatrix} 0 \\ 0 \\ -\text{sgn}\,(z) \end{pmatrix}, & \text{otherwise}, \end{cases}$$

where $a, b, c, d, k, m, e, \delta$ are real parameters.

With $a = 3, b = 20, c = -20, d = -e = 10, k = m = 4$, and $\delta = 1$, the system described in Eq. (5.1) produces a chaotic attractor with two merged basins of attraction: the upper basin and the lower basin, as illustrated in Fig. 5.2a. The controller $U(X) = f_3(X)$ includes three switching planes: S_1, S_2, and $z = 0$. The planes S_1 and S_2 are responsible for forming the upper and lower chaotic attractors, while the plane $z = 0$ works for connecting these two chaotic attractors.

Following this approach, to generate a chaotic attractor with three merged basins of attraction, the controller $U(X)$ can be designed using $\delta(0, 0, -\text{sign}(z-h) - ch/\delta)$, given by

$$f_4(X) = \begin{cases} \text{if} \quad z < h, \\ k\begin{pmatrix} -x \\ -y \\ d \end{pmatrix}, & \text{if} \quad z > 0, z + \sqrt{x^2+y^2} > k, \\ m\begin{pmatrix} -x \\ -y \\ e \end{pmatrix}, & \text{if} \quad z < 0, z - \sqrt{x^2+y^2} < -m, \\ \delta\begin{pmatrix} 0 \\ 0 \\ -\text{sgn}\,(\text{z}) \end{pmatrix}, & \text{otherwise}, \\ \text{if} \quad z \geq h, \\ k\begin{pmatrix} -x \\ -y \\ d - \frac{ch}{k} \end{pmatrix}, & \text{if} \quad z - h + \sqrt{x^2+y^2} > k, \\ \delta\begin{pmatrix} 0 \\ 0 \\ -\text{sgn}\,(\text{z}-\text{h}) - \frac{\text{ch}}{\delta} \end{pmatrix}, & \text{otherwise}, \end{cases}$$

where $a, b, c, d, k, m, e, \delta, h$ are real parameters.

When $a = 3$, $b = 20$, $c = -15$, $d = -e = 10$, $k = m = 4$, $h = 2$, and $\delta = 5$, system (5.1) with the controller $U(X) = f_4(X)$ generates a chaotic attractor with three merged basins of attraction: two upper basins and one lower basin, as shown in Fig. 5.2b.

Similarly, chaotic attractors with n merged basins of attraction can be constructed using the above structured design method, as summarized below.

1. Divide the entire space into n subspaces. For simplicity, this division can be made along the z-axis.
2. Duplicate both the upper and lower attractors in each of the subspaces.
3. Implement a switching controller to connect the n independent attractors, generating a unified chaotic attractor with multiple merged basins of attraction, as shown in Fig. 5.3.

The switching controller can be represented as $\delta\,\text{sgn}(z - h_i)$, where h_i denotes the height between two adjacent subspaces, which must be less than that of an individual chaotic attractor.

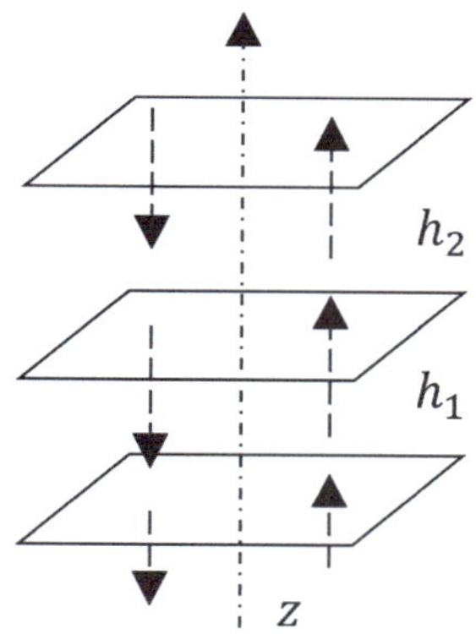

Fig. 5.3 Sketch of the connection of orbit. Reprinted from [3], with permission from World Scientific Pub Co.

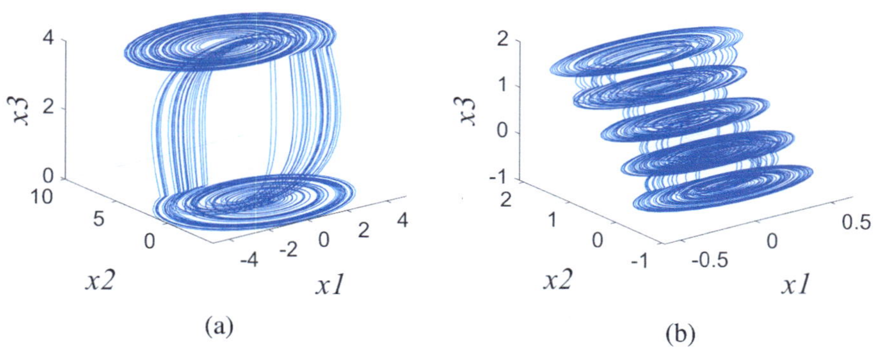

Fig. 5.4 n-Scroll chaotic attractors generated with step-series switching [4]. **a** Double-scroll chaotic attractor. **b** 5-Scroll chaotic attractor

5.2 Generating Multiscroll Chaotic Attractors with Step-Series Switching Control

This section describes of the step-series switching technique for generating multi-scroll chaotic attractors from basic linear systems.

Yang and Li [4] proposed the following PWL function switching system:

$$A = \begin{pmatrix} 0.5 & 10 & 0 \\ -10 & 0.5 & 0 \\ 0 & 0 & -10 \end{pmatrix}, \quad B = \begin{pmatrix} b \\ 4 \\ 4 \end{pmatrix}, \quad c = \begin{pmatrix} 0 \\ 1 \\ 1 \end{pmatrix}, \tag{5.2}$$

where $b = 0$ and $s(\eta) = \begin{cases} 1, & \eta > 0, \\ 0, & \eta \leq 0. \end{cases}$

The system (5.2) has a double-scroll chaotic attractor, as illustrated in Fig. 5.4a.

To generate n-scroll chaotic attractors, Yang and Li [4] modified the function $s(\eta)$ as follows:

$$s(\eta) = \begin{cases} h_1, & \text{if } \eta \le a_1, \\ h_2, & \text{if } a_{i-1} < \eta \le a_i, \text{ for } 2 \le i \le n-1, \\ h_3, & \text{if } \eta > a_n, \end{cases}$$

where $\{a_1, \ldots, a_n\}$ and $\{h_1, \ldots, h_n\}$ are strictly increasing constant sequences. Consequently, system (5.2) can be rewritten as

$$\dot{X} = A\left[X - s\left(C^T X\right) B\right], \tag{5.3}$$

where

$$A = \begin{pmatrix} \alpha & \beta & 0 \\ -\beta & \alpha & 0 \\ 0 & 0 & -\gamma \end{pmatrix}, \quad B = \begin{pmatrix} 0 \\ d \\ d \end{pmatrix}, \quad C = \begin{pmatrix} 0 \\ 1 \\ 1 \end{pmatrix}.$$

With $a_1 = -0.5, a_2 = 0.5, a_3 = 1.5, a_4 = 2.5,\ h_1 = -1, h_2 = 0, h_3 = 1, h_4 = 2, h_5 = 3, \alpha = 0.5,\ \beta = 10,\ \gamma = 5$, and $d = 0$, system (5.3) generates a 5-scroll chaotic attractor, as shown in Fig. 5.4b.

Li et al. [5] designed a circuit for generating multiscroll chaotic attractors, as shown in Fig. 5.5. The equation of the circuit is

$$\begin{pmatrix} \dot{v}_1 \\ \dot{v}_2 \\ \dot{v}_3 \end{pmatrix} = \frac{1}{C} \begin{pmatrix} 0 & \frac{1}{R_{f21}} & 0 \\ -\frac{1}{R_{f12}} & \frac{1}{R_{f22}} & 0 \\ 0 & 0 & -\frac{1}{R_{f33}} \end{pmatrix} \begin{pmatrix} v_1 - w(v_1 + v_3) \\ v_2 - w(v_2 + v_3) \\ v_3 - 0.5w(v_1 + v_3) - 0.5w(v_2 + v_3) \end{pmatrix}, \tag{5.4}$$

where

$$w(\zeta) = \begin{cases} 0, & \text{if } \zeta < \dfrac{V_{CC}}{m}, \\ i\dfrac{V_{CC}}{m}, & \text{if } (2i-1)\dfrac{V_{CC}}{m} \le \zeta \le (2i+1)\dfrac{V_{CC}}{m}, \text{ for } 1 \le i \le n-1, \\ n\dfrac{V_{CC}}{m}, & \text{if } \zeta \ge (2n-1)\dfrac{V_{CC}}{m}. \end{cases}$$

Let $\tau = t/\tau_0, 1/\tau_0 = 1/RC, R = 3k\Omega,\ C = 100pF, V_{BP} = 1V, x = v_1/V_{BP}, y = v_2/V_{BP}$, and $z = v_3/V_{BP}$.

This results in a series of dimensionless state equations, as follows:

$$\begin{aligned} \frac{dx}{d\tau} &= y - w(y+z), \\ \frac{dy}{d\tau} &= -x + ay + w(x+z) - aw(y+z), \\ \frac{dz}{d\tau} &= -z + bw(x+z) + bw(y+z), \end{aligned} \tag{5.5}$$

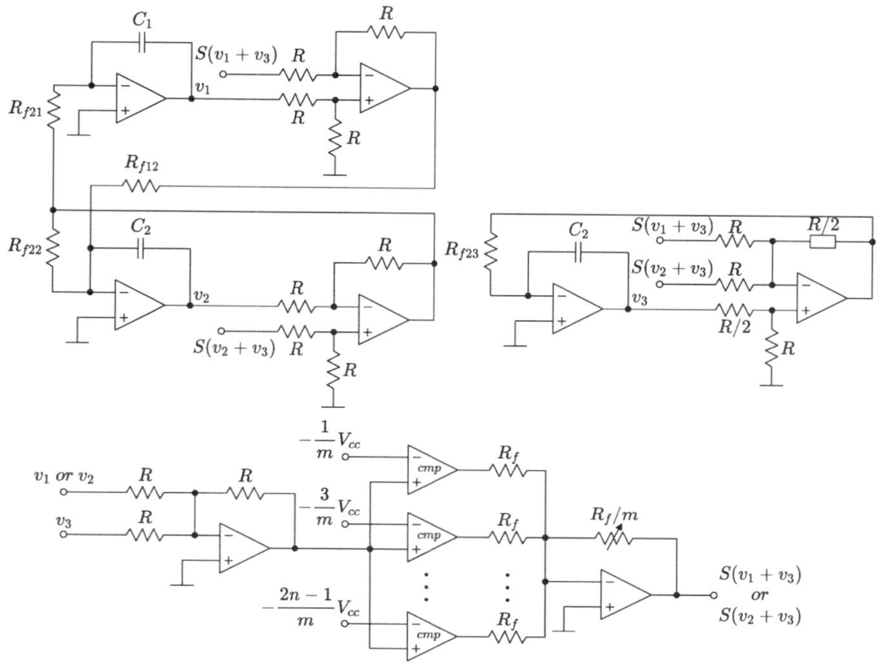

Fig. 5.5 Circuit diagram for 7-scroll chaotic attractors. Reprinted from [3], with permission from World Scientific Pub. Co.

where $w(\zeta) = \frac{1}{2}\sum_{i=0}^{n} u(\zeta - 0.5 - i)$ and $u(\cdot)$ is the unit step function.

With $a = 0.1, b = 0.5$, and $w(\zeta) = 0.5u(\zeta - 0.5) + 0.5u(\zeta - 1.5)$, system (5.5) generates a 7-scroll chaotic attractor, as shown in Fig. 5.6a. The system parameters are set as follows: $V_{CC} = 5V$, $R = R_{f12} = R_{f21} = R_{f33} = 3k\Omega$, $R_{f22} = 30k\Omega$, $R_i = 30k\Omega$, $C_1 = C_2 = C_3 = C = 100pF$, $m = 10$, and $n = 2$. The operational amplifier is powered by $\pm V_{CC}$, and the comparator operates with a supply of $-V_{CC}$ to 0. Figure 5.6b shows the experimental observation of the 7-scroll chaotic attractor.

5.3 Generating Compound Chaotic Attractors with Switching Control

Zhang et al. [6] proposed a method for generating compound chaotic attractors using an autonomous system based on a class of generalized Lorenz systems with switching control.

Consider N 3-D generalized Lorenz systems, each invariant under the transformation $(x, y, z) \to (-x, -y, z)$, which are generically described by

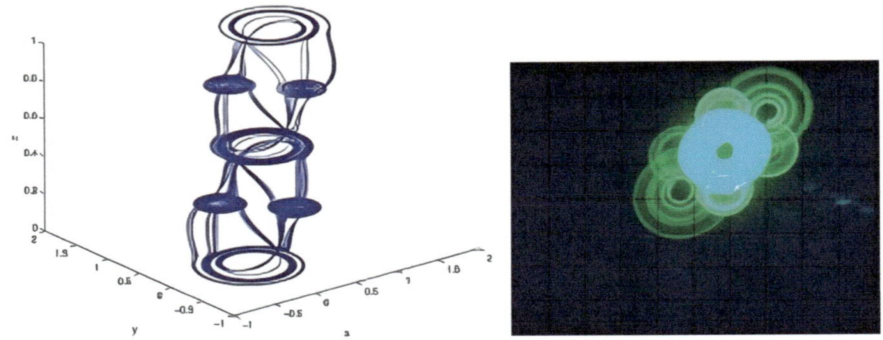

(a) (b)

Fig. 5.6 7-Scroll chaotic attractors. **a** 7-Scroll chaotic attractor obtained via numerical simulation. **b** Experimental observation of 7-scroll chaotic attractor. Reprinted from [3], with permission from World Scientific Pub. Co.

$$\begin{cases} \dot{x} = f_{i1}(x, y, z), \\ \dot{y} = f_{i2}(x, y, z), \\ \dot{z} = f_{i3}(x, y, z), \end{cases} \tag{5.6}$$

where $i = 1, 2, \ldots, N$. The procedure proposed for these N systems to generate a compound chaotic attractor involves the following four steps:

(1) To ensure that all generated attractors of the N systems have comparable sizes in the phase space, apply a state variable scaling transformation to each system, as follows: scale $K_i x$ to x, $K_i y$ to y, and $K_i z$ to z. As a result of this transformation, the system becomes

$$\begin{cases} \dot{x} = K_i\, f_{i1}\left(\dfrac{x}{K_i}, \dfrac{y}{K_i}, \dfrac{z}{K_i}\right), \\ \dot{y} = K_i\, f_{i2}\left(\dfrac{x}{K_i}, \dfrac{y}{K_i}, \dfrac{z}{K_i}\right), \\ \dot{z} = K_i\, f_{i3}\left(\dfrac{x}{K_i}, \dfrac{y}{K_i}, \dfrac{z}{K_i}\right), \end{cases} \tag{5.7}$$

where K_i are state variable scaling factors to be determined, $i = 1, 2, ..., N$.

(2) Translate the coordinates along the z-axis by applying different translation values: z_i (for $i = 1, 2, \ldots, N$) to each system. This adjustment ensures that the chaotic attractors are located in distinct regions defined by $V_i = \{(x, y, z) | x_{i1} \leq x \leq x_{i2}, y_{i1} \leq y \leq y_{i2}, z_{i1} \leq z \leq z_{i2}\}$ (for $i = 1, 2, \ldots, N$) within the same phase space:

$$\begin{cases} \dot{x} = K_i\, f_{i1}\left(\dfrac{x}{K_i}, \dfrac{y}{K_i}, \dfrac{z - z_i}{K_i}\right), \\ \dot{y} = K_i\, f_{i2}\left(\dfrac{x}{K_i}, \dfrac{y}{K_i}, \dfrac{z - z_i}{K_i}\right), \\ \dot{z} = K_i\, f_{i3}\left(\dfrac{x}{K_i}, \dfrac{y}{K_i}, \dfrac{z - z_i}{K_i}\right). \end{cases} \tag{5.8}$$

(3) Choose suitable coordinate translation values z_i (for $i = 1, 2, \ldots, N$) such that each pair of adjacent chaotic attractors intersects in a common connected domain, satisfying the following conditions:

$$\begin{cases} V_j \cap V_{j+1} \neq \emptyset, \\ 0 < z_{i+,1} - z_{i2} \leq \delta_i, \end{cases} \tag{5.9}$$

where $i, j = 1, 2, ..., N$ and δ_i $(i = 1, 2, ..., N)$ are suitably positive constants.

(4) Use these N systems to generate a compound chaotic attractor through switching control, as follows:

$$\begin{pmatrix} \dot{x} \\ \dot{y} \\ \dot{z} \end{pmatrix} = \begin{pmatrix} K_1 f_{11}\left(\frac{x}{K_1}, \frac{y}{K_1}, \frac{z-z_1}{K_1}\right) \cdots K_N f_{N1}\left(\frac{x}{K_N}, \frac{y}{K_N}, \frac{z-z_N}{K_N}\right) \\ K_1 f_{12}\left(\frac{x}{K_1}, \frac{y}{K_1}, \frac{z-z_1}{K_1}\right) \cdots K_N f_{N2}\left(\frac{x}{K_N}, \frac{y}{K_N}, \frac{z-z_N}{K_N}\right) \\ K_1 f_{13}\left(\frac{x}{K_1}, \frac{y}{K_1}, \frac{z-z_1}{K_1}\right) \cdots K_N f_{N3}\left(\frac{x}{K_N}, \frac{y}{K_N}, \frac{z-z_N}{K_N}\right) \end{pmatrix} \begin{pmatrix} S_1 \\ S_2 \\ \vdots \\ S_n \end{pmatrix},$$

where $S_1, S_2, ..., S_N$ are switching controllers to be designed.

The procedure outlined above provides the basic steps in principle, with further technical details discussed below for different case studies.

5.3.1 Compound Lorenz-Chen-Lü Chaotic Attractors

Based on Eq. (5.6) and the above steps (1)–(4), a compound Lorenz-Chen-Lü system is constructed, expressed as

$$\begin{cases} \dot{x} = \left\{ \begin{array}{l} (-a_1 x + a_1 y) \cdot S_1(z) + (-a_2 x + a_2 y) \cdot S_2(z) \\ + (-a_3 x + a_3 y) \cdot S_3(z) \end{array} \right\}, \\ \dot{y} = \left\{ \begin{array}{l} \left[(c_2 - a_2)\, x - \frac{x(z-z_2)}{K_2} + c_2 y\right] \cdot S_2(z) \\ + \left[\frac{-x(z-z_3)}{K_3} + c_3 y\right] \cdot S_3(z) + \left[b_1 x - \frac{x(z-z_1)}{K_1} - y\right] \cdot S_1(z) \end{array} \right\}, \\ \dot{z} = \left\{ \begin{array}{l} \left[\frac{xy}{K_1} - c_1(z - z_1)\right] \cdot S_1(z) + \left[\frac{xy}{K_2} - b_2(z - z_2)\right] \cdot S_2(z) \\ + \left[\frac{xy}{K_3} - b_3(z - z_3)\right] \cdot S_3(z) \end{array} \right\}, \end{cases} \tag{5.10}$$

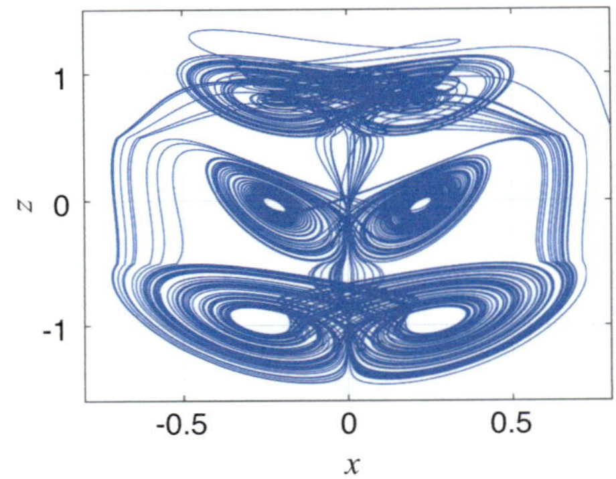

Fig. 5.7 Compound Lorenz-Chen-Lü chaotic attractor [6]

where $a_1 = 10$, $b_1 = 28$, and $c_1 = 2.67$ are the parameters of the Lorenz system [7]; $a_2 = 35$, $b_2 = 2.67$, and $c_2 = 28$ are the parameters of the Chen system [8]; $a_3 = 36$, $b_3 = 3$, and $c_3 = 20$ are the parameters of the Lü system [9]; $K_1 = 1/38$, $K_2 = 1/36$, and $K_3 = 1/30$ are the state variable scale factors for the Lorenz, Chen, and Lü systems, respectively; $z_1 = -0.75$, $z_2 = 0.14$, and $z_3 = -1.65$ are the coordinate translation values along the z-axis for the Lorenz, Chen, and Lü systems, respectively.

The numerical values are determined through trial-and-error simulations. While these values are not unique, a suitable set can be easily found for analysis.

In Eq. (5.10), $S_1(z)$, $S_2(z)$, and $S_3(z)$ are the switching controllers, designed by

$$\begin{cases} S_1(z) = 0.5\,(\mathrm{sgn}\,(z + 0.5) - \mathrm{sgn}\,(z - 0.5)), \\ S_2(z) = 0.5\,(1 + \mathrm{sgn}\,(z - 0.5)), \\ S_3(z) = 0.5\,(1 - \mathrm{sgn}\,(z + 0.5)). \end{cases} \tag{5.11}$$

The constants are also determined through trial-and-error simulations. By combining Eqs. (5.10) and (5.11), the desired compound Lorenz-Chen-Lü chaotic attractor can be generated, as shown in Fig. 5.7.

5.3.2 *Compound Lorenz-Chen-Lü-SM-Rucklidge Chaotic Attractors*

Based on Eq. (5.6) and the above four steps, a compound Lorenz-Chen-Lü-SM-Rucklidge and piecewise-Lorenz system is constructed, as follows:

$$
\left\{
\begin{aligned}
\dot{x} &= \left\{ \begin{array}{l} (-a_1x + a_1y) \cdot S_1(z) + (-a_2x + a_2y) \cdot S_2(z) \\ + (-a_3x + a_3y) \cdot S_3(z) + (a_4y - a_4x) \cdot S_4(z) \\ + y \cdot S_5(z) + y \cdot S_6(z) \end{array} \right\}, \\
\dot{y} &= \left\{ \begin{array}{l} \left[(c_2 - a_2)x - \frac{x(z-z_2)}{K_2} + c_2y\right] \cdot S_2(z) \\ + \left[\frac{-x(z-z_3)}{K_3} + c_3y\right] \cdot S_3(z) + \left[b_1x - \frac{x(z-z_1)}{K_1} - y\right] \cdot S_1(z) \\ + \left[-\mathrm{sgn}(x)(x - z_4)\right] \cdot S_4(z) + \left[-\frac{x(z-z_5)}{K_5} - a_5y + b_5x\right] \cdot S_5(z) \\ + \left[-\frac{x(z-z_6)}{K_6} - a_6y + x\right] \cdot S_6(z) \end{array} \right\}, \\
\dot{z} &= \left\{ \begin{array}{l} \left[\frac{xy}{K_1} - c_1(z - z_1)\right] \cdot S_1(z) + \left[\frac{xy}{K_2} - b_2(z - z_2)\right] \cdot S_2(z) \\ + \left[\frac{xy}{K_3} - b_3(z - z_3)\right] \cdot S_3(z) + (b_4x - c_4) \cdot S_4(z) \\ + \left[\frac{x^2}{K_5} - (z - z_5)\right] \cdot S_5(z) + \left[\frac{x^2}{K_6} - b_6(z - z_6)\right] \cdot S_6(z) \end{array} \right\},
\end{aligned}
\right\} \tag{5.12}
$$

where $a_1 = 10, b_1 = 28$, and $c_1 = 8/3$ are parameters of the Lorenz system; $a_2 = 35$, $b_2 = 8/3$, and $c_2 = 28$ are parameters of the Chen system; $a_3 = 36$, $b_3 = 3$, and $c_3 = 20$ are parameters of the Lü system; $a_4 = 0.5$, $b_4 = 2.5$, and $c_4 = 1/5$ are parameters of the SM system [10]; $a_5 = 2$ and $b_5 = 7.7$ are parameters of the Rucklidge system [11, 12]; $a_6 = 0.8$ and $b_6 = 0.4$ are parameters of the piecewise-Lorenz system [13, 14]. The scaling factors for the state variables of the Lorenz system, Chen system, Lü system, SM system, Rucklidge system, and piecewise-Lorenz system are $K_1 = 1/38$, $K_2 = 1/36$, $K_3 = 1/30$, $K_5 = 1/8$, and $K_6 = 1/2$, respectively. The coordinate translation values in the z-direction for these systems are $z_1 = -0.75$, $z_2 = 0.14$, $z_3 = -1.65$, $z_4 = -2.42$, $z_5 = 1.8$, and $z_6 = 1$, all determined through trial-and-error simulations.

In Eq. (5.12), the functions $S_1(z)$, $S_2(z)$, $S_3(z)$, $S_4(z)$, $S_5(z)$, and $S_6(z)$ represent the switching controllers, designed as follows:

$$
\left\{
\begin{array}{l}
S_1(z) = 0.5(\mathrm{sgn}(z + 0.5) - \mathrm{sgn}(z - 0.5)), \\
S_2(z) = 0.5(\mathrm{sgn}(z - 0.5) - \mathrm{sgn}(z - 1.1)), \\
S_3(z) = 0.5(\mathrm{sgn}(z + 1.5) - \mathrm{sgn}(z + 0.5)), \\
S_4(z) = 0.5(1 - \mathrm{sgn}(z + 1.5)), \\
S_5(z) = 0.5(1 + \mathrm{sgn}(z - 2.18)), \\
S_6(z) = 0.5(\mathrm{sgn}(z - 1.1) - \mathrm{sgn}(z - 2.18)).
\end{array}
\right. \tag{5.13}
$$

By applying Eqs. (5.12) and (5.13), the compound chaotic attractors of the Lorenz-Chen-Lü-SM-Rucklidge and the piecewise-Lorenz systems can be generated, as illustrated in Fig. 5.8.

5.3.3 *Compound Multiwing Piecewise-Lorenz and Chen Chaotic Attractor*

According to Eq. (5.6) and the above four steps, the compound multiwing piecewise-Lorenz and Chen system is constructed, described as follows:

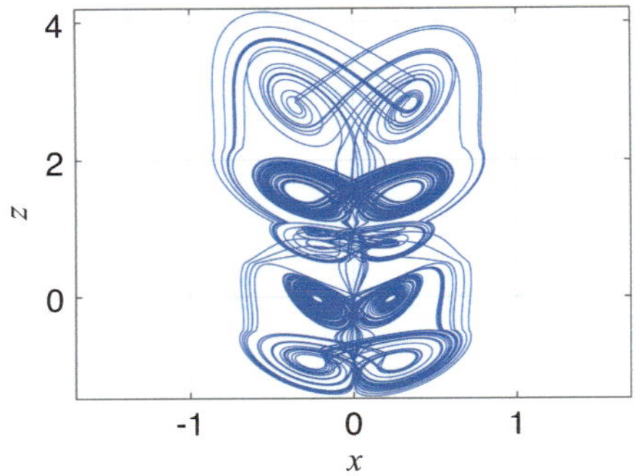

Fig. 5.8 Compound Lorenz-Chen-Lü-SM-Rucklidge and piecewise-Lorenz chaotic attractor [6]

$$\begin{cases} \dot{x} = [-a_1(y-x)] \cdot S_1(z) + [a_2(y-x)] \cdot S_2(z), \\ \dot{y} = \begin{Bmatrix} \left[-\operatorname{sgn}(x) \cdot (z-z_1)\right] \cdot S_1(z) \\ +\left[\operatorname{sgn}(x) \cdot [W_2 \cdot b_2 - (z-z_2) + c_2 y]\right] \cdot S_2(z) \end{Bmatrix}, \\ \dot{z} = (g_1 - W_1) \cdot S_1(z) + [g_2 - d_2(z-z_2)] \cdot S_2(z), \end{cases} \tag{5.14}$$

where $a_1 = 0.5$, $W_1 = 1$, $k_1 = 2.26$, $A_1 = 2.3$, $A_{11} = 2.06$, $A_{21} = 2.06$, $A_{31} = 2.06$, $A_{41} = 2.06$, $H_{11} = 2A_1/A_{11}$, $H_{21} = 2A_1/A_{21}$, $H_{31} = 2A_1/A_{31}$, $H_{41} = 2A_1/A_{41}$, $E_{11} = A_1/k_1$, $E_{21} = 2A_1/k_1$, $E_{31} = 3A_1/k_1$, and $E_{41} = c4A_1/k_1$ are parameters of the multiwing piecewise-Lorenz system [15]. Additionally, $a_2 = 1.18$, $b_2 = 5.82$, $c_2 = 0.7$, $d_2 = 0.168$, $W_2 = 2.5$, $k_2 = 4$, $A_2 = 2$, $A_{12} = 2.5$, $A_{22} = 2.2$, $A_{32} = 2.2$, $A_{42} = 2.2$, $H_{12} = A_2/A_{12}$, $H_{22} = A_2/A_{22}$, $H_{32} = A_2/A_{32}$, $H_{42} = A_2/A_{42}$, $E_{12} = A_2/k_2$, $E_{22} = 2A_2/k_2$, $E_{32} = 3A_2/k_2$, and $E_{42} = 4A_2/k_2$ are parameters of the multiwing Chen system [16]. The coordinate translation values along the z-axis for the multiwing piecewise-Lorenz and multiwing Chen systems are $z_1 = 1.3$ and $z_2 = -21.7$, respectively, which are determined by simulations.

In Eq. (5.14), g_1 represents a nonlinear function used for generating a multiwing piecewise-Lorenz chaotic attractor, defined as follows:

$$\begin{cases} g_{01} = |k_1 x|, \\ g_{11} = W_1 \cdot H_{11} \cdot 0.5\left[\operatorname{sgn}(x + W_1 \cdot E_{11}) - \operatorname{sgn}(x - W_1 \cdot E_{11}) - 2\right], \\ g_{21} = W_1 \cdot H_{21} \cdot 0.5\left[\operatorname{sgn}(x + W_1 \cdot E_{21}) - \operatorname{sgn}(x - W_1 \cdot E_{21}) - 2\right], \\ g_{31} = W_1 \cdot H_{31} \cdot 0.5\left[\operatorname{sgn}(x + W_1 \cdot E_{31}) - \operatorname{sgn}(x - W_1 \cdot E_{31}) - 2\right], \\ g_{41} = W_1 \cdot H_{41} \cdot 0.5\left[\operatorname{sgn}(x + W_1 \cdot E_{41}) - \operatorname{sgn}(x - W_1 \cdot E_{41}) - 2\right], \\ g_1 \ = g_{01} + g_{11} + g_{21} + g_{31} + g_{41}, \end{cases} \tag{5.15}$$

where g_2 represents a nonlinear function used to generate a multiwing Chen chaotic attractor, expressed as

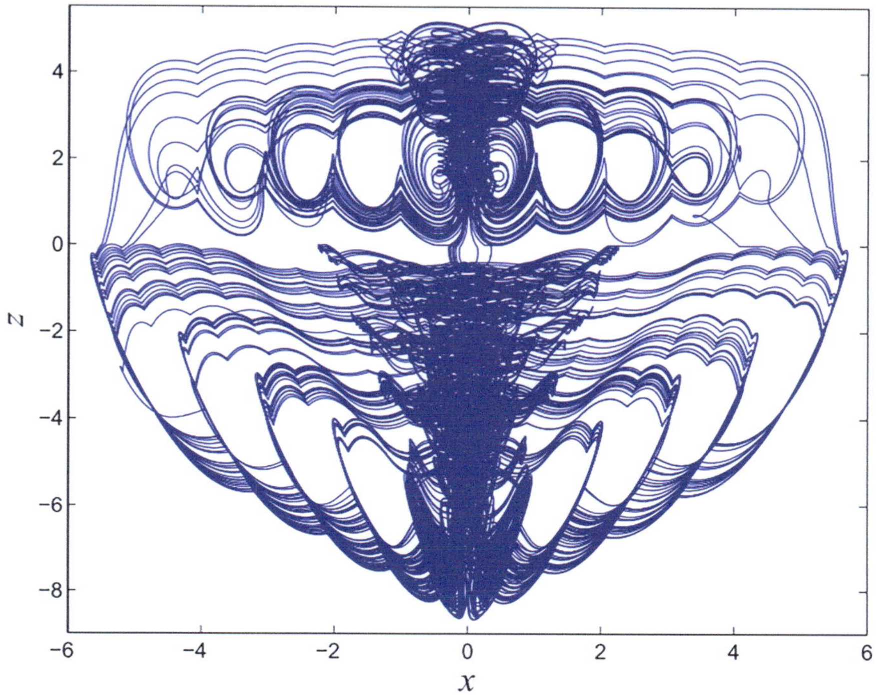

Fig. 5.9 Compound multiwing piecewise-Lorenz and Chen chaotic attractor. Reprinted from [6], with permission from World Scientific Pub. Co.

$$\begin{cases} g_{02} = |k_2 y|, \\ g_{12} = W_2 \cdot H_{12} \cdot 0.5\left[\operatorname{sgn}(x + W_2 \cdot E_{12}) - \operatorname{sgn}(x - W_2 \cdot E_{12}) - 2\right], \\ g_{22} = W_2 \cdot H_{22} \cdot 0.5\left[\operatorname{sgn}(x + W_2 \cdot E_{22}) - \operatorname{sgn}(x - W_2 \cdot E_{22}) - 2\right], \\ g_{32} = W_2 \cdot H_{32} \cdot 0.5\left[\operatorname{sgn}(x + W_2 \cdot E_{32}) - \operatorname{sgn}(x - W_2 \cdot E_{32}) - 2\right], \\ g_{42} = W_2 \cdot H_{42} \cdot 0.5\left[\operatorname{sgn}(x + W_2 \cdot E_{42}) - \operatorname{sgn}(x - W_2 \cdot E_{42}) - 2\right], \\ g_2 \ = g_{02} + g_{12} + g_{22} + g_{32} + g_{42}, \end{cases} \tag{5.16}$$

where $S_1(z)$ and $S_2(z)$ are switching controllers, designed as

$$\begin{cases} S_1(z) = \dfrac{1 + \operatorname{sgn}(z)}{2}, \\ S_2(z) = \dfrac{1 - \operatorname{sgn}(z)}{2}. \end{cases} \tag{5.17}$$

The functional forms and numerical values are determined by simulations.

By applying Eqs. (5.14)–(5.17), the compound multiwing piecewise-Lorenz and Chen chaotic attractor can be generated, as demonstrated in Fig. 5.9.

5.3.4 Compound Multiwing Lorenz-Chen-Lü Chaotic Attractor

Based on Eq. (5.6) and the above four steps, the compound multiwing Lorenz-Chen-Lü system is constructed, as follows:

$$\begin{cases} \dot{x} = \begin{Bmatrix} [-a_1(y-x)] \cdot S_1(z) + \left[\frac{-y(z-z_3)}{K_3} + c_3 x\right] \cdot S_3(z) \\ + \left[-\mathrm{sgn}(y) \cdot [W_2 \cdot b_2 - (z-z_2) + c_2 x]\right] \cdot S_2(z) \end{Bmatrix}, \\ \dot{y} = \begin{Bmatrix} \left[-\mathrm{sgn}(x) \cdot (z-z_1)\right] \cdot S_1(z) \\ + [a_2(x-y)] \cdot S_2(z) + [a_3(x-y)] \cdot S_3(z) \end{Bmatrix}, \\ \dot{z} = \begin{Bmatrix} (g_1 - W_1) \cdot S_1(z) + [g_2 - d_2(z-z_2)] \cdot S_2(z) \\ + [g_3 - b_3(z-z_3)] \cdot S_3(z) \end{Bmatrix}, \end{cases} \tag{5.18}$$

where $a_1 = 0.5$, $W_1 = 1$, $k_1 = 2.5$, $A_1 = 2.6$, $A_{11} = 2.0$, $H_{11} = 2A_1/A_{11}$, and $E_{11} = A_1/k_1$ are parameters of the multiwing Lorenz system [15]; $a_2 = 1.18$, $b_2 = 5.82$, $c_2 = 0.7$, $d_2 = 0.168$, $W_2 = 2.4$, $k_2 = 4$, $A_2 = 2$, $A_{12} = 2.5$, $H_{12} = A_2/A_{12}$, and $E_{12} = A_2/k_2$ are parameters of the multiwing Chen system [16]; $a_3 = 36$, $b_3 = 3$, $c_3 = 20$, $k_3 = 5$, $k_{13} = 0.3$, $k_{23} = 0.25$, $k_{33} = 0.18$, $k_{43} = 0.165$, $A_3 = 30$, $H_{03} = k_3/K_3$, $H_{13} = (A_3 \cdot K_3)/k_{13}$, and $E_{13} = (A_3 \cdot K_3)/k_3$ are parameters of the multiwing Lü system [16]. The state variable scale factor for the multiwing Lü system is $K_3 = 0.24$. The coordinate translation values along the z-axis for the multiwing piecewise-Lorenz system, multiwing Chen system and multiwing Lü system are $z_1 = -0.5$, $z_2 = -8.5$, and $z_3 = -0.95$, respectively. These values were determined by simulations.

In Eq. (5.18), the terms g_1, g_2, and g_3 represent nonlinear functions used for generating the multiwing Lorenz chaotic attractor, the multiwing Chen chaotic attractor and the multiwing Lü chaotic attractor, respectively, which are defined as follows:

$$\begin{cases} g_{01} = |k_1 x|, \\ g_{11} = W_1 \cdot H_{11} \cdot 0.5\left[\mathrm{sgn}(x + W_1 \cdot E_{11}) - \mathrm{sgn}(x - W_1 \cdot E_{11}) - 2\right], \\ g_1 \;= g_{01} + g_{11}, \\ g_{02} = |k_2 x|, \\ g_{12} = W_2 \cdot H_{12} \cdot 0.5\left[\mathrm{sgn}(x + W_2 \cdot E_{12}) - \mathrm{sgn}(x - W_2 \cdot E_{12}) - 2\right], \\ g_2 \;= g_{02} + g_{12}, \\ g_{03} = H_{03} \cdot x^2, \\ g_{13} = H_{13} \cdot \left[\mathrm{sgn}(x + E_{13}) - \mathrm{sgn}(x - E_{13}) - 2\right], \\ g_3 \;= g_{03} + g_{13}. \end{cases} \tag{5.19}$$

The controllers $S_1(z)$, $S_2(z)$, and $S_3(z)$ designed for the system are denoted as

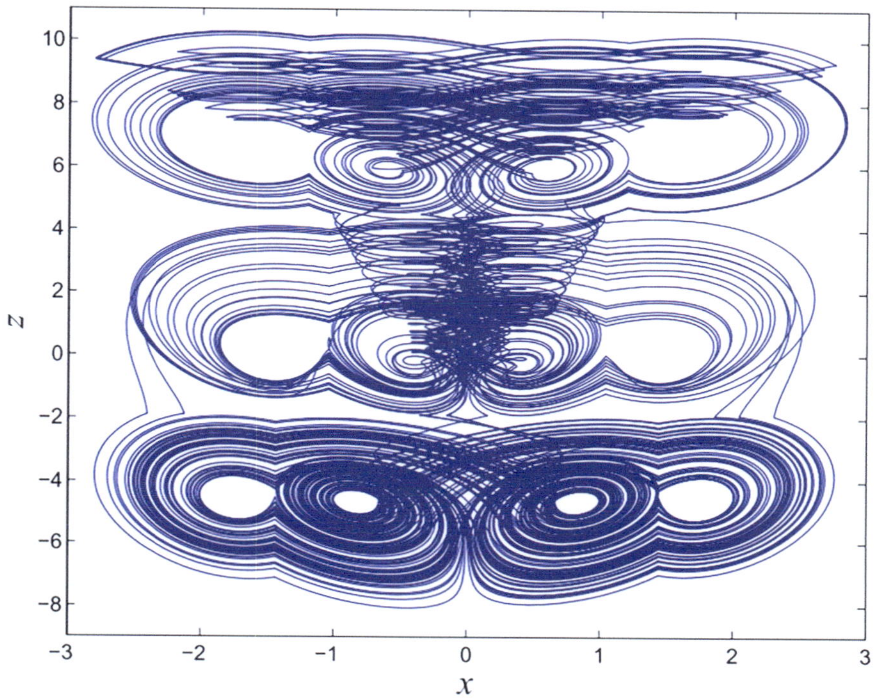

Fig. 5.10 Compound multiwing piecewise-Lorenz and Chen chaotic attractor. Reprinted from [6], with permission from World Scientific Pub. Co.

$$\begin{cases} S_1(z) = \dfrac{\operatorname{sgn}(z+0.9) - \operatorname{sgn}(z-4.5)}{2}, \\ S_2(z) = \dfrac{1+\operatorname{sgn}(z-4.5)}{2}, \\ S_3(z) = \dfrac{1-\operatorname{sgn}(z+1.9)}{2}. \end{cases} \tag{5.20}$$

The functional forms and numerical values are determined through simulations.

By applying Eqs. (5.18)–(5.20), the compound multiwing Lorenz-Chen-Lü chaotic attractor can be produced, as illustrated in Fig. 5.10.

5.3.5 *Circuit Design and Implementation for the Compound Lorenz-Chen-Lü Chaotic Attractor*

This subsection introduces the circuit designed and implemented by Zhang et al. [6] to generate the compound Lorenz-Chen-Lü chaotic attractor, as an example for demonstrating the circuit's efficacy based on the proposed simulation method.

Let $a_1 = 10$, $b_1 = 28$, $c_1 = 2.67$, $a_2 = 35$, $b_2 = 2.67$, $c_2 = 28$, $a_3 = 36$, $b_3 = 3$, $c_3 = 20$, $K_1 = 1/38$, $K_2 = 1/36$, $K_3 = 1/30$, $z_1 = -0.75$, $z_2 = 0.14$, and $z_3 = -1.65$ in Eq. (5.10). Then,

$$\begin{cases} \dot{x} = \begin{Bmatrix} 10(-x+y)\cdot S_1(z) + 35(-x+y)\cdot S_2(z) \\ +36(-x+y)\cdot S_3(z) \end{Bmatrix}, \\ \dot{y} = \begin{Bmatrix} [-7x - 38x(z-z_2) + 28y]\cdot S_2(z) \\ +[-30x(z-z_3) + 20y]\cdot S_3(z) \\ +[28x - 38x(z-z_1) - y]\cdot S_1(z) \end{Bmatrix}, \\ \dot{z} = \begin{Bmatrix} [38xy - 2.67(z-z_1)]\cdot S_1(z) + [36xy - 2.67(z-z_2)]\cdot S_2(z) \\ +[30xy - 3(z-z_3)]\cdot S_3(z) \end{Bmatrix}, \end{cases} \tag{5.21}$$

where $S_1(z)$, $S_2(z)$, and $S_3(z)$ are defined in Eq. (5.11).

Let $\tau = \tau_0 t$, where $\tau_0 = 100$ is the time-scale factor. It follows from Eq. (5.21) that following:

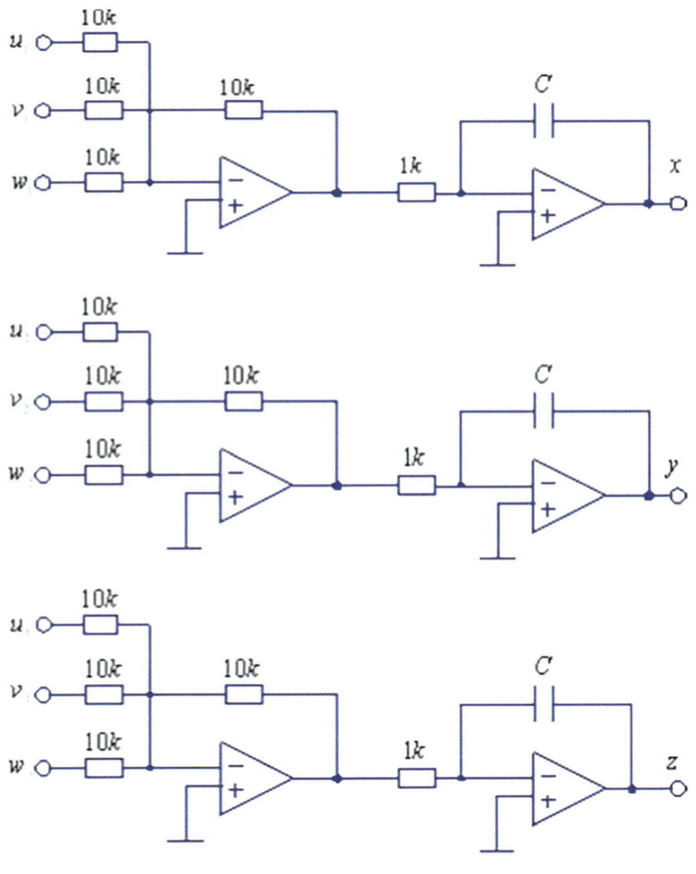

Fig. 5.11 The main circuit diagram for realizing the compound Lorenz-Chen-Lü chaotic attractor. Reprinted from [6], with permission from World Scientific Pub. Co.

$$
\begin{cases}
\dot{x} = \left\{ \begin{array}{l} 1000\,(-x+y)\cdot S_1(z) + 3500\,(-x+y)\cdot S_2(z) \\ +3600\,(-x+y)\cdot S_3(z) \end{array} \right\}, \\
\dot{y} = \left\{ \begin{array}{l} [-700x - 3800x\,(z-z_2) + 2800y]\cdot S_2(z) \\ +[-3000x\,(z-z_3) + 2000y]\cdot S_3(z) \\ +[2800x - 3800x\,(z-z_1) - y]\cdot S_1(z) \end{array} \right\}, \\
\dot{z} = \left\{ \begin{array}{l} [3800xy - 267\,(z-z_1)]\cdot S_1(z) \\ +[3600xy - 267\,(z-z_2)]\cdot S_2(z) \\ +[3000xy - 300\,(z-z_3)]\cdot S_3(z) \end{array} \right\}.
\end{cases}
\tag{5.22}
$$

Using Eq. (5.22), a modular circuit diagram is developed, as shown in Fig. 5.11. From this diagram, the corresponding state equations of the circuit are derived, as follows:

$$
\begin{cases}
\dot{x} = \left[-\frac{1}{R_1C_1}x - \frac{1}{R_2C_1}(-y) \right]\cdot S_1(z) + \left[-\frac{1}{R_3C_1}x - \frac{1}{R_4C_1}(-y) \right]\cdot S_2(z) \\
\quad + \left[-\frac{1}{R_5C_1}x - \frac{1}{R_6C_1}(-y) \right]\cdot S_3(z) \\
\dot{y} = \left[-\frac{1}{R_7C_2}(-x) - \frac{1}{R_8C_2}x\,(z-z_1) - \frac{1}{R_9C_2}y \right]\cdot S_1(z) \\
\quad + \left[-\frac{1}{R_{10}C_2}x - \frac{1}{R_{11}C_2}(-y) - \frac{1}{10R_{12}C_2}x\,(z-z_2) \right] \\
\quad \cdot S_2(z) + \left[-\frac{1}{10R_{13}C_2}x\,(z-z_3) - \frac{1}{R_{14}C_2}(-y) \right]\cdot S_3(z) \\
\dot{z} = \left[-\frac{1}{10R_{15}C_3}(-x)y - \frac{1}{R_{16}C_3}\,(z-z_1) \right]\cdot S_1(z) \\
\quad + \left[-\frac{1}{10R_{17}C_3}x(-y) - \frac{1}{R_{18}C_3}\,(z-z_2) \right]\cdot S_2(z) \\
\quad + \left[-\frac{1}{10R_{19}C_3}x(-y) - \frac{1}{R_{20}C_3}\,(z-z_3) \right]\cdot S_3(z).
\end{cases}
\tag{5.23}
$$

Let $C_1 = C_2 = C_3 = 10\,\text{nF}$ in Fig. 5.11. By comparing Eqs. (5.22) with (5.23), one obtains

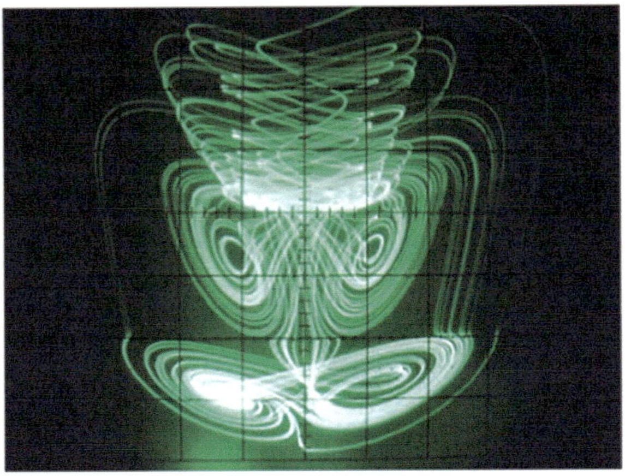

Fig. 5.12 Experimental observation of the compound Lorenz-Chen-Lü chaotic attractor. Reprinted from [6], with permission from World Scientific Pub. Co.

$$\begin{cases} 1000 = \dfrac{1}{R_1 C} \to R_1 = 100\,\text{k}\Omega, & 1000 = \dfrac{1}{R_2 C} \to R_2 = 100\,\text{k}\Omega, \\ 3500 = \dfrac{1}{R_3 C} \to R_3 = 28\,\text{k}\Omega, & 3500 = \dfrac{1}{R_4 C} \to R_4 = 28\,\text{k}\Omega, \\ 3600 = \dfrac{1}{R_5 C} \to R_5 = 27\,\text{k}\Omega, & 3600 = \dfrac{1}{R_6 C} \to R_6 = 27\,\text{k}\Omega, \\ 2800 = \dfrac{1}{R_7 C} \to R_7 = 35\,\text{k}\Omega, & 3800 = \dfrac{1}{10 R_8 C} \to R_8 = 2.63\,\text{k}\Omega, \\ 100 = \dfrac{1}{R_9 C} \to R_9 = 1000\,\text{k}\Omega, & 700 = \dfrac{1}{R_{10} C} \to R_{10} = 143\,\text{k}\Omega, \\ 2800 = \dfrac{1}{R_{11} C} \to R_{11} = 36\,\text{k}\Omega, & 3600 = \dfrac{1}{10 R_{12} C} \to R_{12} = 2.78\,\text{k}\Omega, \\ 3000 = \dfrac{1}{10 R_{13} C} \to R_{13} = 3.3\,\text{k}\Omega, & 2000 = \dfrac{1}{R_{14} C} \to R_{14} = 50\,\text{k}\Omega, \\ 3800 = \dfrac{1}{10 R_{15} C} \to R_{15} = 2.63\,\text{k}\Omega, & 267 = \dfrac{1}{R_{16} C} \to R_{16} = 375\,\text{k}\Omega, \\ 3600 = \dfrac{1}{10 R_{17} C} \to R_{17} = 2.78\,\text{k}\Omega, & 267 = \dfrac{1}{R_{18} C} \to R_{18} = 375\,\text{k}\Omega, \\ 3000 = \dfrac{1}{10 R_{19} C} \to R_{19} = 3.3\,\text{k}\Omega, & 300 = \dfrac{1}{R_{20} C} \to R_{20} = 333.3\,\text{k}\Omega. \end{cases} \tag{5.24}$$

The results from the circuit experiments are presented in Fig. 5.12.

5.4 Design of Grid Multiwing Hyperchaotic Lorenz System Family with Switching Control

Yu et al. [17] presented a systematic method for generating various grid multiwing hyperchaotic attractors by constructing super-heteroclinic loops and using switching control in piece-wise hyperchaotic Lorenz systems.

The family of generalized Lorenz systems, which includes the Lorenz system, Chen system, and Lü system, has two key characteristics:

(1) The system remains unchanged under the linear transformation

$$(x, y, z) \to (-x, -y, z).$$

(2) The system has three equilibrium points: $E(0, 0, 0)$ and $E_{1,2}(\pm x^{(e)}, \pm y^{(e)}, z^{(e)})$. The points $E_{1,2}$ are symmetric with respect to both the x- and y-axes.

Based on these properties, two fundamental 4-D linear systems can be constructed, by linearizing the system around the equilibrium points $E_{1,2}(\pm x^{(e)}, \pm y^{(e)}, z^{(e)})$ and incorporating a linear differential equation with state feedback. In so doing, a fundamental 4-D linear system corresponding to $E_1(\pm x^{(e)}, \pm y^{(e)}, z^{(e)})$ is obtained, as follows:

$$\begin{pmatrix} \dot{x} \\ \dot{y} \\ \dot{z} \\ \dot{u} \end{pmatrix} = J_1 \begin{pmatrix} x \\ y \\ z \\ u \end{pmatrix}, \tag{5.25}$$

where

$$J_1 = \begin{pmatrix} D_x f_1\left(x^{(e)}, y^{(e)}, z^{(e)}\right) & D_y f_1\left(x^{(e)}, y^{(e)}, z^{(e)}\right) & D_z f_1\left(x^{(e)}, y^{(e)}, z^{(e)}\right) & 0 \\ D_x f_2\left(x^{(e)}, y^{(e)}, z^{(e)}\right) & D_y f_2\left(x^{(e)}, y^{(e)}, z^{(e)}\right) & D_z f_2\left(x^{(e)}, y^{(e)}, z^{(e)}\right) & k \\ D_x f_3\left(x^{(e)}, y^{(e)}, z^{(e)}\right) & D_y f_3\left(x^{(e)}, y^{(e)}, z^{(e)}\right) & D_z f_3\left(x^{(e)}, y^{(e)}, z^{(e)}\right) & 0 \\ k_1 & k_2 & k_3 & k_4 \end{pmatrix}.$$

By linearizing the 3-D generalized Lorenz system family at its two symmetric equilibrium points and applying state feedback, one obtains

$$\begin{pmatrix} \dot{x} \\ \dot{y} \\ \dot{z} \\ \dot{u} \end{pmatrix} = \begin{pmatrix} -a & a & 0 & 0 \\ 1 & -1 & -T(x)\sqrt{b(c-1)} & 1 \\ T(x)\sqrt{b(c-1)} & T(x)\sqrt{b(c-1)} & -b & 0 \\ k_1 & 0 & 0 & 0 \end{pmatrix} \times \left(\begin{pmatrix} x \\ y \\ z \\ u \end{pmatrix} - \mathbf{F}(x, y, z, u) \right), \tag{5.26}$$

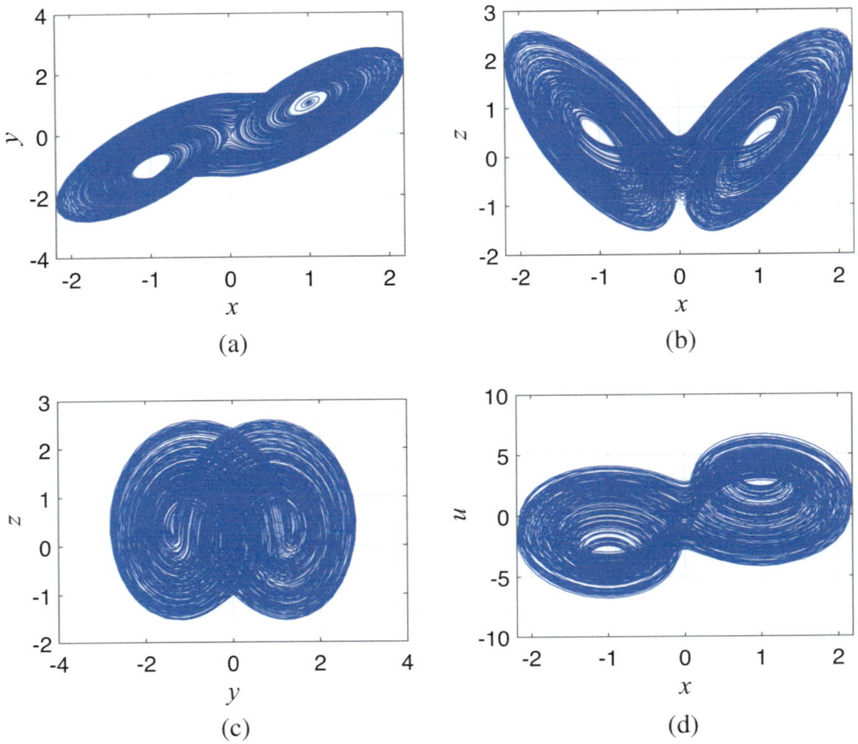

Fig. 5.13 PWL hyperchaotic double-wing Lorenz attractor [17]. **a** x-y plane. **b** x-z plane. **c** y-z plane. **d** x-u plane

where $T(x) = \tanh(Bx)$ and

$$\mathbf{F}(x, y, z, u) = \begin{pmatrix} f_1(x, y, z, u) \\ f_2(x, y, z, u) \\ f_3(x, y, z, u) \\ f_4(x, y, z, u) \end{pmatrix} = \begin{pmatrix} x_0 \tanh(Bx) \\ y_0 \tanh(Bx) \\ z_0 \tanh(Bx) \\ u_0 \tanh(Bx) \end{pmatrix}. \tag{5.27}$$

The system described by Eq. (5.26) represents a PWL double-wing hyperchaotic Lorenz system. The parameters are set as follows: $a = 10$, $b = 2.4$, $c = 40$, and $k_1 = -49$. The projections of the LWL hyperchaotic double-wing Lorenz attractor onto the x-y, x-z, y-z, and x-u planes are illustrated in Fig. 5.13.

Similarly, the PWL double-wing hyperchaotic Chen system is constructed as follows:

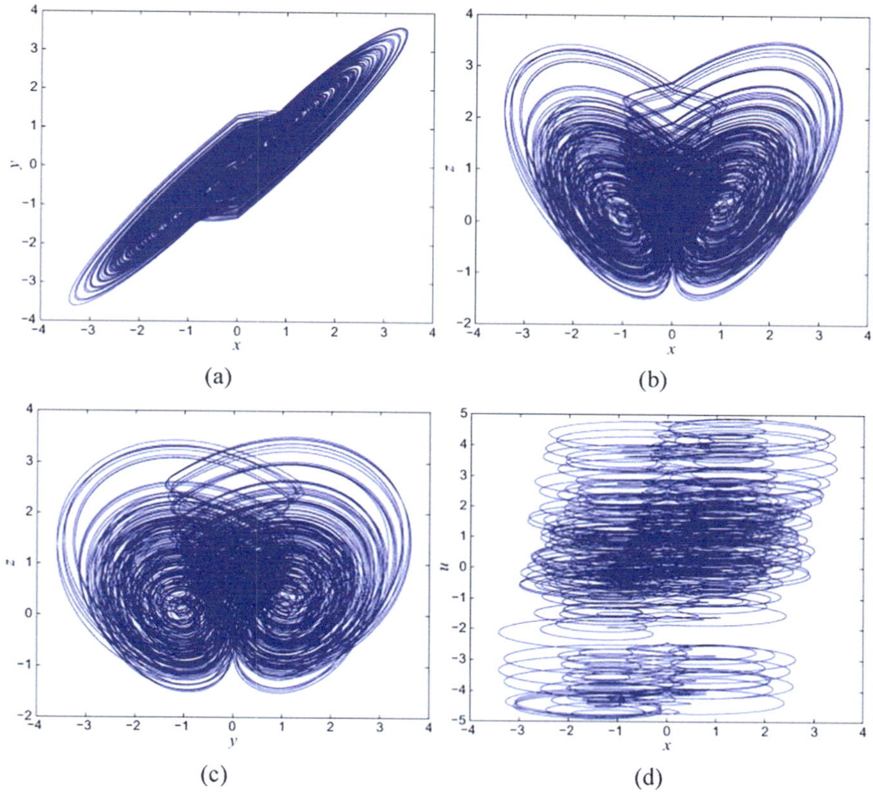

Fig. 5.14 PWL hyperchaotic double-wing Chen attractor. **a** x-y plane. **b** x-z plane. **c** y-z plane. **d** x-u plane. Reprinted from [17], with permission by the IEEE

$$
\begin{pmatrix} \dot{x} \\ \dot{y} \\ \dot{z} \\ \dot{u} \end{pmatrix} = \begin{pmatrix} -a & a & 0 & 0 \\ -c & c & -T(x)\sqrt{b(d+1)} & 1 \\ T(x)\sqrt{b(d+1)} & T(x)\sqrt{b(d+1)} & -b & 0 \\ k_1 & k_2 & 0 & k_4 \end{pmatrix} \times \left(\begin{pmatrix} x \\ y \\ z \\ u \end{pmatrix} - \mathbf{F}(x, y, z, u) \right), \quad (5.28)
$$

where the parameters are $a = 40$, $b = 3$, $c = 28$, $d = -7$, $k_1 = -4$, $k_2 = 0.5$, and $k_4 = 0.2$. Figure 5.14 illustrates the projections of the PWL hyperchaotic double-wing Chen attractor onto the x-y, x-z, y-z, and x-u planes.

By applying the fundamental design principles as discussed above, it is possible to generate various grid multiwing hyperchaotic Lorenz attractors, by constructing a system in the following form:

$$\begin{pmatrix} \dot{x} \\ \dot{y} \\ \dot{z} \\ \dot{u} \end{pmatrix} = \begin{pmatrix} -a & a & 0 & 0 \\ -H & H & -T(x)\cdot G & k \\ T(x)\cdot G & T(x)\cdot G & -b & 0 \\ k_1 & k_2 & k_3 & k_4 \end{pmatrix} \times \left(\begin{pmatrix} x \\ y \\ z \\ u \end{pmatrix} - \mathbf{F}(x,y,z,u) \right), \tag{5.29}$$

where the switching controller is

$$\mathbf{F}(x,y,z,u) = \begin{pmatrix} f_1(x,y,z) \\ f_2(x,y,z) \\ f_3(x,y,z) \\ f_4(x,y,z) \end{pmatrix}$$
$$= \begin{pmatrix} x_0 \tanh(Bx) + \sum_{i=1}^{M} x_0 \left[\tanh\left(B\left(x+2ix_0\right)\right) + \tanh\left(B\left(x-2ix_0\right)\right)\right] \\ y_0 \tanh(Bx) + \sum_{i=1}^{M} y_0 \left[\tanh\left(B\left(x+2ix_0\right)\right) + \tanh\left(B\left(x-2ix_0\right)\right)\right] \\ z_{00} + z_0 \tanh(Bz) + \sum_{j=1}^{N} z_0 \left[\tanh\left(B\left(z+2jz_0\right)\right) + \tanh\left(B\left(x-2jz_0\right)\right)\right] \\ 0 \end{pmatrix}.$$

The corresponding switching function is

$$\begin{aligned} T(x) = {} & \tanh(Bx) + \sum_{i=1}^{M}(-1)^i \left[\tanh\left(B\left(x+2ix_0\right)\right)\right. \\ & \left. + \tanh\left(B\left(x-2ix_0\right)\right)\right], \end{aligned} \tag{5.30}$$

where M, N are positive integers.

Let $a = 10$, $b = 2.4$, $c = 40$, $H = -1$, $G = \sqrt{b(c-1)}$, $k = 1$, $k_1 = -40$, and $k_2 = k_3 = k_4 = 0$. When $z_{00} = z_0 = 0$ and $f_3(x,y,z,u) = 0$, Eqs. (5.29) and (5.30) can generate a $2M+2$-wing hyperchaotic Lorenz attractor. When $z_{00}, z_0 \neq 0$, these equations can generate a $(2M+2) \times (2N+2)$-wing hyperchaotic Lorenz attractor. For instance, for $x_0 = 1$, $y_0 = 0.8714$, $z_{00} = 0$, $z_0 = 0$, and $M = 2$, Eqs. (5.29) and (5.30) generate a 6-wing hyperchaotic Lorenz attractor, as shown in Fig. 5.15a. With $x_0 = 1$, $y_0 = 0.8714$, $z_{00} = -0.33$, $z_0 = 2.1$, $M = 2$, and $N = 1$, these equations generate a 6×4-grid wing hyperchaotic Lorenz attractor, as shown in Fig. 5.15b.

Similarly, for the parameter set $a = 40$, $b = 3$, $c = H = 28$, $d = -7$, $G = \sqrt{b(c+d)}$, $k = 1$, $k_1 = -4$, $k_2 = 0.5$, $k_3 = 0$, and $k_4 = 0.2$, with $z_{00} = z_0 = 0$ and $f_3(x,y,z,u) = 0$, Eqs. (5.29) and (5.30) can generate a $2M+2$-wing hyperchaotic Chen attractor. If $z_{00}, z_0 \neq 0$, these equations can generate a $(2M+2) \times (2N+2)$-wing hyperchaotic Chen attractor. For example, with $x_0 = 1$, $y_0 = 1.0221$, $z_{00} = 0$, $z_0 = 0$, and $M = 2$, Eqs. (5.29) and (5.30) generate a 6-wing hyperchaotic Chen attractor, as shown in Fig. 5.15c. With $x_0 = 1$, $y_0 = 1.0221$, $z_{00} = -0.8$, $z_0 = 2.1$, $M = 2$, and $N = 1$, these equations generate a 6×4-grid wing hyperchaotic Chen attractor, as shown in Fig. 5.15d.

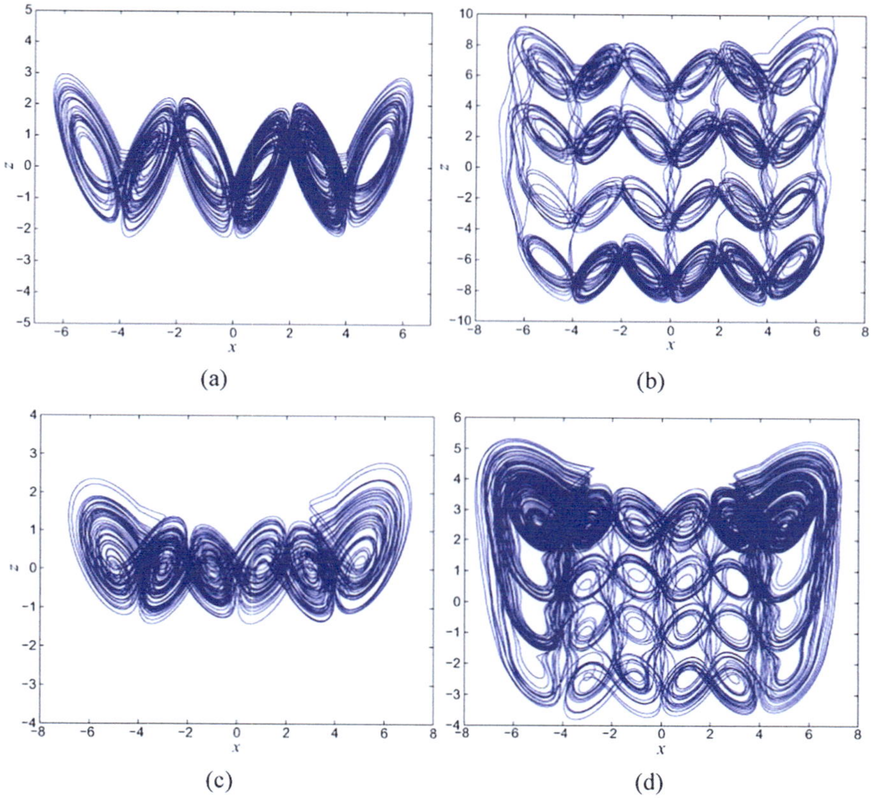

Fig. 5.15 x-y plane projections of multiwing hyperchaotic attractors. **a** 6-wing hyperchaotic Lorenz attractor. **b** 6 × 4-grid wing hyperchaotic Lorenz attractor. **c** 6-wing hyperchaotic Chen attractor. **d** 6 × 4-grid wing hyperchaotic Chen attractor. Reprinted from [17], with permission by the IEEE

According to Eq. (5.28), a modular circuit diagram is designed to implement the PWL grid multiwing hyperchaotic Lorenz and Chen attractors, as illustrated by Fig. 5.16. Furthermore, Fig. 5.16 shows the circuit diagram for the Lorenz system with $w_1 = -(x - f_1)$, $w_2 = (y - f_2)$, and $w_3 = w_4 = 0$. When modified for the Chen system, the diagram is represented as $w_1 = -(x - f_2)$, $w_2 = (y - f_2)$, $w_3 = -(y - f_2)$, and $w_4 = -u$.

In Eq. (5.28), let $a = 10, b = 2.4, c = 40, H = -1, G = \sqrt{b(c-1)} = 9.67, k = 1, k_1 = -40$, and $k_2 = k_3 = k_4 = 0$. The state equation for the PWL grid multiwing hyperchaotic Lorenz system is constructed, expressed as

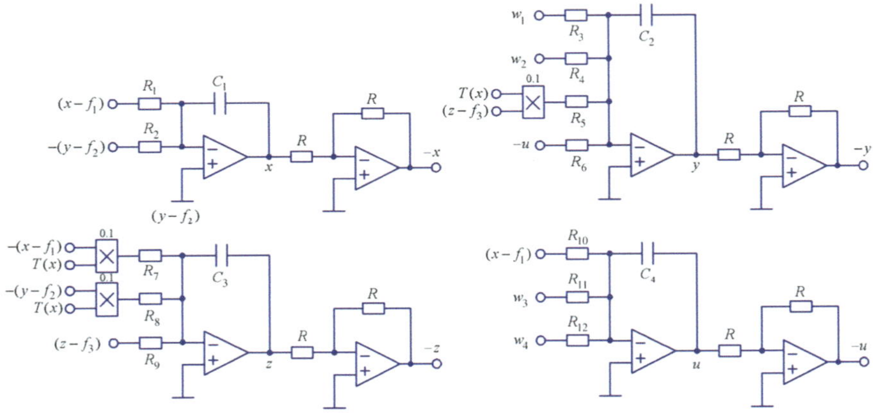

Fig. 5.16 Module-based circuit diagram for realizing PWL grid multiwing hyperchaotic Lorenz and Chen attractors. Reprinted from [17], with permission by the IEEE

$$\begin{cases} dx/d\tau = -10\,(x-f_1) - 10\,[-(y-f_2)], \\ dy/d\tau = -[-(x-f_1)] - (y-f_2) \\ \qquad\qquad -9\cdot 67\cdot T(x)\cdot(z-f_3) - (-u), \\ dz/d\tau = -9.67\cdot T(x)\cdot[-(x-f_1)] \\ \qquad\qquad -9.67\cdot T(x)\cdot[-(y-f_2)] - 2.4\,(z-f_3), \\ du/d\tau = -40\,(x-f_1). \end{cases} \tag{5.31}$$

By applying a time-scale transformation, where $\tau = \tau_0 t$ and $\tau_0 = 100$, the following equation is obtained:

$$\begin{cases} dx/dt = -1000\,(x-f_1) - 1000\,[-(y-f_2)], \\ dy/dt = -100\,[-(x-f_1)] - 100\,(y-f_2) \\ \qquad\qquad -967\cdot T(x)\cdot(z-f_3) - 100(-u), \\ dz/dt = -967\cdot T(x)\cdot[-(x-f_1)] \\ \qquad\qquad -967.T(x)\cdot[-(y-f_2)] - 240\,(z-f_3), \\ du/dt = -4000\,(x-f_1)\ . \end{cases} \tag{5.32}$$

In Fig. 5.16, by setting $w_1 = -(x-f_1)$, $w_2 = (y-f_2)$, $w_3 = 0$, and $w_4 = 0$, the circuit equation for the PWL grid multiwing hyperchaotic Lorenz system is obtained, as follows:

$$\begin{cases} dx/dt = -\dfrac{1}{R_1C_1}(x-f_1) - \dfrac{1}{R_2C_1}[-(y-f_2)], \\ dy/dt = -\dfrac{1}{R_3C_2}[-(x-f_1)] - \dfrac{1}{R_4C_2}(y-f_2) \\ \qquad - \dfrac{1}{10R_5C_2}T(x)(z-f_3) - \dfrac{1}{R_6C_2}(-u), \\ dz/dt = -\dfrac{1}{10R_7C_3}\cdot T(x)\cdot[-(x-f_1)] \\ \qquad - \dfrac{1}{10R_8C_3}\cdot T(x)\cdot[-(y-f_2)] - \dfrac{1}{R_9C_3}(z-f_3), \\ du/dt = -\dfrac{1}{R_{10}C_1}(x-f_1). \end{cases} \tag{5.33}$$

By setting the capacitances $C_1 = C_2 = C_3 = C_4 = 20\,\text{nF}$ and by comparing Eqs. (5.32) with (5.33), one can obtain the values of the corresponding resistance in Fig. 5.16 for the hyperchaotic Lorenz system, as follows:

$$\begin{cases} \dfrac{1}{R_1C_1} = 1000 \rightarrow R_1 = 50\,\text{k}\Omega, \\ \dfrac{1}{R_2C_1} = 1000 \rightarrow R_2 = 50\,\text{k}\Omega, \\ \dfrac{1}{R_3C_2} = 100 \rightarrow R_3 = 500\,\text{k}\Omega, \\ \dfrac{1}{R_4C_2} = 100 \rightarrow R_4 = 500\,\text{k}\Omega, \\ \dfrac{1}{10R_5C_2} = 967 \rightarrow R_5 = 5.17\,\text{k}\Omega, \\ \dfrac{1}{R_6C_2} = 100 \rightarrow R_6 = 500\,\text{k}\Omega, \\ \dfrac{1}{10R_7C_3} = 967 \rightarrow R_7 = 5.17\,\text{k}\Omega, \\ \dfrac{1}{10R_8C_3} = 967 \rightarrow R_8 = 5.17\,\text{k}\Omega, \\ \dfrac{1}{R_9C_3} = 240 \rightarrow R_9 = 208\,\text{k}\Omega, \\ \dfrac{1}{R_{10}C_1} = 4000 \rightarrow R_{10} = 12.5\,\text{k}\Omega\,. \end{cases} \tag{5.34}$$

Similarly, by setting $a = 40, b = 5, d = -7, c = H = -1$, and $G = \sqrt{b(c+d)}$, along with $k = 1, k_1 = -4, k_2 = 0.5, k_3 = 0$, and $k_4 = 0.2$, the state equation for the PWL grid multiwing hyperchaotic Chen system can be established. Following the same procedure, all the required resistance values for the hyperchaotic Chen system shown in Fig. 5.16 can be obtained, as follows:

$$\begin{cases} \dfrac{1}{R_1C_1} = 4000 \to R_1 = 12.5\,\mathrm{k}\Omega, \\ \dfrac{1}{R_2C_1} = 4000 \to R_2 = 12.5\,\mathrm{k}\Omega, \\ \dfrac{1}{R_3C_2} = 2800 \to R_3 = 17.9\,\mathrm{k}\Omega, \\ \dfrac{1}{R_4C_2} = 2800 \to R_4 = 17.9\,\mathrm{k}\Omega, \\ \dfrac{1}{10R_{r_5C_2}} = 794 \to R_5 = 6.3\,\mathrm{k}\Omega, \\ \dfrac{1}{R_6C_2} = 100 \to R_6 = 500\,\mathrm{k}\Omega, \\ \dfrac{1}{10R_7C_3} = 794 \to R_7 = 6.3\,\mathrm{k}\Omega, \\ \dfrac{1}{10R_8C_3} = 794 \to R_8 = 6.3\,\mathrm{k}\Omega, \\ \dfrac{1}{R_9C_3} = 300 \to R_9 = 166.7\,\mathrm{k}\Omega, \\ \dfrac{1}{R_{10}C_1} = 400 \to R_{10} = 125\,\mathrm{k}\Omega, \\ \dfrac{1}{R_{11}C_3} = 50 \to R_{11} = 1000\,\mathrm{k}\Omega, \\ \dfrac{1}{R_{12}C_4} = 20 \to R_{12} = 2500\,\mathrm{k}\Omega\,. \end{cases} \tag{5.35}$$

Figure 5.17 displays the circuit diagrams used to implement the switching function $T(x)$ and the nonlinear functions $\pm(x - f_1)$, $\pm(y - f_2)$, and $\pm(z - z_0 - f_3)$.

By setting the parameters $a = 40$, $b = 5$, $d = -7$, $c = H = -1$, and $G = \sqrt{b(c+d)}$, along with $k = 1$, $k_1 = -4$, $k_2 = 0.5$, $k_3 = 0$, and $k_4 = 0.2$, one can derive the state equation for the PWL grid multiwing hyperchaotic Chen system. From this approach, all the necessary resistance values for the hyperchaotic Chen system illustrated in Fig. 5.16 can be calculated.

With $x_0 = 1$, $y_0 = 0.8714$, $z_{00} = 0$, and $z_0 = 0$, the circuits in Figs. 5.16 and 5.17 can be realized to generate a 6-wing hyperchaotic Lorenz attractor, as shown in Fig. 5.18.

With $x_0 = 1$, $y_0 = 0.8714$, $z_{00} = -0.33$, and $z_0 = 2.1$, the circuits in Figs. 5.16 and 5.17 can be realized to generate a 6×4-wing hyperchaotic Lorenz attractor, as shown in Fig. 5.19.

In Fig. 5.16, let $a = 40$, $b = 3$, $d = -7$, $c = H = -1$, $G = \sqrt{b(c+d)}$, $k = 1$, $k_1 = -4$, $k_2 = 0.5$, $k_3 = 0$, $k_4 = 0.2$, $w_1 = -(x - f_2)$, $w_2 = (y - f_2)$, $w_3 = -(y - f_2)$, and $w_4 = -u$. All resistance values can be calculated using Eq. (5.35).

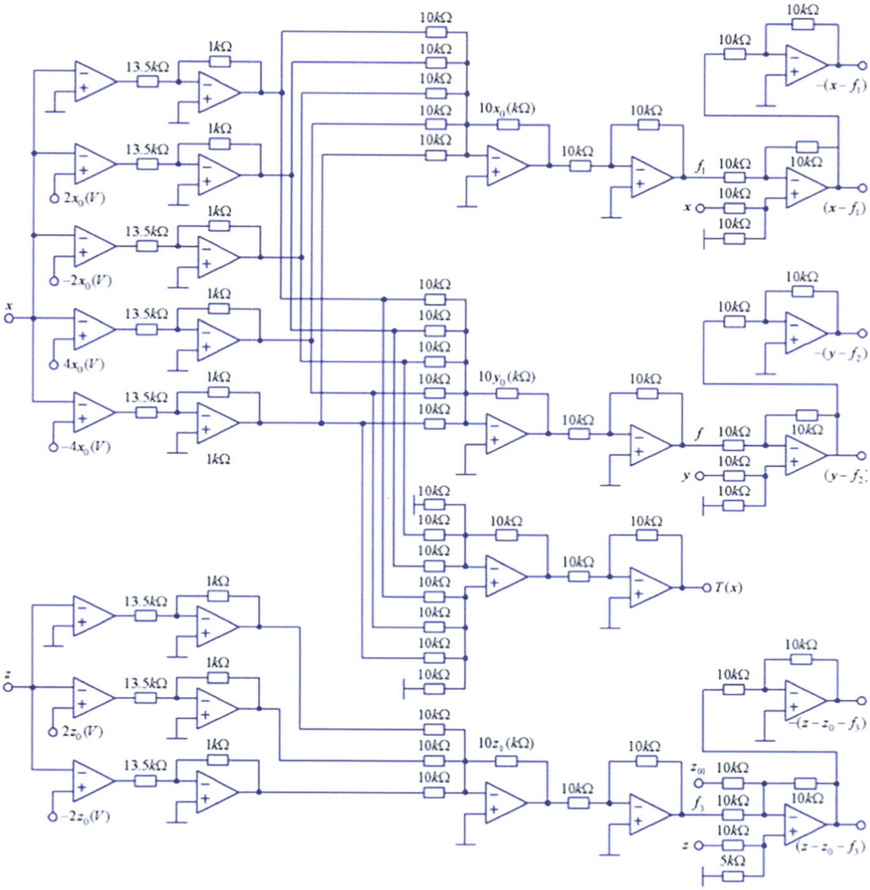

Fig. 5.17 Circuit diagrams for realizing the switching function $T(x)$ and nonlinear functions $\pm(x - f_1)$, $\pm(y - f_2)$, $\pm(z - z_0 - f_3)$. Reprinted from [17], with permission by the IEEE

In Fig. 5.17, with $x_0 = 1$, $y_0 = 1.0221$, and $z_{00} = z_0 = 0$, both circuits shown in Figs. 5.16 and 5.17 can generate a 6-wing hyperchaotic Chen attractor, as shown in Fig. 5.20.

With $x_0 = 1$, $y_0 = 1.0221$, $z_{00} = 0$, and $z_0 = 0.8$, the circuits shown Figs. 5.16 and 5.17 can generate a 6×4-wing hyperchaotic Chen attractor, as shown in Fig. 5.21.

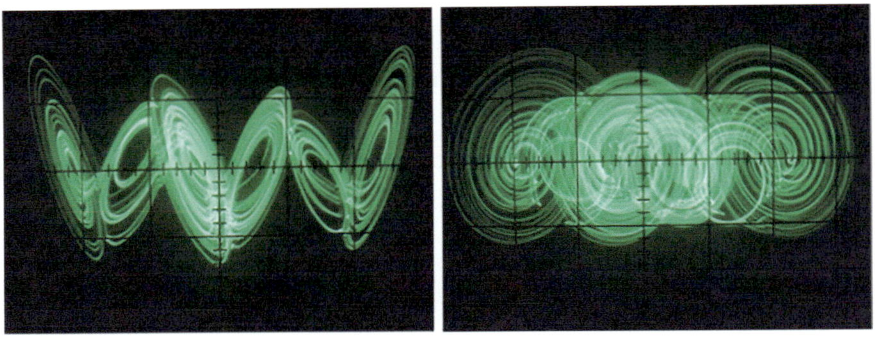

(a) (b)

Fig. 5.18 Experimental observations of the 6-wing hyperchaotic Lorenz attractor, where $x = 2.3$ V/div, $y = 2.0$ V/div, and $z = 2.0$ V/div. **a** x-z plane. **b** y-z plane. Reprinted from [17], with permission by the IEEE

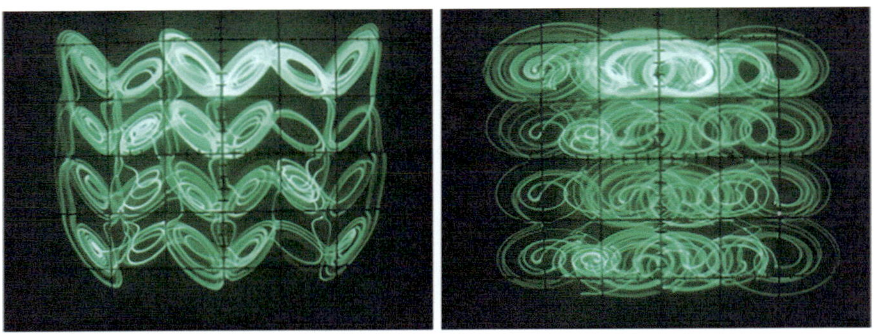

(a) (b)

Fig. 5.19 Experimental observations of 6×4-wing hyperchaotic Lorenz attractor, where $x = 2.3$ V/div, $y = 2.0$ V/div, and $z = 4.0$ V/div. **a** x-z plane. **b** y-z plane. Reprinted from [17], with permission by the IEEE

5.5 Conclusion

This chapter presents some systematic methods for designing controlled systems with multiscroll chaotic attractors using switching controls. By utilizing piecewise linear control, step-series switching, and state variable transformations, these approaches enable the generation of n-scroll chaotic attractors and compound chaotic systems. These methods enable the generation of complex chaotic dynamics by integrating multiple attractors under a single framework. Furthermore, the successful implementation of hardware circuits confirms the practicality of the proposed designs, demonstrating their applicability in engineering and technology. The

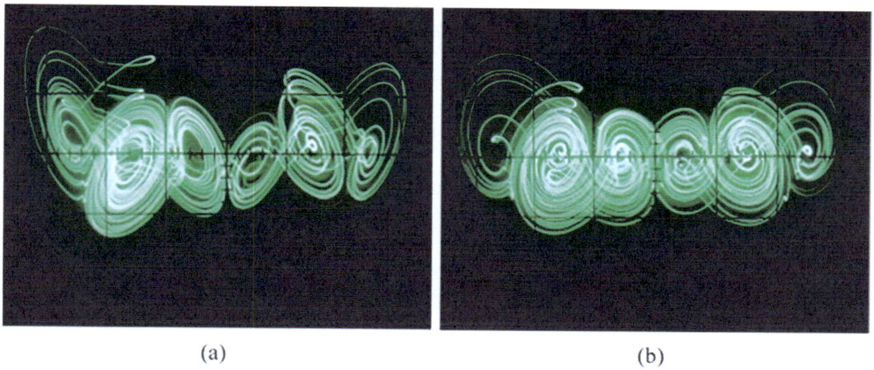

Fig. 5.20 Experimental observations of 6-wing hyperchaotic Chen attractor, where $x = 2.6\,\text{V/div}$, $y = 2.2\,\text{V/div}$, and $z = 2.5\,\text{V/div}$. **a** x-z plane. **b** y-z plane. Reprinted from [17], with permission by the IEEE

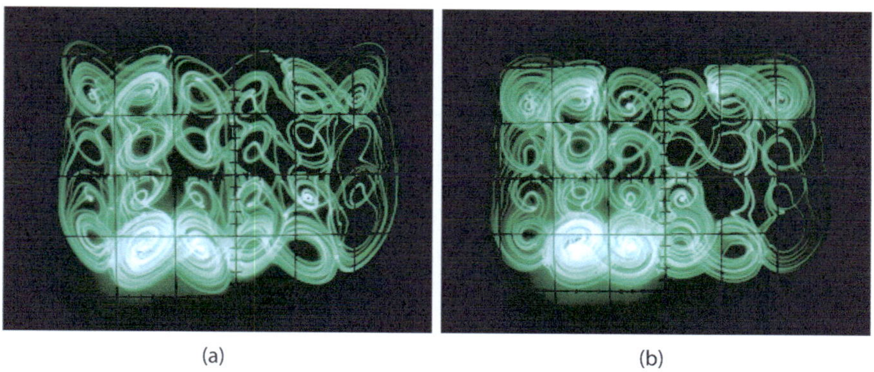

Fig. 5.21 Experimental observations of 6×4-grid wing hyperchaotic Chen attractor, where $x = 2.6\,\text{V/div}$, $y = 2.2\,\text{V/div}$, and $z = 2.25\,\text{V/div}$. **a** x-z plane. **b** y-z plane. Reprinted from [17], with permission by the IEEE

findings provide a solid foundation for further advancing chaos-based systems theory and real engineering applications, including secure communications, signal processing, and nonlinear control systems.

References

1. Lü, J., Yu, X., Chen, G.: Generating chaotic attractors with multiple merged basins of attraction: a switching piecewise-linear control approach. IEEE Trans. Circuits Syst.-I **50**, 198–207 (2003)
2. Lü, J., Zhou, T., Chen, G., Yang, X.: Generating chaos with a switching piecewise-linear controller. Chaos **12**, 344–349 (2002)

3. Lü, J., Chen, G.: Generating multiscroll chaotic attractors: theories, methods and applications. Int. J. Bifur. Chaos **16**, 775–858 (2006)
4. Yang, X.S., Li, Q.: Generate n-scroll attractor in linear system by scalar output feedback. Chaos Solit. Fract. **18**, 25–29 (2003)
5. Li, Q., Yang, X.S., Yang, F.: Multiple-scrolls chaotic attractor and circuit implementation. Electron. Lett. **39**, 1 (2003)
6. Zhang, C., Yu, S., Chen, G.: Design and implementation of compound chaotic attractors. Int. J. Bifur. Chaos **22**, 1250120 (2012)
7. Lorenz, E.N.: Deterministic nonperiodic flow. J. Atmos. Sci. **20**, 130–41 (1963)
8. Chen, G., Ueta, T.: Yet another chaotic attractor. Int. J. Bifur. Chaos **9**, 1465–1466 (1999)
9. Lü, J., Chen, G.: A new chaotic attractor coined. Int. J. Bifur. Chaos **12**, 659–661 (2002)
10. Shimizu, T., Morioka, N.: On the bifurcation of a symmetric limit cycle to an asymmetric one in a simple model. Phys. Lett. A **76**, 201–204 (1980)
11. Rucklidge, A.M.: Chaos in models of double convection. J. Fluid Mech. **237**, 209–229 (1992)
12. Zhou, T., Chen, G.: Classification of chaos in 3-D autonomous quadratic systems-I: basic framework and methods. Int. J. Bifur. Chaos **16**, 2459–2479 (2006)
13. Elwakil, A. S., Kennedy, M. P.: Construction of classes of circuit-independent chaotic oscillators using passive-only nonlinear devices. IEEE Trans. Circuits Syst.-I **48**, 289–307 (2005)
14. Ozoguz, S., Elwakil, A.S., Kennedy, M.P.: Experimental verification of the butterfly attractor in a modified Lorenz system. Int. J. Bifur. Chaos **12**, 1627–1632 (2002)
15. Yu, S., Tang, W.K.S., Lü, J., Chen, G.: Design and implementation of multi-wing butterfly chaotic attractors via Lorenz-type systems. Int. J. Bifur. Chaos **20**, 29–41 (2010)
16. Yu, S., Tang, W.K.S., Lü, J., Chen, G.: Generating $2n$-wing attractors from Lorenz-like systems. Int. J. Circuit Theory Appl. **38**, 243–258 (2010)
17. Yu, S., Lü, J., Yu, X., Chen, G.: Design and implementation of grid multiwing hyperchaotic Lorenz system family via switching control and constructing super-heteroclinic loops. IEEE Trans. Circuits Syst.-I **59**, 1015–1028 (2012)

Chapter 6
Generating Grid Multiscroll and Multiwing Chaotic Attractors

Abstract This chapter presents three methods for transforming systems to generate grid multiscroll chaotic attractors: parallel transformations, mirror transformations, and rotation transformations. Additionally, an approach is introduced for constructing complex grid multiwing hyperchaotic attractors based on 4-D hyperchaotic systems. The effectiveness of these methods is demonstrated through their applications to various systems, including the Lorenz system and the Chen system, allowing for the generation of n-wing and $n \times m$-wing hyperchaotic attractors. Circuit designs and experimental results support the theoretical design and analysis, highlighting their practicality in engineering applications.

6.1 Generating Grid Multiscroll Chaotic Attractors with Transformations

Ai et al. [1] proposed three transformation methods for generating grid multiscroll chaotic attractors.

6.1.1 Transformation Approach

There are three primary methods for designing grid-based multiscroll chaotic attractors: parallel transformation (PT), mirror transformation (MT), and rotation transformation (RT) [1].

The parallel, mirror, and rotation transformations introduce distinct operations to achieve the same goal of generating grid multiscroll attractors, each using a certain transformation–parallel, mirror symmetry, or rotation–to generate 1-D multiscroll attractors.

A chaotic system can generate a 1-D multiscroll attractor in the m-D space, as shown in Eq. (6.1). With the same range of the 1-D variable x_i in various transformation functions $f_i(x_i)$, to replace the variable x_i, it is possible to generate a grid of multiscroll attractors. The transformation

J. Lü et al., *Chaotic Systems with Multiscroll Attractors*,
https://doi.org/10.1007/978-3-032-09854-2_6

$$\begin{cases} \dot{x}_1 = f_1\left(x_1, x_2, \cdots, x_m\right), \\ \dot{x}_2 = f_2\left(x_1, x_2, \cdots, x_m\right), \\ \quad\vdots \\ \dot{x}_m = f_m\left(x_1, x_2, \cdots, x_m\right), \end{cases} \tag{6.1}$$

will be specified for different considerations below.

If the equilibrium point of the system (6.1) is a saddle-focus with an index of 2, it can generate a scroll attractor in the phase plane. Here, the step size is set to $h = 0.001$ and the initial values for the iteration are chosen to be $(x, y, z) = (0.1, 0.1, 0.3)$.

6.1.2 Parallel Transformation

The dimensionless Chua system [2] is described by

$$\begin{cases} \dot{x} = \alpha\left(y - h\left(x\right)\right), \\ \dot{y} = x - y + z, \\ \dot{z} = -\beta y, \end{cases} \tag{6.2}$$

where $\alpha = 10$, $\beta = 16$, and $h(x)$ is a nonlinear function: to generate $(2N + 2)$-scroll attractors,

$$h\left(x\right) = kx - pk\left\lfloor -\text{sgn}(x) + \sum_{i=0}^{N} \text{sgn}\left(x + 2ip\right) + \sum_{i=0}^{N} \text{sgn}\left(x - 2ip\right)\right\rfloor; \tag{6.3}$$

to generate $(2N + 3)$-scroll attractors,

$$h\left(x\right) = kx - pk\left\lfloor \sum_{i=0}^{N} \text{sgn}\left(x + (1 + 2ip)\right) + \sum_{i=0}^{N} \text{sgn}\left(x - (1 + 2i)\,p\right)\right\rfloor, \tag{6.4}$$

where $\lfloor\cdot\rfloor$ is the floor function, $k = 0.3$, $p = 0.5$, and

$$\text{sgn}(x) = \begin{cases} 1, & \text{if } x > 0, \\ 0, & \text{if } x = 0, \\ -1, & \text{if } x < 0. \end{cases}$$

Figure 6.1 shows the 5-scroll and 6-scroll chaotic attractors.

To generate grid multiscroll chaotic attractors from the 1-D multiscroll attractor in system (6.2), a parallel transformation function $y - f_1(y)$ is applied to the y-dimension of the system. This transforms system (6.2) to the following form:

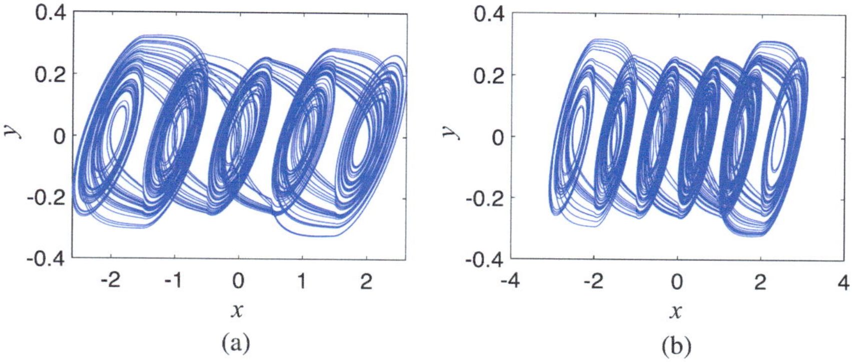

Fig. 6.1 Multiscroll chaotic attractors on the x-y plane [1]. **a** 5-scroll attractor. **b** 6-scroll attractor

$$\begin{cases} \dot{x} = \alpha\,(y - f_1\,(y) - h\,(x)), \\ \dot{y} = x - (y - f_1\,(y)) + z, \\ \dot{z} = -\beta\,(y - f_1\,(y)), \end{cases} \tag{6.5}$$

where

$$f_1\,(y) = \sum_{i=1}^{M_1} A\left[1 + \text{sgn}(y - y_i)\right] \pm \sum_{j=1}^{M_2} A\left[1 \pm \text{sgn}(y - y_j)\right], \tag{6.6}$$

with $A = 0.25$, $\alpha = 10$, $\beta = 16$, $h(x)$ is defined as above, M_1 and M_2 are non-negative integers.

System (6.5) is capable of producing chaotic attractors with either $(2N + 2) \times (M_1 + M_2 + 1)$ or $(2N + 3) \times (M_1 + M_2 + 1)$ scrolls. It is worth noting that the values of A, y_i and y_j are determined through a trial-and-error approach, whereas A affects the attractor's size while y_i and y_j control its position. For instance, if $f_1(y) = A(1 + \text{sgn}(y - 2.5))$, $y_1 = 2.5$, then a 5×2 scroll attractor is generated, as shown in Fig. 6.2a. Similarly, if $f_1(y) = A(1 + \text{sgn}(y - 2.5)) - A(1 - \text{sgn}(y + 2.5)) - A(1 - \text{sgn}(y + 7.5))$, $y_1 = 2.5$, $y_2 = -2.5$, and $y_3 = -7.5$, then a 6×4 scroll attractor is generated, as displayed in Fig. 6.2b.

The chaotic attractors are uniformly distributed with excellent symmetry, and the span along the y-axis remains consistent. This demonstrates the validity of this method for designing multiscroll chaotic attractors, and additional attractors with varied scroll patterns can be generated by adjusting the relevant parameters.

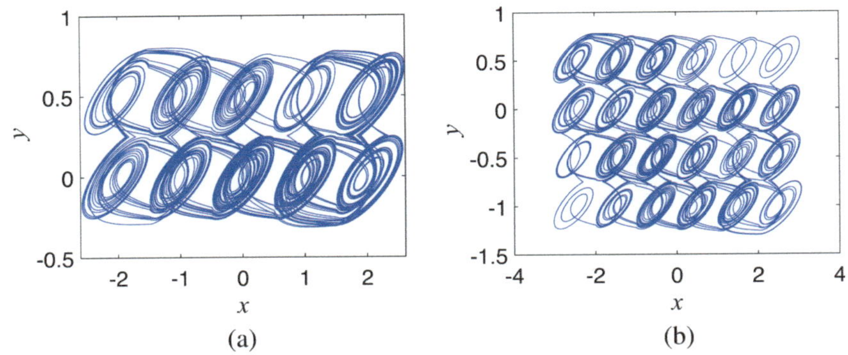

Fig. 6.2 Grid multiscroll chaotic attractors generated by PT [1]. **a** 5 × 2-scroll attractor. **b** 6 × 4-scroll attractor

6.1.3 *Mirror Transformation*

By substituting the variable y into Eq. (6.2) with a mirror transformation $\mathrm{sgn}(y - y_0)(y - y_0) + y_0 - f_2(y)$, the system is transformed to

$$\begin{cases} \dot{x} = \alpha\left(\mathrm{sgn}\left(y - y_0\right)\left(y - y_0\right) + y_0 - f_2\left(y\right) - h\left(x\right)\right), \\ \dot{y} = -\left(y - y_0\right) - \mathrm{sgn}\left(y - y_0\right) \times \left(-x + y_0 - f_2\left(y\right) - z\right), \\ \dot{z} = -\beta \mathrm{sgn}\left(y - y_0\right)\left(y - y_0\right) + y_0 - f_2\left(y\right), \end{cases} \tag{6.7}$$

where

$$f_2(y) = \sum_{i=1}^{M_3} A\left[1 + \mathrm{sgn}\left(\mathrm{sgn}(y - y_0)(y - y_0) - (y_i - y_0)\right)\right],$$

with $\alpha = 10$, $\beta = 16$, and $h(x)$ remaining the same as previously defined; M_3 is a non-negative integer, while A, y_0, and y_i are determined through a trial-and-error approach. The constant A affects the attractor's size, while y_0 and y_i control its positioning.

Using the mirror transformation, chaotic attractors with either $(2N + 2) \times (2M_3 + 2)$- or $(2N + 3) \times (2M_3 + 2)$-scrolls can be generated from system (6.7). With $A = 0.25$, $y_0 = -0.27$, $y_1 = 0.25$, $y_2 = 0.75$, and $y_3 = 1.25$, the phase diagrams on the x-y plane are displayed in Fig. 6.3, which correspond to $M_3 = 1$ and $M_3 = 2$ in Figs. 6.3a and b, respectively.

The above simulation confirms that grid multiscroll chaotic attractors can be generated from a 1-D multiscroll chaotic system using mirror transformation. These grid multiscroll attractors exhibit mirror symmetry along the y-axis and are uniformly distributed, with a consistent span in the y-axis.

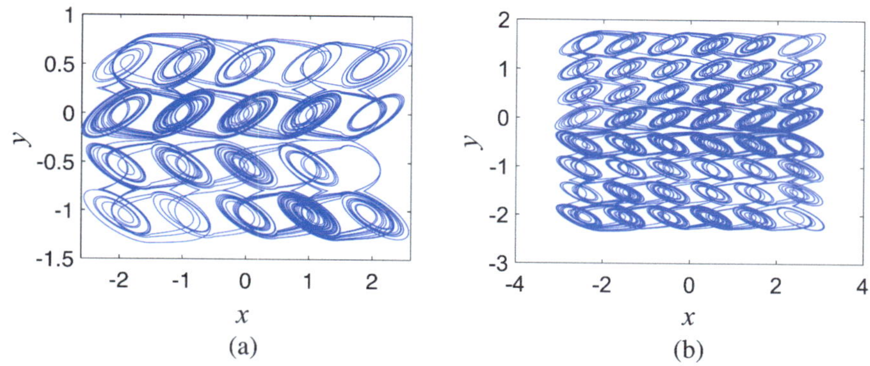

Fig. 6.3 Grid multiscroll chaotic attractors generated by MT [1]. **a** 5 × 4-scroll attractor. **b** 6 × 5-scroll attractor

6.1.4 Rotation Transformation

Applying a rotation transformation to system (6.2) leads to

$$\begin{cases} \dot{x} = H_x\left(H_1 \cos(\theta) + H_2 \sin(\theta)\right), \\ \dot{y} = H_y\left(-H_1 \sin(\theta) + H_2 \cos(\theta)\right), \\ \dot{z} = -\beta\left((|x - x_0| + x_0)\sin(\theta) + (|y - y_0| + y_0)\cos(\theta)\right), \end{cases} \tag{6.8}$$

where the rotation angle $\theta = \pi/4$. The components H_x and H_y are defined as follows:

$$\begin{cases} H_x = \dfrac{1}{2 \times 0.04}\left(|x - x_0 + 0.04| - |x - x_0 - 0.04|\right), \\ H_y = \dfrac{1}{2 \times 0.04}\left(|y - y_0 + 0.04| - |y - y_0 - 0.04|\right), \end{cases}$$

with

$$\begin{cases} H_1 = \alpha \begin{bmatrix} (|x - x_0| + x_0)\sin(\theta) + (|y - y_0| + y_0)\cos(\theta) \\ -h\left((|x - x_0| + x_0)\cos(\theta) - (|y - y_0| + y_0)\sin(\theta)\right) \end{bmatrix}, \\ \\ H_2 = \begin{bmatrix} \left((|x - x_0| + x_0)\cos(\theta) - (|y - y_0| + y_0)\sin(\theta)\right) \\ -\left((|x - x_0| + x_0)\sin(\theta) + (|y - y_0| + y_0)\cos(\theta)\right) + z \end{bmatrix}, \end{cases}$$

in which α, β, and $h(x)$ are as previously defined. It should be noted that the turning points x_0 and y_0 are determined through trial-and-error in simulation.

The general rotation transformation method follows a three-step procedure. First, the 1-D multiscroll attractor is rotated by an angle θ. Second, the attractor is expanded to be a circular grid multiscroll attractor using H_1, H_2, H_x, and H_y. Finally, the chaotic attractor's position determines the values of the turning points x_0 and y_0, and

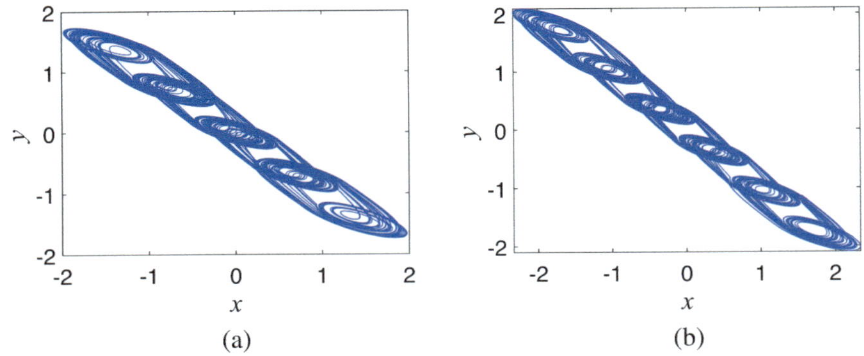

Fig. 6.4 Multiscroll chaotic attractors generated by RT [1]. **a** 5-scroll attractor. **b** 6-scroll attractor

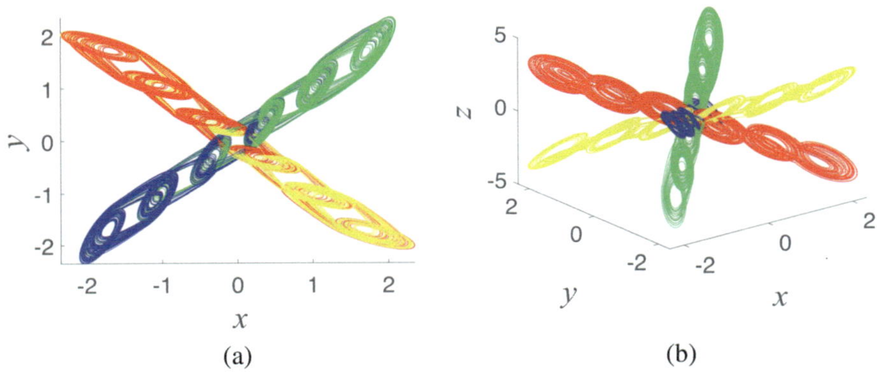

Fig. 6.5 4×6 circular grid multiscroll chaotic attractors generated by RT [1]. **a** x-y plane. **b** x-y-z space

the circular grid multiscroll attractor is centered along the x-direction on the $x = x_0$ plane and along the y-direction on the $y = y_0$ plane. The phase diagram on the x-y plane is shown in Fig. 6.4. In Fig. 6.4a, $x_0 = -14$ and $y_0 = -9$, while in Fig. 6.4b, $x_0 = -23$ and $y_0 = -21$. If the parameters are allowed to remain constant, the value of the parameter θ will be $\pi/4$, $3\pi/4$, $-\pi/4$, and $-3\pi/4$, respectively. This will result in the generation of the attractor shown in Fig. 6.5, where (a) presents the 4×6 circular grid multiscroll chaotic attractors on the x-y plane generated by RT, and (b) presents that in the 3-D space.

This demonstrates that system (6.8) produces circular grid multiscroll chaotic attractors through the RT applied to a multiscroll chaotic system. The resulting chaotic attractors exhibit a well-symmetrical distribution.

6.2 Generating Grid Multiwing Hyperchaotic Attractors with Transformations

Zhang and Yu [3] explored an approach for generating grid multwing hyperchaotic attractors from 4-D quadratic hyperchaotic systems.

6.2.1 *A Modified Double-Wing Hyperchaotic Lü System*

Zhang and Yu [3] developed a modified double-wing hyperchaotic Lü system using state-feedback control and a non-uniform variable scaling compression transformation. This method enables the construction of an n-wing hyperchaotic Lü system.

First, recall the 3-D Lü system [4]:

$$\begin{cases} \dot{u} = -au + av, \\ \dot{v} = -uw + cv, \\ \dot{w} = uv - bw, \end{cases} \tag{6.9}$$

where $a = 36$, $b = 3$, and $c = 20$.

To generate n-wing and $n \times m$-wing hyperchaotic attractors, a modified double-wing hyperchaotic Lü system is constructed. The third state equation of system (6.9) is modified to be $\dot{w} = kv^2 - bw$, with $k = 5$. Since the state variables in system (6.9) exceed the linear dynamic range of the active devices and are limited by their saturation values, a non-uniform variable scaling compression transformation is adopted for easier circuit implementation. This transformation is defined as $x = p_1 u$, $y = p_2 v$, and $z = p_3 w$, where $p_1 = 1/4$, $p_2 = 1/5$, and $p_3 = 1/7$ are non-uniform variable scaling compression factors. As a result, system (6.9) becomes

$$\begin{cases} \dot{x} = -a_{11}x + a_{12}y, \\ \dot{y} = -a_{21}xz + a_{22}y, \\ \dot{z} = a_{31}y^2 - a_{31}z, \end{cases} \tag{6.10}$$

where $a_{11} = a = 36$, $a_{12} = ap_1/p_2 = 45$, $a_{21} = p_2/(p_1 p_3) = 5.6$, $a_{22} = c = 20$, $a_{31} = kp_3/(p_2)^2 = 17.9$, and $a_{32} = b = 3$.

Based on system (6.10), by introducing a linear feedback controller u into the first equation and applying a first-order nonlinear differential state equation for u, the modified double-wing hyperchaotic Lü system is obtained as follows:

$$\begin{cases} \dot{x} = -a_{11}x + a_{12}y + a_{13}u, \\ \dot{y} = -a_{21}xz + a_{22}y, \\ \dot{z} = a_{31}y^2 - a_{32}z, \\ \dot{u} = a_{41}xz + a_{42}u, \end{cases} \tag{6.11}$$

Fig. 6.6 Modified double-wing hyperchaotic Lü attractor on the y-z plane [3]

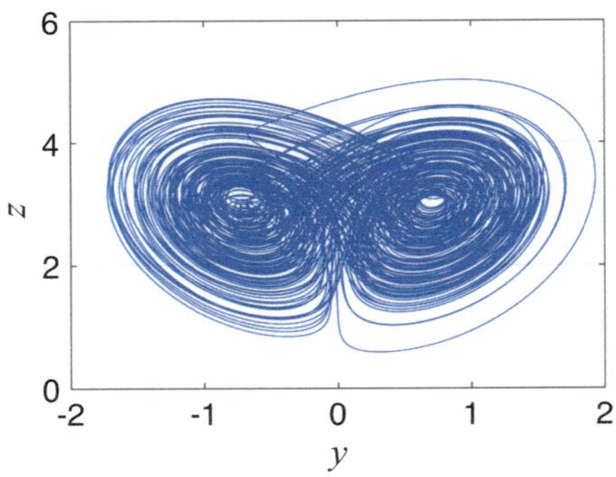

where $a_{11} = 36, a_{12} = 45, a_{13} = 2, a_{21} = 5.6, a_{22} = 20, a_{31} = 17.9, a_{32} = 3, a_{41} = 0.5$, and $a_{42} = 0.3$. Based on system (6.11), the double-wing hyperchaotic attractor obtained from numwerical simulation is presented in Fig. 6.6.

6.2.2 *The n-Wing Hyperchaotic Lü System*

In this section, based on system (6.10), an n-wing hyperchaotic Lü system is constructed by incorporating a duality-symmetric multi-segment quadratic function, allowing for the generation of n-wing hyperchaotic Lü attractors.

To construct the n-wing hyperchaotic system, the square term $a_{31}y_2$ in the third equation of system (6.10) is replaced by a function $f(y)$, which will be determined below, resulting in

$$\begin{cases} \dot{x} = -a_{11}x + a_{12}y + a_{13}u, \\ \dot{y} = -a_{21}xz + a_{22}y, \\ \dot{z} = f(y) - a_{32}z, \\ \dot{u} = a_{41}xz + a_{42}u, \end{cases} \tag{6.12}$$

where $f(y)$ is a duality-symmetric multi-segment quadratic function, given by

$$f(y) = H_0 y^2 + \sum_{i=1}^{N} H_i \left[\operatorname{sgn}(y + C_i) - \operatorname{sgn}(y - C_i) - 2\right], \tag{6.13}$$

in which $i = 1, 2, 3, \ldots, N$, and $N \geq 1$ is a positive integer. The parameters H_0, H_i and C_i are obtained as

$$\begin{cases} H_0 = k, \\ H_i = A/k_i, \\ C_i = 0.5\,(n + 1)\,A/k. \end{cases}$$

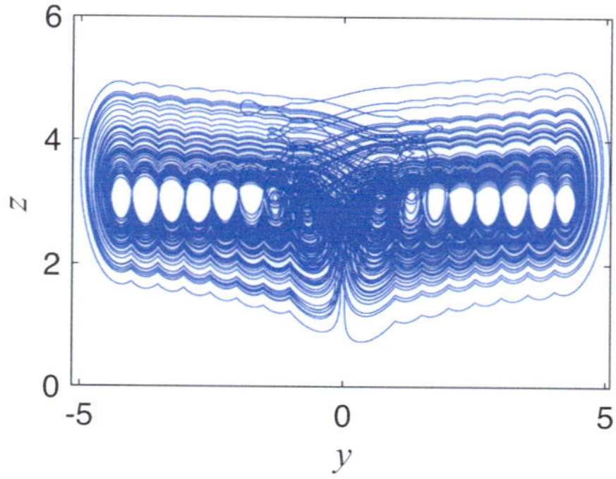

Fig. 6.7 16-wing hyperchaotic Lü attractor on the y-z plane [3]

Let $N = 7$, $A = 18$, $k = 17.9$, $k_1 = 1.7$, $k_2 = 1.45$, $k_3 = 1$, $k_4 = 0.8$, $k_5 = 0.65$, $k_6 = 0.57$, and $k_7 = 0.55$. It follows that $H_0 = 17.9$, $H_1 = 10.59$, $H_2 = 12.41$, $H_3 = 18$, $H_4 = 22.5$, $H_5 = 27.69$, $H_6 = 31.58$, and $H_7 = 32.73$, along with $C_1 = 1.0056$, $C_2 = 1.5084$, $C_3 = 2.0112$, $C_4 = 2.514$, $C_5 = 3.0168$, $C_6 = 3.5196$, and $C_7 = 4.0223$. Using Eq. (6.12), a $(2N + 2)$-wing hyperchaotic attractor can be generated. For instance, by setting $N = 7$, a 16-wing hyperchaotic Lü attractor is generated as displayed in Fig. 6.7.

6.2.3 Construction of $n \times m$-Wing Hyperchaotic Systems with Switching Control

In this section, based on system (6.12), a switching control in the z-axis is applied to construct a class of $n \times m$-wing hyperchaotic systems, referred to as the first class as compared to the sent class to be discussed below, which generate $n \times m$-wing hyperchaotic attractors. By introducing a switching controller $g_1(z)$ to system (6.12), one obtains

$$\begin{cases} \dot{x} = -a_{11}x + a_{12}y + a_{13}u, \\ \dot{y} = -a_{21}x\left(z - g_1(z)\right) + a_{22}y, \\ \dot{z} = f(y) - a_{32}\left(z - g_1(z)\right), \\ \dot{u} = a_{41}x\left(z - g_1(z)\right) + a_{42}u, \end{cases} \tag{6.14}$$

where $f(y)$ is a duality-symmetric multi-segment quadratic function defined in (6.13), and the switching controller $g_1(z)$ is

$$g_1(z) = \sum_{i=1}^{M} \left[\pm B\left(1 \pm \text{sgn}(z - z_i)\right)\right].$$

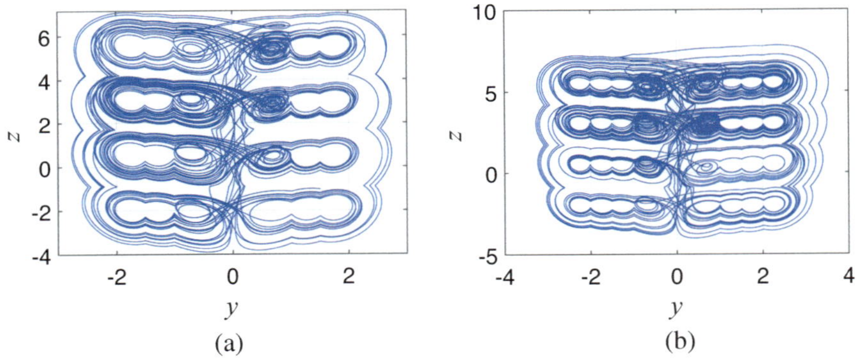

Fig. 6.8 The first class hyperchaotic Lü attractor. **a** 6 × 4-wing attractor on the y-z plane. **b** 8 × 4-wing attractor on the y-z plane [3]

Basd on Eq. (6.14), this first class of grid $(2N+2)(M+1)$-wing hyperchaotic attractor can be generated. For example, with $N=2$, $B=1.25$, $z_1=4.2$, $z_2=1.7$, $z_3=0.9$, $g_1(z)=B(1+\text{sgn}(z-z_1))-B(1-\text{sgn}(z-z_2))-B(1-\text{sgn}(z+z_3))$, and by setting $M=2$ and $M=3$, an 8 × 4-wing and 6 × 4-wing hyperchaotic attractors can be generated are shown in Fig. 6.8.

6.2.4 *Construction of n × m-Wing Hyperchaotic Systems with Mirror Symmetry Conversion*

In this section, based on system (6.12), a mirror symmetry conversion-based approach is introduced to generate the second class of $n \times m$-wing hyperchaotic systems, which can generate $n \times m$-wing hyperchaotic attractors.

System (6.12) is further modified, as follows:

$$\begin{cases} \dot{x} = -a_{11}x + a_{12}y + a_{13}u, \\ \dot{y} = -a_{21}x \times (\text{sgn}\,(z-z_0) \times (z-z_0) - g_2\,(z)) + a_{22}y, \\ \dot{z} = \text{sgn}\,(z-z_0) \times f\,(y) - a_{32}\,(z - z_0 - \text{sgn}\,(z-z_0) \times g_2\,(z)), \\ \dot{u} = a_{41}x \times (\text{sgn}\,(z-z_0) \times (z-z_0) - g_2\,(z)) + a_{42}u, \end{cases} \tag{6.15}$$

where $f(y)$ is a duality-symmetric multi-segment quadratic function defined in (6.13), and $g_2(z)$ is the mirror symmetry conversion function given by

$$g_2\,(z) = \sum_{i=1}^{M} \left[\pm B\,(1 \pm \text{sgn}\,(\text{sgn}\,(z-z_0) \times (z-z_0) - (z_i - z_0)))\right].$$

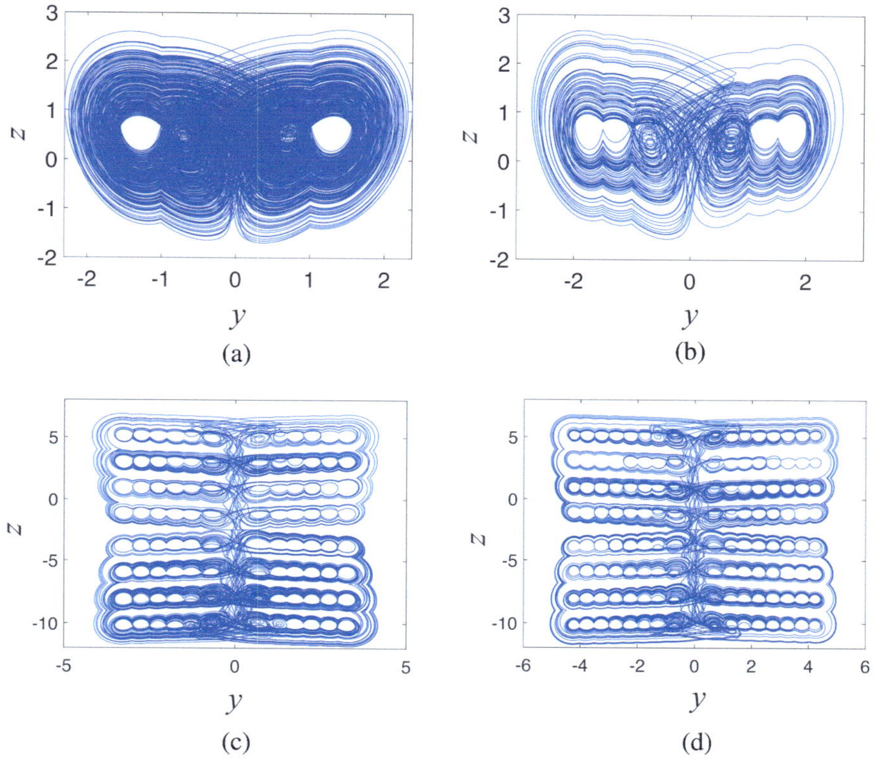

Fig. 6.9 The second class hyperchaotic Lü attractor [3]. **a** 4 × 1-wing attractor on the y-z plane. **b** 6 × 1-wing attractor on the y-z plane. **c** 12 × 8-wing attractor on the y-z plane. **d** 16 × 8-wing attractor on the y-z plane

Based on Eq. (6.14), the second class of grid $n \times m$-wing hyperchaotic attractors can be generated. For example, by setting $N = 1, 2$, $M = 0$, and $g_2(z) = 0$, a 4 × 1-wing and a 6 × 1-wing hyperchaotic attractor are generated, as shown in Fig. 6.9a and b. Additionally, by setting $N = 7, 5$, $M = 3$, $B = 1.05$, $z_0 = -2.5$, $z_1 = 4$, $z_2 = 1.8$, $z_3 = -0.3$, and

$$\begin{aligned} g_2(z) = {} & B\left(1 + \operatorname{sgn}\left(\operatorname{sgn}(z - z_0) \times (z - z_0) - (z_1 - z_0)\right)\right) \\ & - B\left(1 - \operatorname{sgn}\left(\operatorname{sgn}(z - z_0) \times (z - z_0) - (z_2 - z_0)\right)\right) \\ & - B\left(1 - \operatorname{sgn}\left(\operatorname{sgn}(z - z_0) \times (z - z_0) - (z_3 - z_0)\right)\right) - z_0, \end{aligned}$$

the system can generate 12 × 8-wing hyperchaotic attractors as shown in Fig. 6.9c, and 16 × 8-wing hyperchaotic attractors as shown in Fig. 6.9d.

6.2.5 Circuit Design and Experimental Observation

In this section, based on the dimensionless state equations and using the enhanced module-based method, systems (6.14) and (6.15) are implemented. Experimental results are then observed from the electronic device.

Using the above-defined parameters in system (6.14), one obtains

$$\begin{cases} \dfrac{dx}{d\tau} = -36x + 45y + 0.2u, \\ \dfrac{dy}{d\tau} = -5.6x\,(z - g_1(z)) + 20y, \\ \dfrac{dz}{d\tau} = f(y) - 3\,(z - g_1(z)), \\ \dfrac{du}{d\tau} = 0.5x\,(z - g_1(z)) + 0.3u. \end{cases} \tag{6.16}$$

By letting $\tau = \tau_0 t$, where $\tau_0 = 1000$ is the time-scale factor, system (6.16) is transformed to

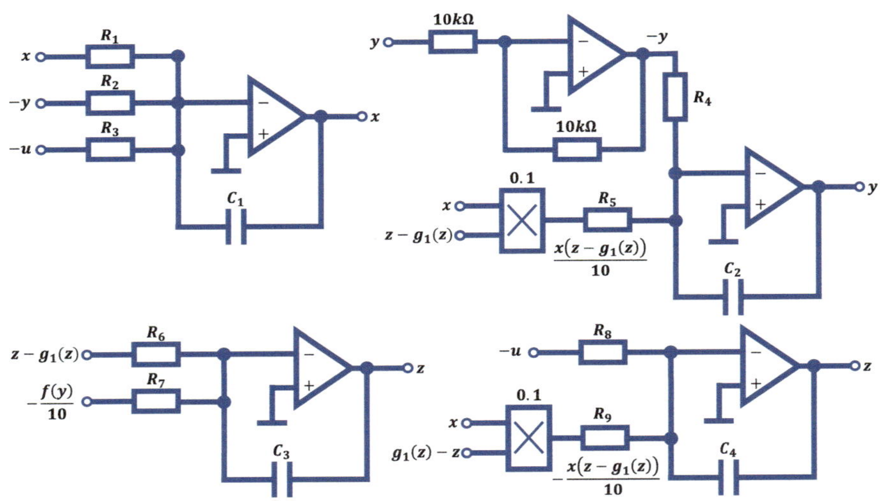

Fig. 6.10 Circuit design for the first class of $n \times m$-wing hyperchaotic system [3]

$$\begin{cases} \dfrac{dx}{d\tau} = -36000x + 45000y + 2000u, \\ \dfrac{dy}{d\tau} = -56000x\,(z - g_1\,(Z)) + 20000y, \\ \dfrac{dz}{d\tau} = 1000f\,(y) - 3000\,(z - g_1\,(z)), \\ \dfrac{du}{d\tau} = 500x\,(z - g_1\,(z)) + 300u. \end{cases} \tag{6.17}$$

Figure 6.10 shows an improved module-based circuit diagram for system (6.16). The circuit design diagrams for both $z - g_1(z)$ and $-f(y)/10$ are presented in Fig. 6.11.

From Fig. 6.10, one gets

$$\begin{cases} \dfrac{dx}{d\tau} = -\dfrac{1}{R_1C_1}x - \dfrac{1}{R_2C_1}(-y) - \dfrac{1}{R_3C_1}(-u), \\ \dfrac{dy}{d\tau} = -\dfrac{1}{10R_5C_2}x\,(z - g_1\,(Z)) + \dfrac{1}{R_4C_2}(-y), \\ \dfrac{dz}{d\tau} = -\dfrac{1}{R_6C_3}(-0.1f\,(y)) - \dfrac{1}{10R_6C_3}(z - g_1\,(z)), \\ \dfrac{du}{d\tau} = -\dfrac{1}{10R_9C_4}x\,(z - g_1\,(z)) + \dfrac{1}{R_8C_4}(-u). \end{cases} \tag{6.18}$$

Let $C_1 = C_2 = C_3 = C_4 = 10\,\text{nF}$, and by comparing Eqs. (6.17) with (6.18), the resistance values indicated in Fig. 6.10 are determined as follows: $R_1 = 2.8\,\text{k}\Omega$, $R_2 = 2.2\,\text{k}\Omega$, $R_3 = 50\,\text{k}\Omega$, $R_4 = 5\,\text{k}\Omega$, $R_5 = 1.8\,\text{k}\Omega$, $R_6 = 33.3\,\text{k}\Omega$, $R_7 = 10\,\text{k}\Omega$, $R_8 = 333.3\,\text{k}\Omega$, and $R_9 = 20\,\text{k}\Omega$.

Referring to Figs. 6.10 and 6.11, the circuit experimental observations of the first class of multiwing hyperchaotic attractors are displayed in Fig. 6.12. When K_1 and K_2 shown in Fig. 6.11b are switched on, and K_3 is switched off, a 6×4-wing butterfly hyperchaotic attractor is generated, as shown in Fig. 6.12a. When K_1, K_2, and K_3 are all switched on, an 8×4-wing butterfly hyperchaotic attractor is produced, as shown in Fig. 6.12b.

Using the above-defined parameter values in system (6.15), one obtains

$$\begin{cases} \dot{x} = -36x + 45y + 2u, \\ \dot{y} = -5.6x \times (\text{sgn}\,(z - z_0) \times (z - z_0) + z_0) + 20y, \\ \dot{z} = \text{sgn}\,(z - z_0) \times f\,(y) - 3\,(z - z_0 + \text{sgn}\,(z - z_0) \times z_0), \\ \dot{u} = 0.5x \times (\text{sgn}\,(z - z_0) \times (z - z_0) + z_0) + 0.3u. \end{cases} \tag{6.19}$$

Based on Eq. (6.19), an enhanced module-based circuit diagram is shown in Fig. 6.13a. Similarly, the circuit design for $-f(y)/10$ is illustrated in Fig. 6.11b, while the circuit designs for $z - z_0$, $\text{sgn}(z - z_0)$, and $\text{sgn}(z - z_0) \times (z - z_0) + z_0$ are shown in Fig. 6.13b.

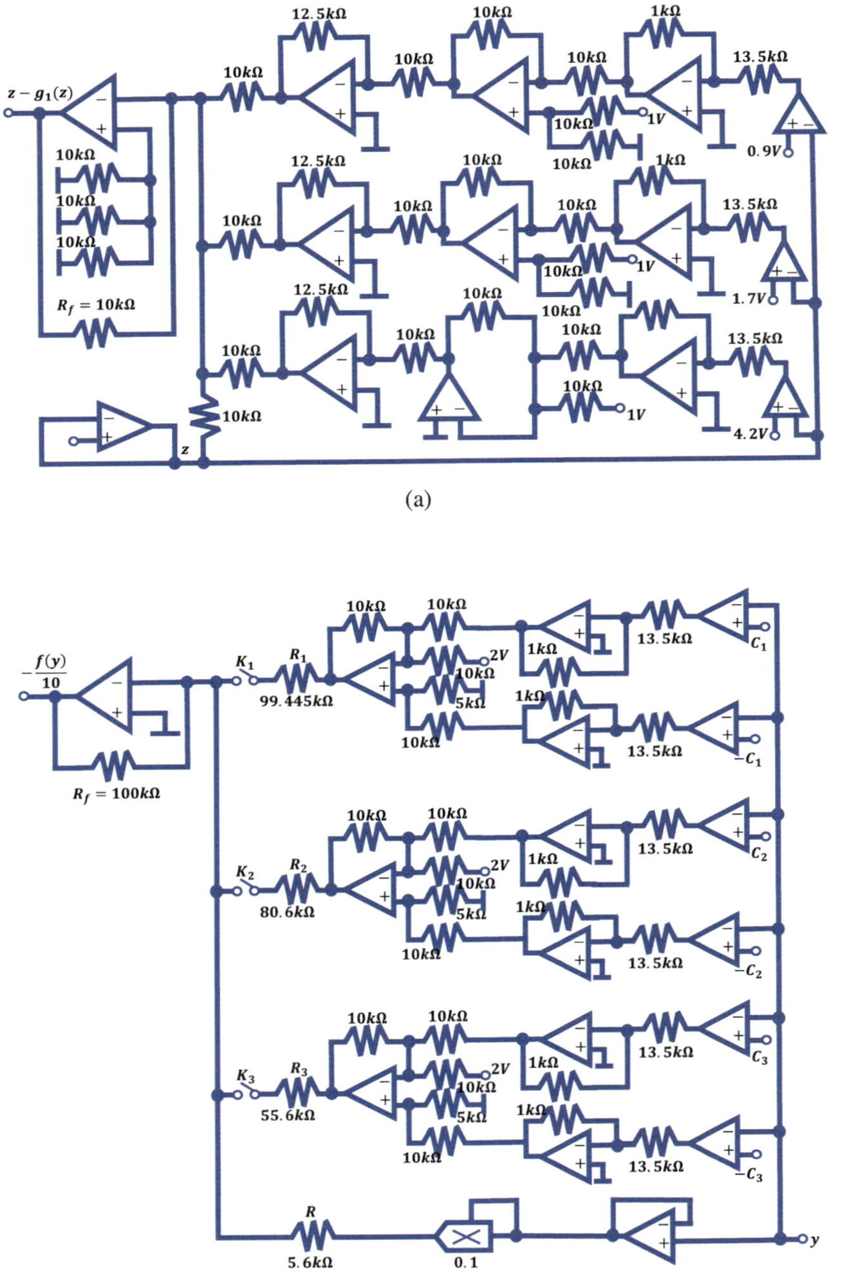

Fig. 6.11 Circuit design for $z - g_1(z)$ and $-f(y)/10$ [3]. **a** Circuit design for $z - g_1(z)$. **b** Circuit design for $-f(y)/10$

Referring to Figs. 6.11b, 6.13, and 6.14, the circuit experimental observations for the second class of multiwing hyperchaotic attractors are presented in Fig. 6.15. According to Fig. 6.11b, when K_1 is switched on and K_2, K_3 are switched off, a 4×2-wing butterfly hyperchaotic attractor is generated, as shown in Fig. 6.15a. When K_1 and K_2 are activated while K_3 is deactivated, a 6×2-wing butterfly hyperchaotic attractor is generated, as depicted in Fig. 6.15b.

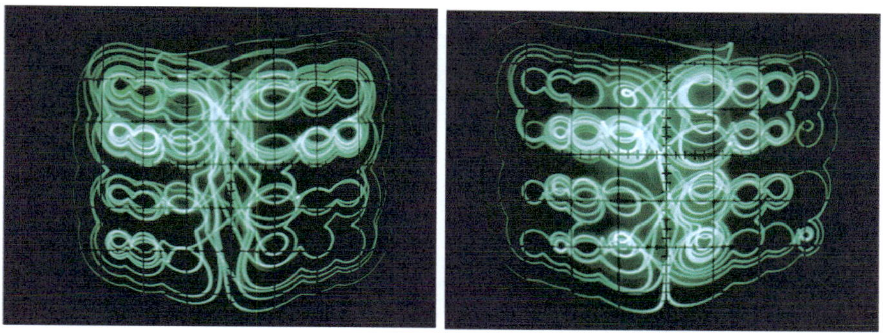

(a) (b)

Fig. 6.12 Experimental observations. **a** 6×4-wing hyperchaotic attractor. **b** 8×4-wing hyperchaotic attractor. Reprinted from [3], with permission from Wiley

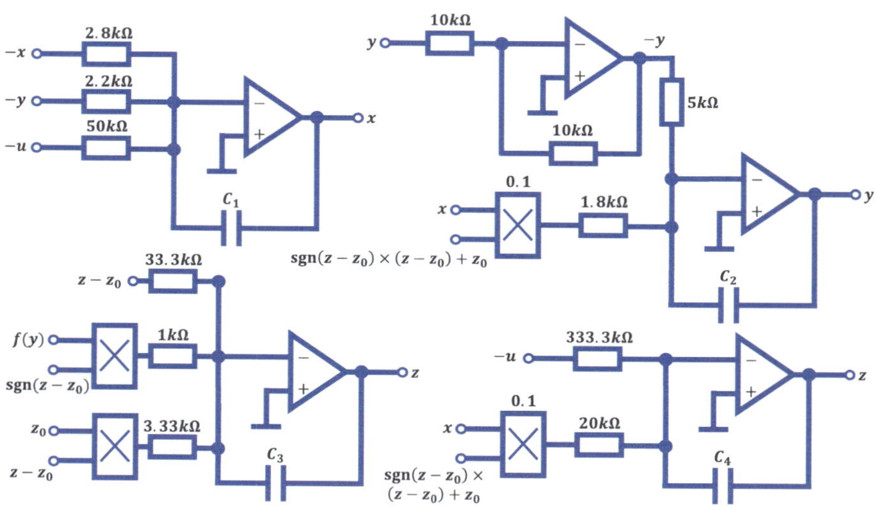

Fig. 6.13 Circuit design for the second class of $n \times m$-wing hyperchaotic system [3]

6.3 Conclusion

This chapter provides a systematic exploration of transformation-based methods for generating multiscroll chaotic attractors and grid multiwing hyperchaotic attractors. Parallel, mirror, and rotation transformations are introduced as effective techniques for constructing grid multiscroll chaotic attractors with high symmetry and scalability. Additionally, an approach is formulated for designing n-wing and $n \times m$-wing hyperchaotic attractors using 4-D hyperchaotic systems and switching control strategies. The feasibility and robustness of these methods are demonstrated through circuit designs and experimental observations, offering a practical means for applications in secure communication, nonlinear control, and signal processing. These findings pave the way for future advancements in chaos-based systems design and relevant real-world applications.

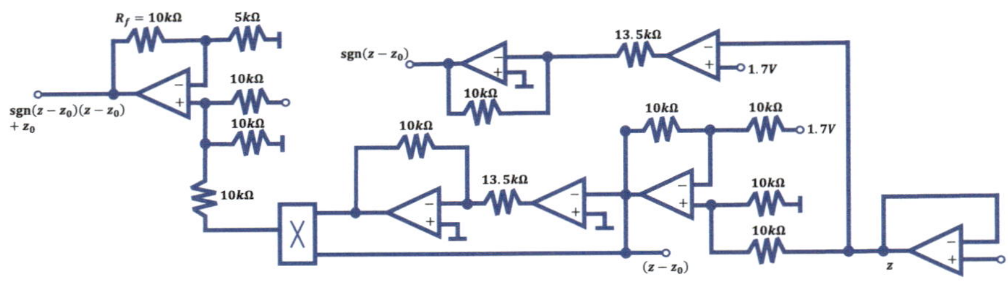

Fig. 6.14 Circuit design of $z - z_0$, $\mathrm{sgn}(z - z_0)$, and $\mathrm{sgn}(z - z_0)(z - z_0) + z_0$ [3]

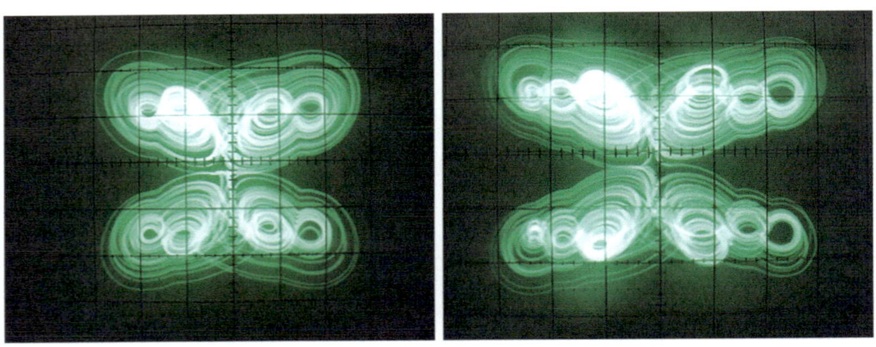

(a) (b)

Fig. 6.15 Experimental observations. **a** 4×2-wing hyperchaotic attractor. **b** 6×2-wing hyperchaotic attractor. Reprinted from [3], with permission from Wiley

References

1. Ai, X., Sun, K., He, S., Wang, H.: Design of grid multiscroll chaotic attractors via transformations. Int. J. Bifur. Chaos **25**, 1530027 (2015)
2. Chua, L.O., Komuro, M., Matsumoto, T.: The double scroll family. IEEE Trans. Circuits Syst. **33**, 1072–1118 (1986)
3. Zhang, C., Yu, S.: On constructing complex grid multi-wing hyperchaotic system: theoretical design and circuit implementation. Int. J. Circuit Theory and Appl. **41**, 221–237 (2013)
4. Lü, J., Chen, G.: A new chaotic attractor coined. Int. J. Bifur. Chaos **12**, 659–661 (2002)
5. Lü, J., Murali, K., Sinha, S., et al.: Generating multi-scroll chaotic attractors by thresholding. Phys. Lett. A **372**, 3234–3239 (2008)

Chapter 7
Generating Multiscroll Chaotic Attractors from Fractional-Fractal Systems

Abstract This chapter examines the generation of multiscroll chaotic attractors from fractional and fractal systems. Fractional-order systems are modified by incorporating fractal-fractional derivatives to produce self-similar chaotic dynamics. Special transformations, such as parabolic and triangular mappings, are utilized to construct various multiscroll attractors. Additionally, the chapter investigates hidden chaotic attractors within fractional-order systems, including the generalized Lorenz system, the Rabinovich-Fabrikant system, and non-smooth Chua systems. It reveals their unique properties and the coexistence of multiple attractors. These methodologies provide new insights into chaos theory and its applications in nonlinear dynamics.

7.1 Multiscroll Chaotic Attractors from Fractional-Order Systems

In this section, a simple fractional-order system is constructed to generate n-scroll chaotic attractors.

Ahmad [1] proposed a simple fractional-order chaotic oscillator, described by

$$\begin{cases} \dfrac{dx_1^\alpha}{dt^\alpha} = x_2, \\ \dfrac{dx_2}{dt} = x_3, \\ \dfrac{dx_3}{dt} = -a\,(x_1 + x_2 + x_3 - f(x_1)), \\ y = x_1, \end{cases} \tag{7.1}$$

where dx_1^α/dt^α is the fractional derivative of x_1 with order α, a is the control parameter for this oscillator, $f(x_1) = \mathrm{sgn}(x_1)$, and y is the system output.

It is evident that the order of system (7.1) is $2+\alpha$. In the integer-order case, system (7.1) can generate double-scroll chaotic attractors, for $0.49 < a < 1$. In the fractional-order case, the range of the parameter a for generating chaos depends on the specific fractional-order α.

J. Lü et al., *Chaotic Systems with Multiscroll Attractors*,
https://doi.org/10.1007/978-3-032-09854-2_7

To generate multiscroll chaotic attractors from the fractional system (7.1), the nonlinearity $f(x_1)$ is slightly modified as follows:

$$f(x_1) = \text{sgn}(x_1) + \sum_{k=1}^{N} (\text{sgn}(x_1 - k) + \text{sgn}(x_1 + k)). \tag{7.2}$$

When $\alpha = 1$ and $a = 0.4$, system (7.1) with (7.2) of an integer order can generate a 2-scroll chaotic attractor, as shown in Fig. 7.1. When $\alpha = 0.5$ and $a = 0.01$, system (7.1) with (7.2) of a fractional order 2.5 can also generate a 2-scroll chaotic attractor, as shown in Fig. 7.2.

7.2 Multiscroll Chaotic Attractors from Fractal Transformations

This section combines fractal and chaotic systems to generate multi-scroll chaotic attractors.

Yan et al. [2] introduced a no-equilibrium chaotic system, which possesses a wide range of complex behaviors, including chaos, hyperchaos and multiscroll attractors.

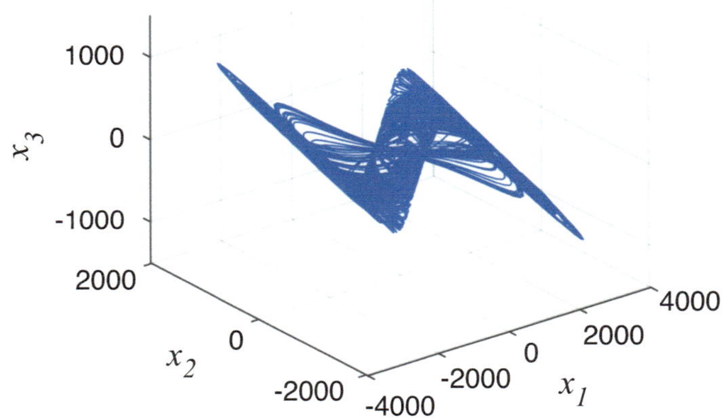

Fig. 7.1 2-scroll chaotic attractor observed from the integer-order system (7.1) with $\alpha = 1, a = 0.4$

7.2.1 A Chaotic System Without Equilibrium

Based on a 3-D chaotic system without equilibrium [2], a 4-D system is introduced as follows:

$$\begin{cases} \dot{x} = ay + bxz, \\ \dot{y} = -cx + yzw, \\ \dot{z} = d - |x|, \\ \dot{w} = gz, \end{cases} \tag{7.3}$$

where a, b, c, d, g are real constants. This model consists of three nonlinear terms and four linear terms, whereas x, y, z, and w represent the state variables, while b, d and g are three positive parameters fixed at $b = 15, d = 1$, and $g = 5$. Additionally, the parameters a and c are variable parameters used to generate multiscroll attractors.

From a practical standpoint, the absolute-valued term in system (7.3) can be easily implemented by circuits using electronic components such as resistors, diodes and operational amplifiers. Furthermore, the symmetry of the system provides the potential for generating coexisting multiple attractors.

The system (7.3) can be written in vector form as

$$\frac{d\theta}{dt} = f(\theta) = \begin{bmatrix} f_1(x, y, z, w) \\ f_2(x, y, z, w) \\ f_3(x, y, z, w) \\ f_4(x, y, z, w) \end{bmatrix}, \tag{7.4}$$

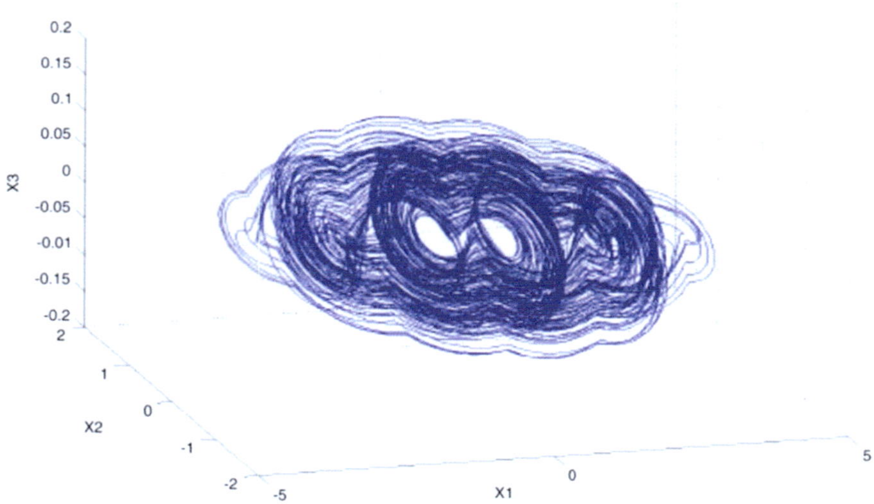

Fig. 7.2 6-scroll chaotic attractor observed from the fractional-order system (7.1) with $\alpha = 0.5, a = 0.01$. Reprinted from [1], with permission from the Elsevier

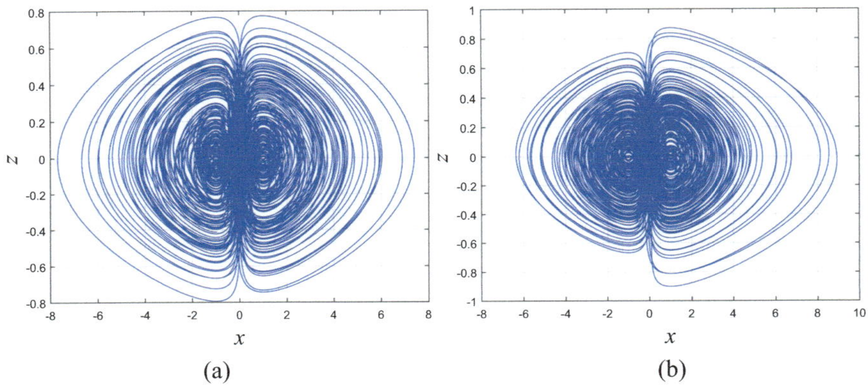

Fig. 7.3 Chaotic and hyper-chaotic orbits on the x-z plane [2]. **a** Chaotic ($a = 0.9$). **b** Hyper-chaotic ($a = 0.98$)

where $\theta \in R^4$ is the state vector and $f(\theta)$ is a smooth function given by

$$\begin{cases} f_1(x, y, z, w) = ay + bxz, \\ f_2(x, y, z, w) = -cx + yzw, \\ f_3(x, y, z, w) = d - |x|, \\ f_4(x, y, z, w) = gz. \end{cases} \tag{7.5}$$

The divergence of the vector field ($div(f) = \nabla \cdot f$) for the nonlinear system (7.4) can be computed, yielding

$$div(f) = \nabla \cdot f = \frac{\partial f_1}{\partial x} + \frac{\partial f_2}{\partial y} + \frac{\partial f_3}{\partial z} + \frac{\partial f_4}{\partial w} = z(b + w). \tag{7.6}$$

With $b = 15$, the divergence of the vector field satisfies $\nabla \cdot f = z(b + w) < 0$, implying that system (7.3) is dissipative.

The number of scrolls in the attractor can be easily adjusted by modifying the parameters of system (7.3). The initial values of the system are fixed at (-0.1, -0.1, -0.1, -0.1), and the parameter c is set to be 1. Figure 7.3 illustrates the hyperchaotic attractor obtained when $a = 0.9$ and the chaotic attractor obtained when $a = 0.98$, respectively.

The hyperchaotic system (7.3) not only can generate double-scroll chaotic attractors but also can produce multiscroll attractors. By adjusting the parameters a and c, two typical multiscroll chaotic attractors are obtained, as shown in Fig. 7.4a–d.

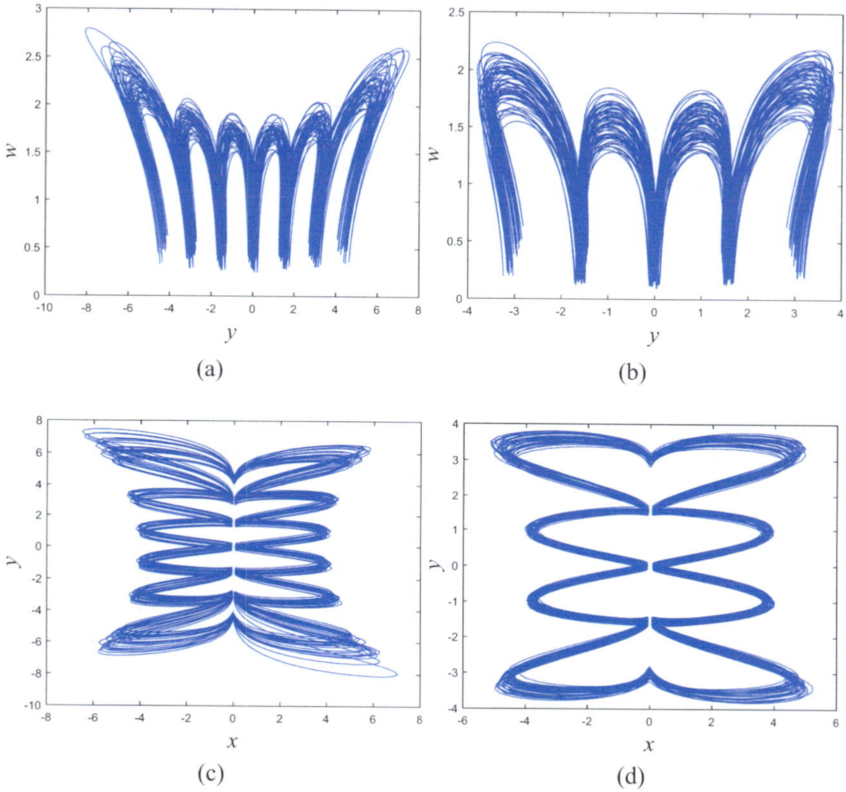

Fig. 7.4 Multiscroll chaotic attractors generated by varying parameter a or c [2]. **a** y-w plane with parameters $a = 0.02$ and $c = 1$. **b** y-w plane with parameters $a = 0.05$ and $c = 1$. **c** x-y plane with parameters $a = 0.02$ and $c = 1$. **d** x-y plane with parameters $a = 0.05$ and $c = 1$

7.2.2 Hidden Coexisting Attractors

Generally, multi-stability refers to the coexistence of multiple attractors, which is a fundamental property of many nonlinear dynamical systems. The attractors generated by system (7.3) are all hidden, in the sense that the equilibrium points are not located on the boundary of the system stability region; otherwise, it is self-excited. While hidden coexisting chaotic attractors have been observed in various nonlinear circuits, the phenomenon of multiple coexisting hidden multiscroll chaotic attractors is quite rare.

By selecting several typical initial conditions, one can realize the coexistence of multiple hidden multiscroll attractors, as shown in Fig. 7.5a.

The basin of attraction is a useful concept for examining coexisting hidden attractors. It is defined as the set of initial points in the phase space from which the system's trajectory will converge to a particular attractor, namely stable limit

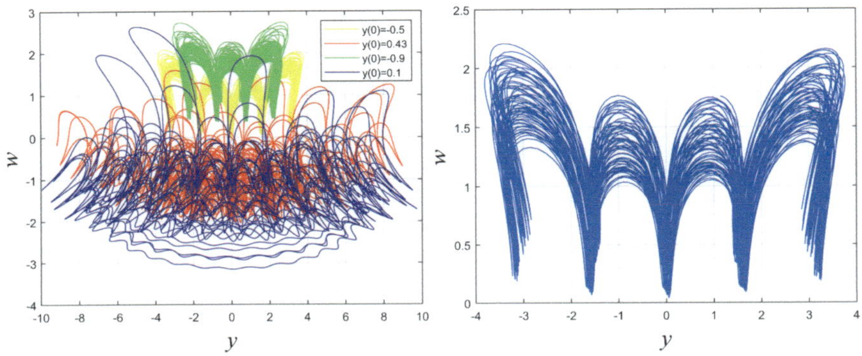

(a) (b)

Fig. 7.5 Coexistence of hidden multiple-scroll attractors with different initial conditions and the corresponding basin of attraction on the $x(0)$-$y(0)$ plane [2]. **a** Coexistence of hidden multiple-scroll attractors. **b** Basin of attraction on the $x(0)$-$y(0)$ plane, with $a = 0.08$, $b = 15$, $c = 1$, $d = 1$, $g = 5$, and $z(0) = w(0) = -0.1$

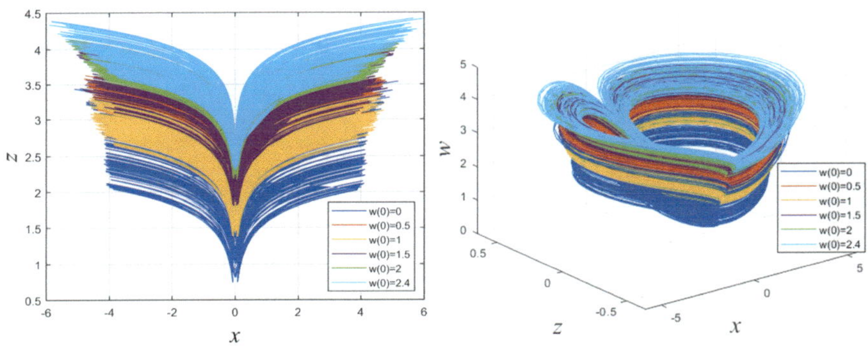

(a) (b)

Fig. 7.6 Initial offset-boosting behavior with different initial value $x(0)$. **a** Coexisting six hidden chaotic attractors in $x - w$ plane, **b** Coexisting six hidden chaotic attractors in $x - z - w$ plane

points. For system (7.3), the basins of attraction in the $x(0)$–$y(0)$ initial plane, with $z(0) = w(0) = -0.1$, are shown in Fig. 7.5b.

Now, fix parameters $a = 0.2$ and $c = 1$, and the initial condition $(-0.1, -0.1, -0.1, w(0))$, and choose $w(0)$ as the so-called offset-boosted controller, in such a way that it gradually increases. From the initial values $w(0) = (0, 0.5, 1, 1.5, 2)$, and 2.4, the phase diagrams of the six hidden chaotic attractors generated are shown in Fig. 7.6a and b.

7.3 Hidden Chaotic Attractors in Fractional-Order Systems

Danca [3] proposed a method to reveal hidden chaotic attractors from fractional-order nonlinear systems.

7.3.1 *Hidden Chaotic Attractor of a Generalized Fractional-Order Lorenz System*

The generalized fractional-order Lorenz system is a fractional version of the generalized integer-order Lorenz system, derived from the Rabinovich system [4] as follows:

$$\begin{aligned} lx_1^q &= -\sigma\,(x_1 - x_2) - ax_2x_3, \\ x_2^q &= rx_1 - x_2 - x_1x_3, \\ x_3^q &= -x_3 + x_1x_2, \end{aligned} \tag{7.7}$$

where q is the fractional order, $\sigma = -ar$ and $a < 0$. With $q = 1$ and $a = 0$, the system (7.7) coincides with the classical Lorenz system.

Due to the symmetry of the following transformation

$$T\,(x_1, x_2, x_3) \to (-x_1, -x_2, x_3), \tag{7.8}$$

each trajectory has a corresponding symmetrical (twin) trajectory with respect to the x_3-axis.

Let $q = 0.995$, a value at which the system is chaotic. To confirm that the generalized Lorenz system (7.7) exhibits a hidden attractor, it is necessary to verify the existence of small neighborhoods around the unstable equilibrium X_0^*, from which trajectories are attracted to the stable equilibria $X_{1,2}^*$ shown in Fig. 1 of [3]. As seen from Fig. 1a of [3], trajectories originating from a small neighborhood of X_0^* either approach X_1^* (red plots) or X_2^* (blue plots). The plots shown in Fig. 1b of [3] are 50 trajectories starting from a neighborhood of radius $\delta = 0.3$ centered at X_0^*, showing how they are attracted to either X_1^* or X_2^*. The hidden chaotic attractor H is colored in green.

7.3.2 *Hidden Chaotic Attractor of the Rabinovich-Fabrikant System*

The fractional-order Rabinovich-Fabrikant (RF) system [5] is modeled by

$$\begin{aligned} lx_1^q &= x_2\left(x_3 - 1 + x_1^2\right) + ax_1, \\ x_2^q &= x_1\left(3x_3 + 1 - x_1^2\right) + ax_2, \\ x_3^q &= -2x_3\left(b + x_1x_2\right), \end{aligned} \tag{7.9}$$

where $a > 0$ and b are the bifurcation parameters. This fractional-order system exhibits highly complex dynamics, such as multistability, which is a key feature for the potential existence of various hidden attractors.

Considering the instability of the equilibrium X_0^*, with $q = 0.998$, in addition to the hidden attractor H shown in Fig. 2a (green plots) of [3], the system displays some unusual behaviors such as having "unbounded self-excited attractors" $Y_{1,2}^*$, as shown in Fig. 2a of [3]. To visualize that the chaotic attractor H is hidden, Figs. 2b and c of [3] show that all trajectories starting from the neighborhoods of X_0^* either diverge to infinity (gray plots) or converge to the stable equilibria $X_{1,2}^*$ (dotted blue and dotted red plots, respectively).

7.3.3 Hidden Chaotic Attractor in a Non-smooth Fractional-Order Chua System

Consider the non-smooth Chua's system

$$\begin{aligned} lx_1^q &= \alpha\left(x_2 - x_1 - m_1x_1 - \psi\left(x_1\right)\right), \\ x_2^q &= x_1 - x_2 + x_3, \\ x_3^q &= -\left(\beta x_2 + \gamma x_3\right), \end{aligned} \tag{7.10}$$

where $\psi(x_1) = (m_0 - m_1)sat(x_1)$, $sat(\cdot)$ is the saturation function, $m_0 = -0.1768$, $m_1 = -1.1468$, $\alpha = 8.4562$, $\beta = 12.0732$, and $\gamma = 0.0052$.

With $q = 0.9998$, the numerical results are shown in Fig. 3a of [3]. Figure 3b of [3] shows that all trajectories starting from a small neighborhood of the unstable equilibrium either converge to the stable equilibrium X_0^* (red plots) or diverge to infinity (blue plots).

7.4 Conclusion

This chapter presents a study of the design and analysis of multiscroll chaotic attractors in fractional and fractal dynamical systems. Based on fractional-order dynamics and using fractal-fractional derivatives, a range of self-similar chaotic behaviors is generated, demonstrating the effectiveness of fractional-order systems in generating complex attractors. Transformation techniques, including parabola and triangle mappings, facilitate the generation of multiscroll attractors with diverse topologies.

Furthermore, the exploration of hidden chaotic attractors in fractional-order systems, such as the generalized Lorenz system, Rabinovich-Fabrikant system, and a non-smooth Chua system, highlights the coexistence of multiple attractors and unveils some new phenomena in nonlinear dynamical systems. These findings enhance the understanding of chaotic systems and pave the way for potential applications in secure communication, signal processing, and system control.

References

1. Ahmad, W.M.: Generation and control of multi-scroll chaotic attractors in fractional order systems. Chaos Solit. Fract. **25**, 727–735 (2005)
2. Yan, D., Ji'e, M., Wang, L., et al.: Generating novel multi-scroll chaotic attractors via fractal transformation. Nonlinear Dyn. **107**, 3919–3944 (2022)
3. Danca, M.F.: Hidden chaotic attractors in fractional-order systems. Nonlinear Dyn. **89**, 577–586 (2017)
4. Danca, M.F., Feckan, M., Kuznetsov, N., Chen, G.: Looking more closely at the Rabinovich-Fabrikant system. Int. J. Bifur. Chaos **26**, 1650038 (2016)
5. Pikovskii, A.S., Rabinovich, M.I., Trakhtengerts, V.Y.: Onset of stochasticity in decay confinement of parametric instability. Sov. Phys.-JETP. **47**, 715–719 (1978)

Chapter 8
Generating Multifold Torus Chaotic Attractors

Abstract This chapter investigates the design and generation of multifold torus chaotic attractors using PWL systems. A general PWL function is constructed to extend the folded torus circuit, enabling the creation of multifold torus chaotic attractors with increased complexity. A systematic method for generating multidirectional multitorus (MDMT) chaotic attractors is developed, utilizing higher-dimensional PWL systems. Circuit-based implementations verify these theoretical designs, demonstrating their feasibility for practical applications. This study contributes to chaos theory by introducing new approaches for constructing intricate complex dynamical systems with potential engineering applications.

8.1 Multifold Torus Chaotic Attractors from PWL Systems

Yu et al. [1] proposed a systematic method for generating multifold torus chaotic attractors using a simple 3-D PWL system.

Recall the Chua's double-fold torus chaotic circuit, known as the folded torus circuit, described by [2]

$$\begin{cases} \dot{x} = -\alpha g\,(y - x), \\ \dot{y} = -g\,(y - x) - z, \\ \dot{z} = \beta y, \end{cases} \tag{8.1}$$

where α and β are real parameters and

$$g\,(y - x) = m_1\,(y - x) + \frac{m_0 - m_1}{2}\,(|y - x + x_1| - |y - x - x_1|). \tag{8.2}$$

Here, $g\,(y - x)$ is a PWL odd function satisfying $g(x - y) = -g(y - x)$. With $\alpha = 15, \beta = 1, m_0 = 0.1, m_1 = -0.07$, and $x_1 = 1$, system (8.1) generates a double-fold torus chaotic attractor, as shown in Fig. 8.1, with a maximum Lyapunov exponent of 0.0270.

J. Lü et al., *Chaotic Systems with Multiscroll Attractors*,
https://doi.org/10.1007/978-3-032-09854-2_8

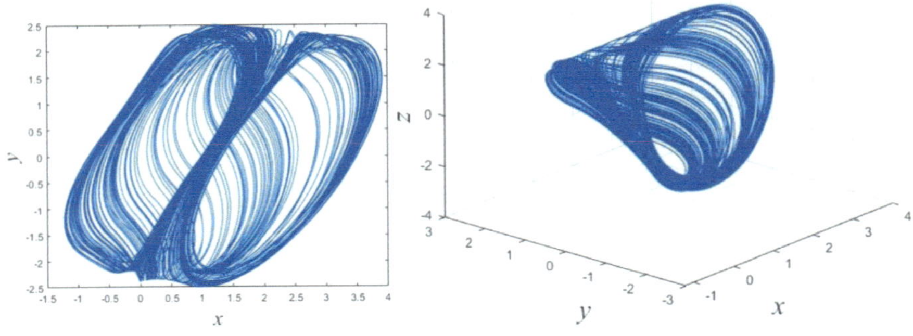

Fig. 8.1 Double-fold torus chaotic attractor. **a** x-y plane. **b** x-y-z space

To produce multifold torus chaotic attractors from the folded circuit described by system (8.1), a new PWL odd function is introduced to replace the original characteristic function (8.2), which is described by

$$g\left(y-x\right)=m_{N-1}\left(y-x\right)+\sum_{i=1}^{N-1}\frac{m_{i-1}-m_i}{2}\left(\left|y-x+x_i\right|-\left|y-x-x_i\right|\right), \tag{8.3}$$

where $g(x-y)$ satisfies the property $g(x-y)=-g(y-x)$, α and β are real parameters.

It is noted that the PWL characteristic function $g(y-x)$ can be reformulated as follows:

$$g\left(y-x\right)=\begin{cases} m_0\left(y-x\right), m_0\left(y-x\right), \text{if } \left|y-x\right|<x_1, \\ m_i\left(y-x\right)+\sum\limits_{j=1}^{i}\left(m_{j-1}-m_j\right)\operatorname{sgn}\left(y-x\right)x_j, \\ \text{if } x_i\leq\left|y-x\right|\leq x_{i+1},\quad 1\leq i\leq N-2, \\ m_{N-1}\left(y-x\right)+\sum\limits_{j=1}^{N-1}\left(m_{j-1}-m_j\right)\operatorname{sgn}\left(y-x\right)x_j, \\ \text{if } \left|y-x\right|>x_{n-1}, \end{cases} \tag{8.4}$$

where m_i $(0\leq i\leq N-1)$ represent the slopes of the segments, or radii, across different piecewise-defined regions, or radii within various piecewise-defined regions, and $\pm x_i$ $(x_i\geq 0, 1\leq i\leq N-1)$ denote the switching points. Given that $g(\cdot)$ is an odd function, only the positive switching points x_i, $1\leq i\leq N-1$, are needed to be specified.

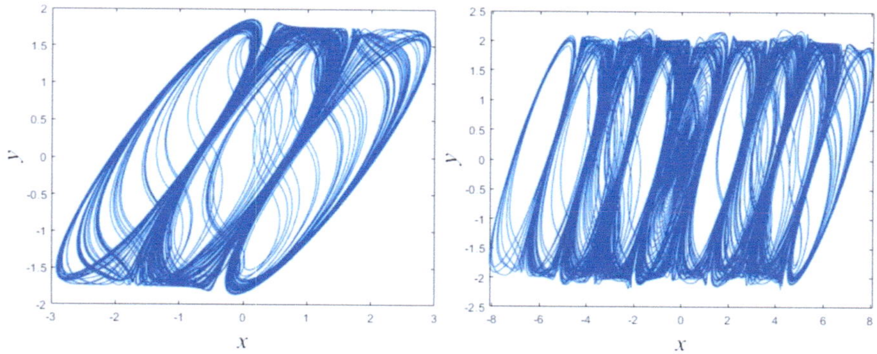

(a) (b)

Fig. 8.2 Multifold torus chaotic attractors [1]. **a** 3-fold torus. **b** 9-fold torus

Numerical simulations indicate that the system (8.1) with function (8.3) has a broad parameter range that supports chaos generation. Below, set $\alpha = 14.5$, $\beta = 1.25$, and $x_1 = 0.75$.

For illustration, two typical cases are examined, where $N = 2$ and $N = 5$. With $N = 2$, $g(y - x)$ can be expressed as $g(y - x) = m_1(y - x) + \frac{1}{2}(m_0 - m_1)(|y - x + x_1| - |y - x - x_1|)$. Let $m_0 = 0.15$ and $m_1 = -0.17$. With these parameters, system (8.1) along with Eq. (8.3) generates a three-fold torus chaotic attractor, as shown in Fig. 8.2a.

With $N = 5, m_0 = -0.17, m_1 = 0.15, m_2 = -0.17, m_3 = 0.15$, and $m_4 = -0.17$, the system (8.1) with Eq. (8.3) generates a 9-fold torus chaotic attractor, as shown in Fig. 8.2b.

Building on the above theoretical analysis, a block circuit is constructed to generate multifold torus chaotic attractors, as depicted in Fig. 8.3. This circuit is described by

$$\begin{cases} C_1 \dfrac{dv_{C1}}{dt} = -f(v_{C2} - v_{C1}), \\ C_2 \dfrac{dv_{C2}}{dt} = -f(v_{C2} - v_{C1}) - i_L, \\ L \dfrac{di_L}{dt} = v_{C2}, \end{cases} \tag{8.5}$$

where

$$\begin{aligned} & f(v_{C2} - v_{C1}) \\ & = G_{N-1}(v_{C2} - v_{C1}) + \tfrac{1}{2} \sum_{i=1}^{N-1} (G_{i-1} - G_i)(|v_{C2} - v_{C1} + E_i| - |v_{C2} - v_{C1} - E_i|). \end{aligned}$$

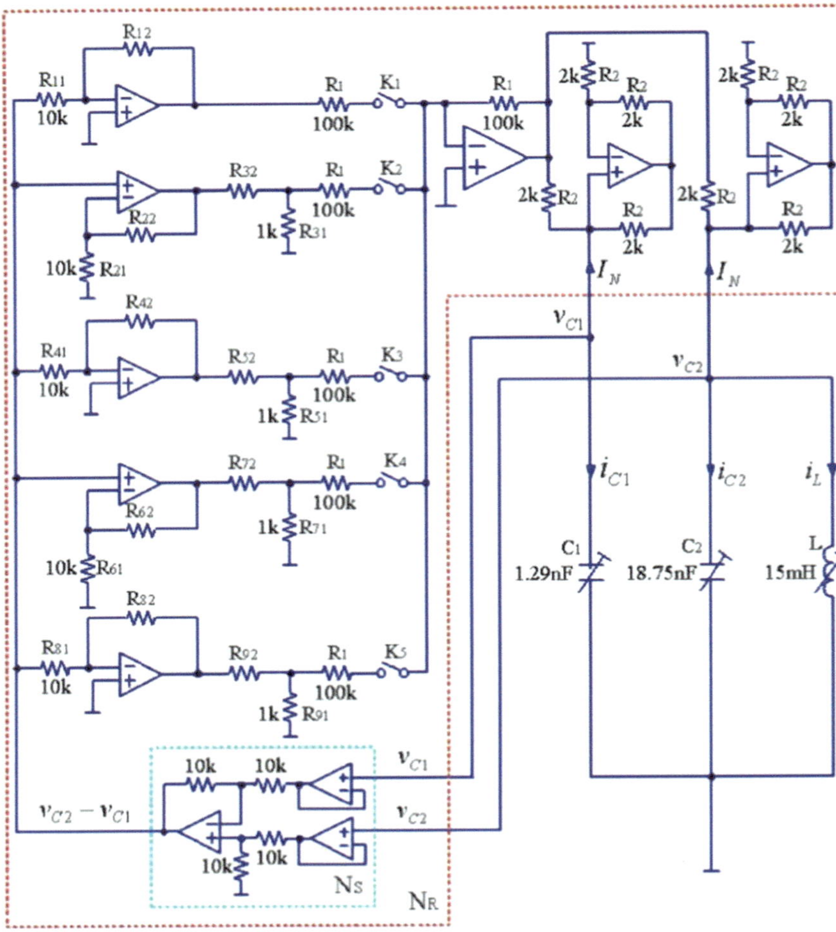

Fig. 8.3 Circuit diagram of multifold torus chaotic attractors. Reprinted from [1], with permission by the AIP

It follows from Eq. (8.5) that

$$
\begin{cases}
C_1 \dfrac{dv_{C1}}{dt} = -\dfrac{1}{RC_2} \dfrac{C_2}{C_1} Rf\,(v_{C2} - v_{C1}), \\
C_2 \dfrac{dv_{C2}}{dt} = -\dfrac{1}{RC_2} \left[Rf\,(v_{C2} - v_{C1}) + Ri_L \right], \\
L \dfrac{di_L}{dt} = \dfrac{1}{RC_2} \dfrac{R^2 C_2}{L} v_{C2}.
\end{cases}
\tag{8.6}
$$

By comparing Eq. (8.1) with Eq. (8.6), the following equivalent relationships can be derived:

$$\begin{cases} \tau_0 = RC_2, \tau = \dfrac{t}{t_0}, \alpha = \dfrac{C_2}{C_1} = 14.5, \\ \beta = \dfrac{R^2 C_2}{L} = 1.25, x = \dfrac{v_{C1}}{v_{BP}}, y = \dfrac{v_{C2}}{v_{BP}}, z = \dfrac{R_{iL}}{v_{BP}}, \\ x_i = \dfrac{E_i}{v_{BP}}, G = \dfrac{1}{R}, G_i = m_i G \; (0 \le i \le N-1), \\ f\,(v_{C2} - v_{C1}) = \dfrac{1}{R} g\,(y - x). \end{cases} \tag{8.7}$$

Here, $V_{BP} = 1\,\mathrm{V}$ and $1/\tau_0 = 1/RC_2$ represent the time-scale transformation. Let $R = 1\,\mathrm{k\Omega}$. Then, it follows from Eq. (8.7) that $C_1 = 1.29\,\mathrm{nF}$, $C_2 = 18.75\,\mathrm{nF}$, and $L = 15\,\mathrm{mH}$.

The subcircuit N_S in Fig. 8.3 functions as the subtraction generator, with output given by $v_{C2} - v_{C1}$. The subcircuit N_R in Fig. 8.3 generates the PWL characteristic function $f(v_{C2} - v_{C1})$, and its input and output satisfy the relation $I_N = f(v_{C2} - v_{C1})$. Additionally, the theoretical values of all resistors in NR can be precisely calculated by using recursive formulas. The operational amplifier chosen here is the TL082, and the power supply voltage for the circuit is $\pm E_C = \pm 15\,\mathrm{V}$. Consequently, the saturation voltage of the operational amplifier is obtained as $E_{\text{sat}} = 14.3\,\mathrm{V}$.

In the following, the theoretical values of all resistors in NR will be calculated to experimentally verify the 3-, 5-, 7-, and 9-fold torus chaotic attractors.

To generate a three-fold torus chaotic attractor, the theoretical values of the resistors in N_R are calculated as follows:

$$\begin{cases} G_0 = \dfrac{m_0}{R} = 0.15\text{mS}, G_1 = \dfrac{m_1}{R} = -0.17\text{mS}, \\ E_1 = x_1 V_{BP}, r_1 = \dfrac{R_{12}}{R_{11}} = -G_1 R_2 = 0.34, \\ r_2 = \dfrac{R_{22}}{R_{21}} = \dfrac{E_{\text{sat}}}{E_1} - 1 = 18.07, \\ r_3 = \dfrac{R_{32}}{R_{31}} = -\dfrac{1 + r_2}{R_2\,(G_1 - G_0)} - 1 = 28.79\,. \end{cases} \tag{8.8}$$

Similarly, the theoretical values of the resistors in N_R for the 5-, 7-, and 9-fold torus chaotic attractors are obtained as follows.

For the 5-fold torus chaotic attractor: $G_0 = -0.17\,\mathrm{mS}$, $G_1 = 0.15\,\mathrm{mS}$, $G_2 = -0.17\,\mathrm{mS}$, $E_i = x_i V_{BP}$ (for $i = 1, 2$), $r_1 = 0.34$, $r_2 = 4.84$, $r_3 = 8.12$, $r_4 = 19.1$, and $r_5 = 28.79$.

For the 7-fold torus chaotic attractor: $G_0 = 0.15\,\mathrm{mS}$, $G_1 = -0.17\,\mathrm{mS}$, $G_2 = 0.15\,\mathrm{mS}$, $G_3 = -0.17\,\mathrm{mS}$, $E_i = x_i V_{BP}$ (for $i = 1, 2, 3$), $r_1 = 0.34$, $r_2 = 3.00$, $r_3 = 5.25$, $r_4 = 6.90$, $r_5 = 9.78$, $r_6 = 18.07$, and $r_7 = 28.79$.

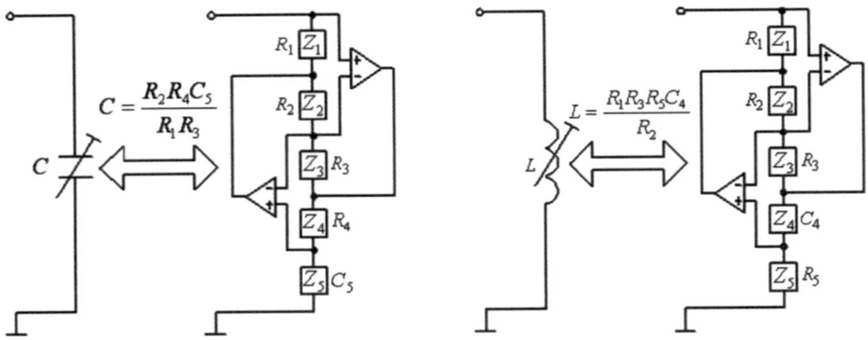

Fig. 8.4 Equivalent circuit of adjustable inductance and capacitance. Reprinted from [1], with permission by the AIP

For the 9-fold torus chaotic attractor: $G_0 = -0.17\,\text{mS}$, $G_1 = 0.15\,\text{mS}$, $G_2 = -0.17\,\text{mS}$, $G_3 = 0.15\,\text{mS}$, $G_4 = -0.17\,\text{mS}$, $E_i = x_i V_{BP}$ (for $1 \leq i \leq 4$), $r_1 = 0.34$, $r_2 = 1.53$, $r_3 = 2.95$, $r_4 = 3.62$, $r_5 = 4.66$, $r_6 = 4.84$, $r_7 = 8.12$, $r_8 = 19.1$, and $r_9 = 28.79$.

In the circuit design, all operational amplifiers shown in Fig. 8.3 are selected to be of the TL082 type. The supply voltage for the electrical source is $\pm E_C = \pm 15\,\text{V}$, with the operational amplifiers having a saturation voltage of $E_{\text{sat}} = 14.3\,\text{V}$. Additionally, all resistors shown in Fig. 8.3 are precisely adjustable, either as resistors or potentiometers. It is noted that the capacitances C_1, C_2, and the inductance L are adjustable, as illustrated in Fig. 8.4. Thus, the real parameter values of α and β can be fine-tuned by adjusting the capacitances C_1, C_2, and the inductance L.

Let $R_1 = 100\,\text{k}\Omega$, $R_2 = 2\,\text{k}\Omega$, $R_{31} = R_{51} = R_{71} = R_{91} = 1\,\text{k}\Omega$, and $R_{11} = R_{21} = R_{41} = R_{61} = R_{81} = 10\,\text{k}\Omega$. Using a formula similar to Eq. (8.8) and $R_{n2} = r_n R_{n1}$, for $1 \leq n \leq 9$, one can calculate all resistors in N_R.

Figure 8.5 presents the experimental observation results for the generated 3-, 5-, 7-, and 9-fold torus chaotic attractors.

8.2 Generating Multi-directional Multitorus Chaotic Attractors

Yu et al. [3] introduced a fourth-order double-torus chaotic circuit. Expanding on this basic circuit, they proposed a theoretical design approach for generating multi-directional multitorus (MDMT) chaotic attractors.

This chaotic system is described by

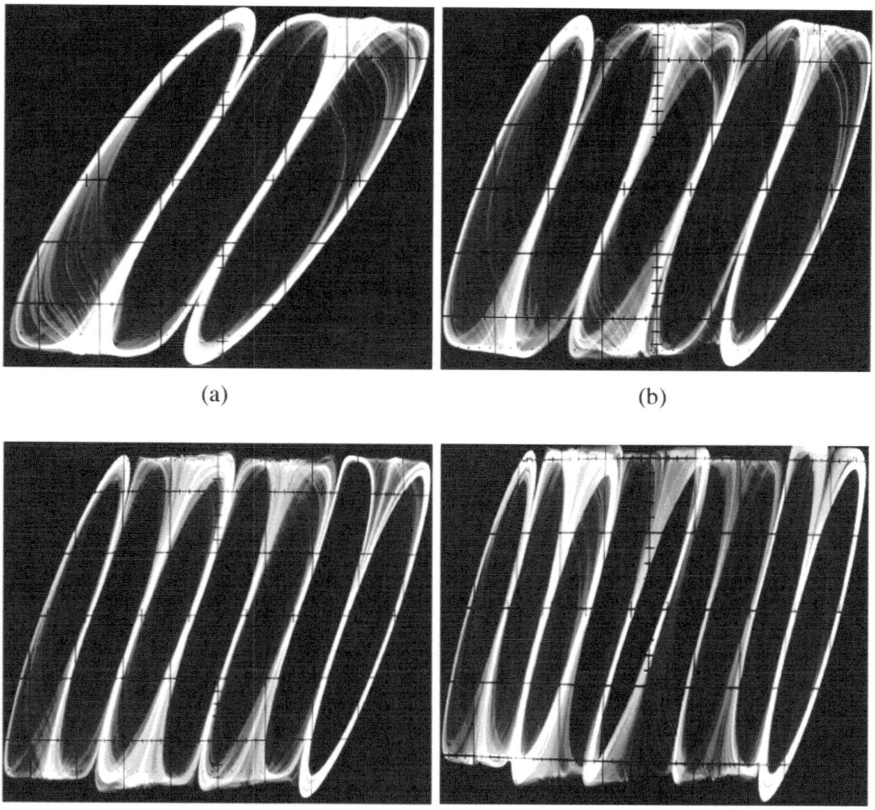

(a) (b) (c) (d)

Fig. 8.5 Experimental observations of multifold torus chaotic attractors. **a** 3-fold torus, with $x = 0.86$ V/div and $y = 0.6$ V/div. **b** 5-fold torus, with $x = 1.25$ V/div and $y = 0.64$ V/div. **c** 7-fold torus, with $x = 1.25$ V/div and $y = 0.64$ V/div. **d** 9-fold torus, with $x = 1.6$ V/div and $y = 0.9$ V/div. Reprinted from [1], with permission from the AIP

$$\begin{cases} \dfrac{dx}{d\tau} = y, \\ \dfrac{dy}{d\tau} = z, \\ \dfrac{dz}{d\tau} = w, \\ \dfrac{dw}{d\tau} = -\alpha x - \beta y - \gamma z - \delta w + \alpha f_1(x), \end{cases} \tag{8.9}$$

where α, β, γ, δ are system parameters and $f_1(x) = A_1 \operatorname{sgn}(x)$ with $A_1 > 0$.

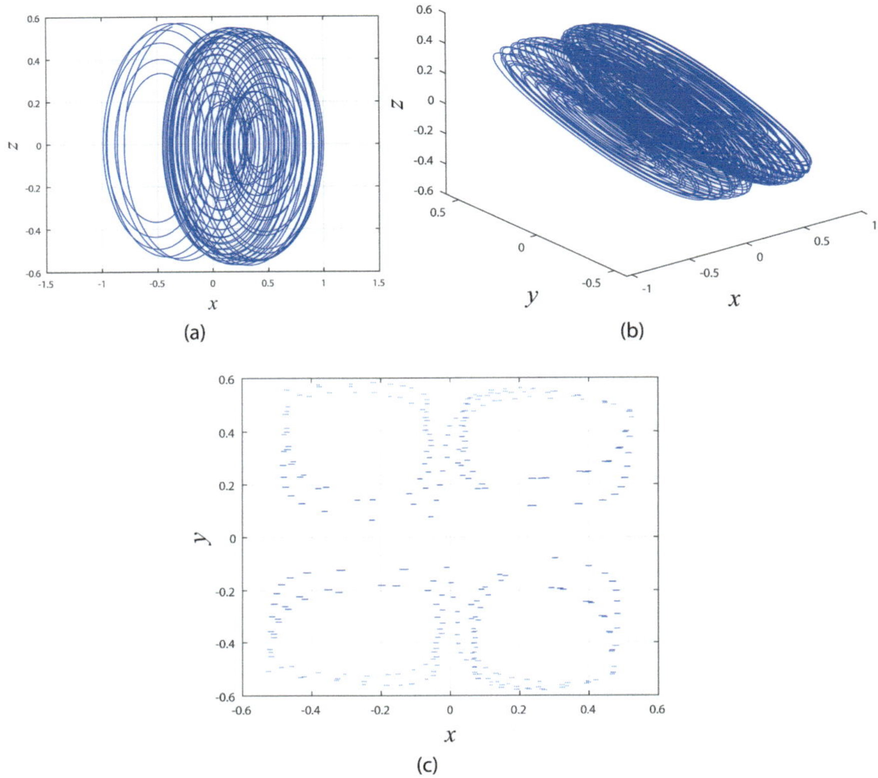

Fig. 8.6 Double-torus chaotic attractor [3]. **a** x-y plane. **b** x-y-z space. **c** Poincaré map on the x-z plane, with $y = 0$

With $\alpha = 0.1$, system (8.9) generates a double-torus chaotic attractor as shown in Fig. 8.6a, and as shown in the 3-D x-y-z space in Fig. 8.6b. Figure 8.6c presents the Poincaré map of the double-torus chaotic attractor on the x-y plane.

A fourth-order circuit is designed to experimentally verify the above-discussed double-torus chaotic attractor. Figure 8.7 presents the circuit diagram. It consists of two main sections: the first represents the basic fourth-order double-torus circuit, denoted as N_0, while the second is the sign function generator in the y-direction, labeled as N_1. The supply voltages for the circuit are set to ± 15 V, and the saturation voltages of the operational amplifiers are set to $\pm 13.5V$. All resistances are precisely adjustable, using resistors or potentiometers. The time-scale transformation is given by $\tau = t/R_0C_0$, where $1/R_0C_0$ is the integration constant of the integrator N_0. Additionally, $\alpha = R/R_1 = R/R_5$. By adjusting the resistors R_1 and R_5, the exact value of α can be obtained.

It is noted that a high-capacity capacitance, C_5, is added into Fig. 8.7, so as to improve the coupling between the subcircuits. The experiments demonstrate that, with the assistance of C_5, the signal trajectory can quickly and uniformly traverse the

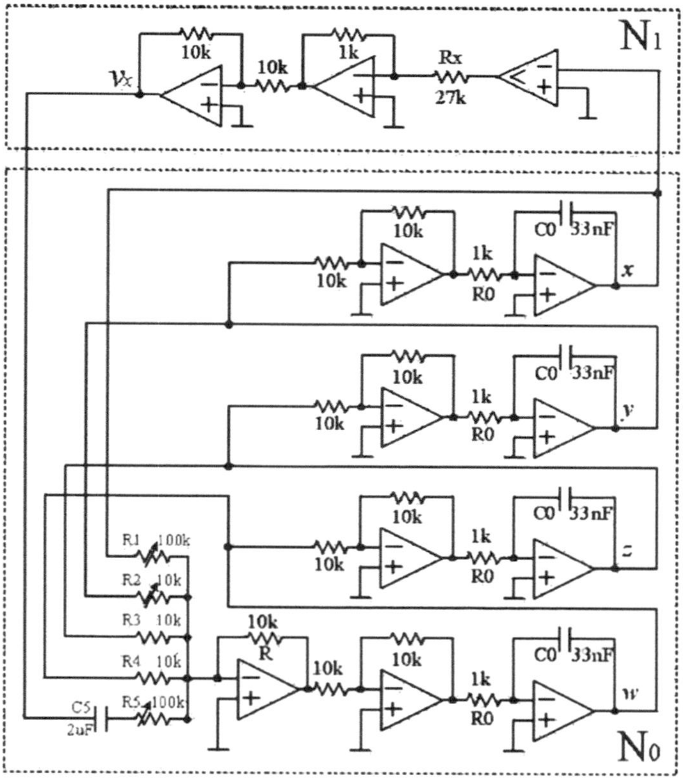

Fig. 8.7 Equivalent circuit of adjustable inductance and capacitance. Reprinted from [3], with permission by the IEEE

entire region of the double tori. The primary role of C_5 is to enhance the uniformity of the current distribution and prevent the current from staying in any localized region for too long. However, similar phenomena can still be observed without C_5 for observation in long-enough time.

Figure 8.8 shows the experimental results of the double-torus chaotic attractor.

Yu et al. [3] introduced a fourth-order MDMT chaotic system, defined by

$$\begin{cases} \dfrac{dx}{d\tau} = y - f_2(y), \\ \dfrac{dy}{d\tau} = z - f_3(z), \\ \dfrac{dz}{d\tau} = w - f_4(w), \\ \dfrac{dw}{d\tau} = \begin{bmatrix} -\alpha x - \beta y - \gamma z - \delta w + \alpha f_1(x) \\ +\beta f_2(y) + \gamma f_3(z) + \delta f_4(w) \end{bmatrix}, \end{cases} \tag{8.10}$$

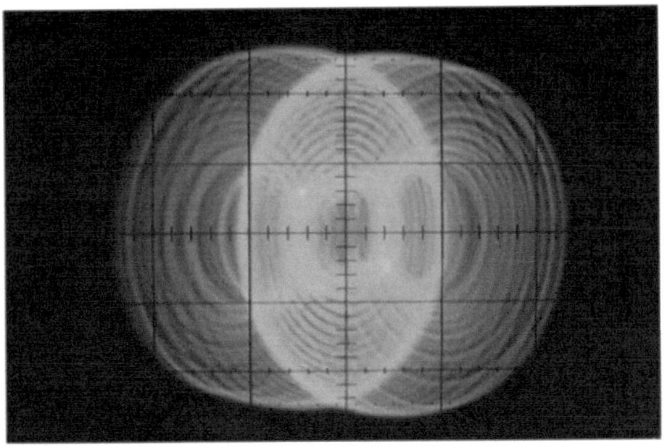

Fig. 8.8 Experimental observation of double-torus chaotic attractors, where $x = 0.2\,\mathrm{V/div}$, $y = 0.1\,\mathrm{V/div}$. Reprinted from [3], with permission by the IEEE

where $\alpha, \beta, \gamma, \delta$ are system parameters, $f_1(x), f_2(y), f_3(z), f_4(w)$ are controllers, which determine the number of tori in the x, y, z, and w directions, respectively.

To generate MDMT chaotic attractors from system (8.10), introduce a series of step functions or saturated functions into system (8.10). To do so, let $f_1(x), f_2(y), f_3(z), f_4(w)$ represent the step function series; that is,

$$f_i(\xi_i) = A_i \sum_{j=1}^{N_i} \{\mathrm{sgn}\,[\xi_i - (2j-1)A_i] + \mathrm{sgn}\,[\xi_i + (2j-1)A_i]\}, \tag{8.11}$$

or

$$f_i(\xi_i) = A_i \sum_{j=1}^{N_i} \{\mathrm{sgn}\,[\xi_i - 2jA_i] + \mathrm{sgn}\,[\xi_i + 2jA_i]\}, \tag{8.12}$$

where $1 \le i \le 4$, $\xi_1 = x, \xi_2 = y, \xi_3 = z, \xi_4 = w$, N_i, and M_i, $1 \le i \le 4$, are non-negative integers, $\mathrm{sgn}(\cdot)$ is the sign function, and A_i, $1 \le i \le 4$, are control parameters of the corresponding step function series with $A_1 > 0$ and $A_2 = A_3 = A_4 = A_1/4$.

Similarly, let $f_1(x), f_2(y), f_3(z), f_4(w)$ represent a series of saturation functions; that is,

$$f_i(\xi_i) = \frac{A_i}{2\eta} \sum_{\substack{j=-P_i,\\ j\neq 0}}^{P_i} \left\{ \left| \left[\xi_i - \left(2j - \frac{|j|}{j}\right) A_i\right] + \eta \right| - \left| \left[\xi_i - \left(2j - \frac{|j|}{j}\right) A_i\right] - \eta \right| \right\}, \tag{8.13}$$

or

$$f_i(\xi_i) = \frac{A_i}{2\eta} \sum_{j=-Q_i}^{Q_i} \{ | [\xi_i - 2jA_i] + \eta | - | [\xi_i - 2jA_i] - \eta | \}. \tag{8.14}$$

where $1 \le i \le 4$, $\xi_1 = x$, $\xi_2 = y$, $\xi_3 = z$, $\xi_4 = w$, N_i, P_i, and Q_i, $1 \le i \le 4$, are non-negative integers. Here, $\eta = 0.001$ and A_i (for $1 \le i \le 4$) are control parameters of the respective step function series, with $A_1 > 0$ and $A_2 = A_3 = A_4 = A_1/4$. Additionally, the chaotic regions for the α parameter depend on the values of N_i and M_i (for $1 \le i \le 4$). Specifically, as N_i and M_i increase, the parameters α, β, γ, and δ also increase. Bifurcation diagrams and Lyapunov exponent spectra can be further analyzed, which are not detailed here for brevity.

With $\alpha = 0.1$, $\beta = \gamma = \delta = 1$, $A_1 = 0.5$, and $N_1 = 2$, system (8.10) with Eq. (8.11) generates a 1-D 8-torus chaotic attractor, as shown in Fig. 8.9a. Figure 8.9b shows the Poincaré map of the 8-torus chaotic attractor on the x-y plane.

System (8.10) with Eq. (8.11), or alternatively with Eqs. (8.12), (8.13), or (8.14), is capable of generating various 2-D, 3-D, or 4-D multitorus chaotic attractors. For simplicity, a brief overview of the 2-D and 3-D cases are discussed here, before delving into a more detailed discussion of the 4-D case.

When $f_1(x) \ne 0$, $f_2(y) \ne 0$, and $f_3(z) = f_4(w) = 0$, system (8.10) with Eq. (8.11), or alternatively with Eqs. (8.12), (8.13), or (8.14), can produce 2-D multitorus chaotic attractors. Generally, system (8.10) with Eq. (8.11) or (8.12) can generate various 2-D $\pi_1 \times \pi_2$-torus chaotic attractors when $f_1(x) \ne 0$, $f_2(y) \ne 0$, and $f_3(z) = f_4(w) = 0$, where $\pi_i = 2N_i + 1$ or $\pi_i = 2M_i + 2$, $i = 1, 2$. For instance, when $\alpha = 0.1$, $\beta = \gamma = \delta = 1$, $A_1 = 0.5$, $A_2 = A_1/4$, and $N_1 = N_2 = 6$, system (8.10) with Eq. (8.11) produces a 2-D 13×13-torus chaotic attractor, as shown in Fig. 8.10a. Figure 8.10b and c display the 2-D 13×13-torus chaotic attractor on the x-y plane and the y-z plane, respectively.

When $f_1(x) \ne 0$, $f_2(y) \ne 0$, and $f_3(z) \ne 0$, $f_4(w) = 0$, system (8.10) with Eq. (8.11), or alternatively with Eqs. (8.12), (8.13), or (8.14), can produce 2-D multitorus chaotic attractors. Generally, system (8.10) with Eq. (8.11) or (8.12) can generate various 3-D $\pi_1 \times \pi_2 \times \pi_3$-torus chaotic attractors, when $f_1(x) \ne 0$, $f_2(y) \ne 0$, and $f_3(z) \ne 0$, $f_4(w) = 0$, where $\pi_i = 2N_i + 1$ or $\pi_i = 2M_i + 2$, $i = 1, 2, 3$. For instance, when $\alpha = 0.1$, $\beta = \gamma = \delta = 1$, $A_1 = 0.5$, $A_2 = A_3 = A_1/4$, and $N_1 = N_2 = 6$, $N_3 = 3$, system (8.10) with Eq. (8.11) produces a 3-D $13 \times 13 \times 7$ -torus chaotic attractor, as shown in Fig. 8.11a–d. Figure 8.11b–d displays the 3-D $13 \times 13 \times 13$-torus chaotic attractor on the x-y plane, the y-z plane, and the z-w plane, respectively.

When $f_1(x) \ne 0$, $f_2(y) \ne 0$, and $f_3(z) \ne 0$, $f_4(w) \ne 0$, system (8.10) with Eq. (8.11), or alternatively with Eqs. (8.12), (8.13), or (8.14), can produce 2-D multitorus chaotic attractors. Generally, system (8.10) with Eqs. (8.11) or (8.12) can generate various 4-D $\pi_1 \times \pi_2 \times \pi_3$-torus chaotic attractors when $f_1(x) \ne 0$, $f_2(y) \ne 0$, and $f_3(z) \ne 0$, $f_4(w) \ne 0$, where $\pi_i = 2N_i + 1$ or $\pi_i = 2M_i + 2$, $i = 1, 2, 3, 4$. For instance, when $\alpha = 0.1$, $\beta = \gamma = \delta = 1$, $A_1 = 0.5$, $A_2 = A_3 = A_4 = A_1/4$,

Fig. 8.9 1-D 8-torus chaotic attractor [3]. **a** x-y-z space. **b** Poincaré map on the x-z plane, with $y = 0$

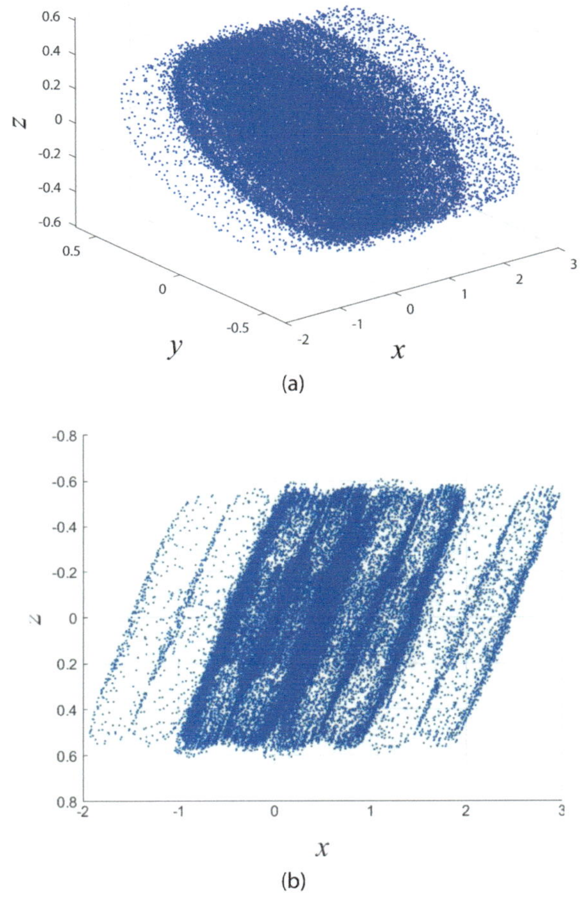

$N_1 = N_2 = 6$, and $N_3 = N_4 = 3$, system (8.10) with (8.11) generates a 4-D $13 \times 13 \times 7 \times 7$-torus chaotic attractor, as shown in Fig. 8.12a–c. Figure 8.12d shows the Poincaré map of this $13 \times 13 \times 7 \times 7$-torus chaotic attractor on the x-y plane.

Based on the operational principles of MDMT chaotic attractors described by Eq. (8.10) and Eq. (8.11), a circuit is designed to physically generate various MDMT chaotic attractors.

This diagram, shown in Fig. 8.13, comprises 6 distinct sections: Part I is the foundational fourth-order multitorus circuit N_0; Part II is the step function series generator N_1 along the x-direction; Part III is the step function series generator N_2 along the y-direction; Part IV is the step function series generator N_3 along the z-direction; Part V is the step function series generator N_4 along the w-direction;

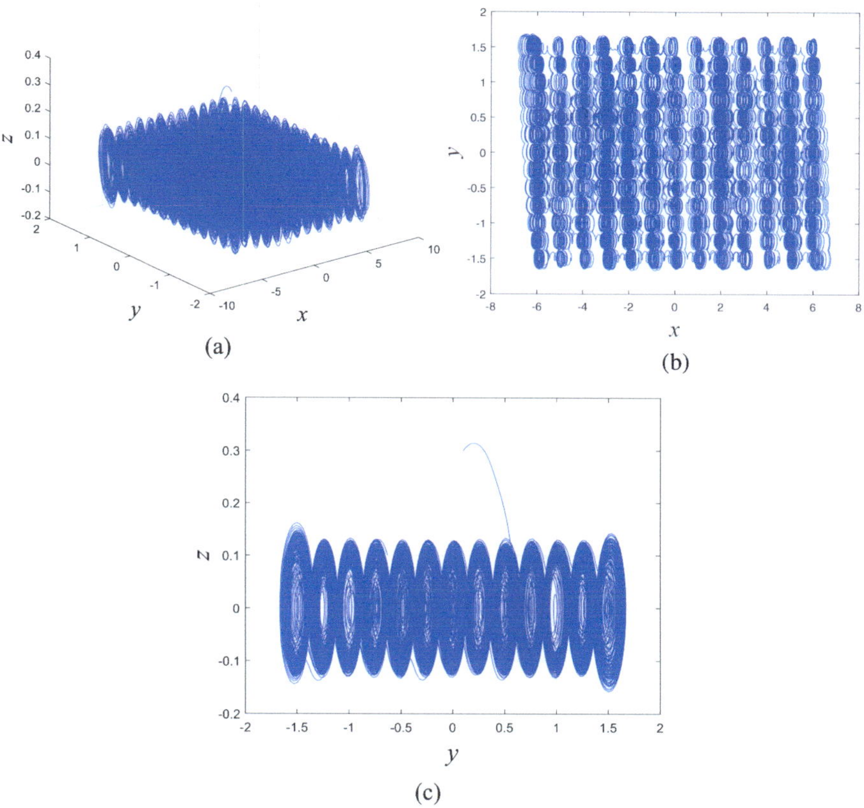

Fig. 8.10 2-D 13 × 13-torus chaotic attractor [3]. **a** x-y-z space. **b** x-y plane. **c** y-z plane

Part VI includes the switching components, such as K_1, K_2, K_3, K_{10}, K_{11}, K_{12}, K_{13}, K_{14}, K_{15}, K_{22} .

In this diagram, the operational amplifiers are powered by supply voltages of $V = \pm 15\text{V}$ with saturation voltages $V_{\text{sat}} = \pm 13.5\text{V}$. The switches K_1, K_2 and K_3 control the dimensions of the MDMT chaotic attractors by interacting with the basic fourth-order multitorus circuit N_0 and the capacitance coupling. Meanwhile, switches K_{10}, K_{11}, K_{12}, K_{13}, K_{14}, and K_{15} manage the number of tori in the x-direction, and switch K_{22} adjusts the number of tori in the y-direction by configuring generators N_1 and N_2, respectively.

Based on the circuit diagram shown in Fig. 8.13, the state equation of the circuit can be written in the integral form, as follows:

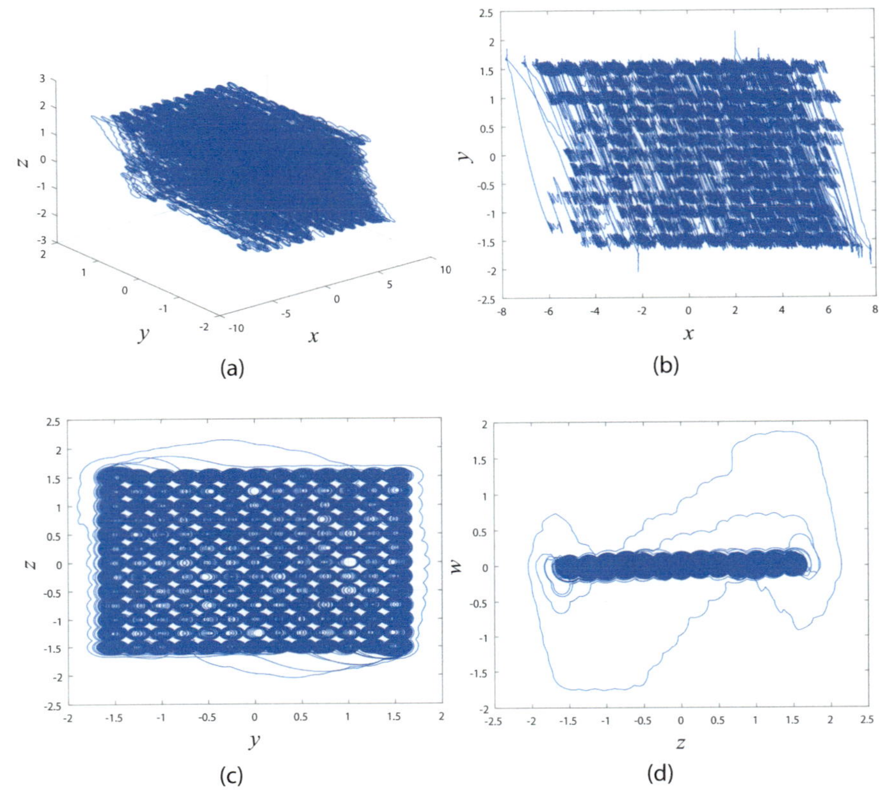

Fig. 8.11 3-D 13 × 13 × 13-torus chaotic attractor [3]. **a** x-y-z space. **b** x-y plane. **c** y-z plane. **d** z-w plane

$$
\begin{cases}
x = \dfrac{1}{R_0 C_0} \displaystyle\int \left(y + K_1 v_y\right) dt, \\
y = \dfrac{1}{R_0 C_0} \displaystyle\int (z + K_2 v_z)\, dt, \\
z = \dfrac{1}{R_0 C_0} \displaystyle\int (w + K_3 v_w)\, dt, \\
w = \dfrac{1}{R_0 C_0} \displaystyle\int \Bigg(-\dfrac{R}{R_1} x - \dfrac{R}{R_2} y - \dfrac{R}{R_3} z - \dfrac{R}{R_4} w - \dfrac{R}{R_5} v_x \\
\qquad - \dfrac{R}{R_6} K_1 v_y - \dfrac{R}{R_7} K_2 v_z - \dfrac{R}{R_8} K_3 v_w \Bigg) dt,
\end{cases}
\tag{8.15}
$$

where v_x, v_y, v_z, v_w are the step function series in the x-, y-, z-, w-direction, respectively, given by

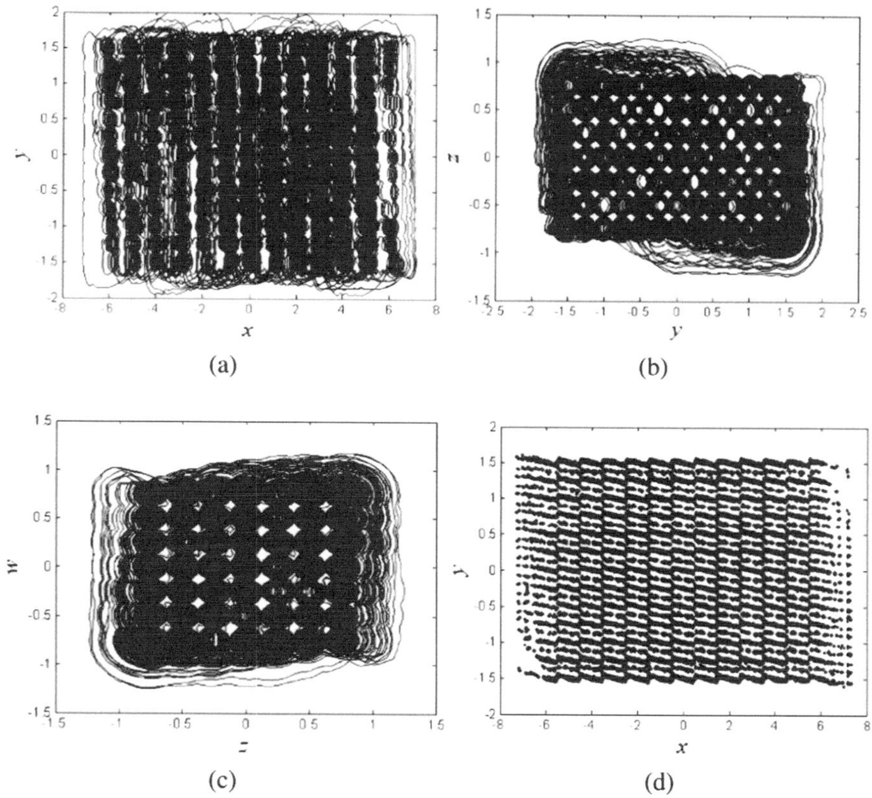

Fig. 8.12 2-D 13 × 13 × 7 × 7-torus chaotic attractor. **a** the x-y plane. **b** the y -z plane. **c** the z-w plane. **d** Poincaré map on the y-z plane with $w = 0$. Reprinted from [3], with permission by the IEEE

$$
\begin{cases}
v_x = -\dfrac{|V_{sat}|}{R_x} \left\{ K_{10}\,\mathrm{sgn}(x) \displaystyle\sum_{j=1}^{5} K_{1j} \left[\mathrm{sgn}\left(x - E_{1j}\right) + \mathrm{sgn}\left(x + E_{1j}\right)\right] \right\}, \\
v_y = -\dfrac{|V_{sat}|}{R_y} \displaystyle\sum_{j=1}^{2} K_{2j} \left[\mathrm{sgn}\left(y - E_{2j}\right) + \mathrm{sgn}\left(y + E_{2j}\right)\right], \\
v_z = -\dfrac{|V_{sat}|}{R_z} K_{31} \left[\mathrm{sgn}\left(z - E_{31}\right) + \mathrm{sgn}\left(z + E_{31}\right)\right], \\
v_w = -\dfrac{|V_{sat}|}{R_w} K_{41} \left[\mathrm{sgn}\left(w - E_{41}\right) + \mathrm{sgn}\left(w + E_{41}\right)\right].
\end{cases}
\tag{8.16}
$$

In this configuration, $R_0 = 1\,\mathrm{k}\Omega$ and $C_0 = 33\,\mathrm{nF}$ are the integral resistor and capacitor, respectively. The resistors R_x, R_y, R_z, and R_w serve for voltage-to-current conversion. All operational amplifiers have saturation voltages of $V_{sat} = \pm 13.5\,\mathrm{V}$.

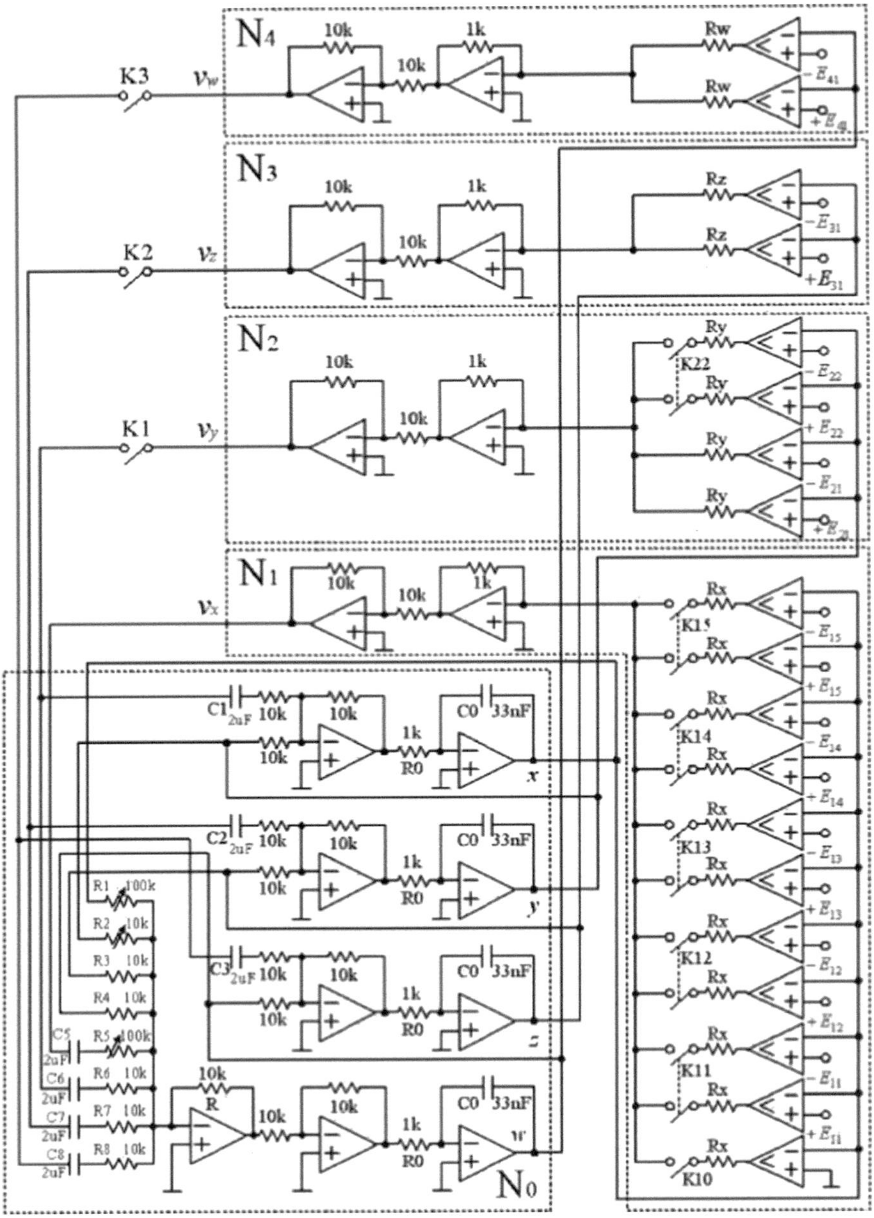

Fig. 8.13 Circuit diagram for implementing multi-directional grid multitorus chaotic attractors. Reprinted, from [3], with permission by the IEEE

The step voltages for the step function series are given by E_{1j} (for $1 \leq j \leq 5$), E_{2k}, $k = 1, 2$, E_{31}, and E_{41}.

The transformation coefficients of the anti-summator are represented by (R/R_i), $1 \leq j \leq 8$, with $R = 10\,\mathrm{k}\Omega$ and the resistors R_i, $1 \leq j \leq 8$, as specified in Fig. 8.13. Additionally, $K_{21} = K_{31} = K_{41} = 1$. The switching functions K_{1j}, $1 \leq j \leq 5$, and K_j, $1 \leq j \leq 3$, are defined such that their value is 1 when the switch is on and 0 when it is off.

From Eqs. (8.15) to (8.16), one has

$$\begin{cases} \dfrac{dx}{d\tau} =y - K_1 f_2(y), \\ \dfrac{dy}{d\tau} =z - K_2 f_3(z), \\ \dfrac{dz}{d\tau} =w - K_3 f_4(w), \\ \dfrac{dw}{d\tau} = -\dfrac{R}{R_1}x - \dfrac{R}{R_2}y - \dfrac{R}{R_3}z - \dfrac{R}{R_4}w + \dfrac{R}{R_5}f_1(x) \\ \qquad + \dfrac{R}{R_6}K_1 f_2(y) + \dfrac{R}{R_7}K_2 f_3(z) + \dfrac{R}{R_8}K_3 f_4(w), \end{cases} \tag{8.17}$$

where $R/R_1 = R/R_5 = \alpha$, $R/R_2 = R/R_6 = \beta$, $R/R_3 = R/R_7 = \gamma$, $R/R_4 = R/R_8 = \delta$, and

$$\begin{cases} f_1(x) =\dfrac{|V_{sat}|}{R_5}\left\{K_{10}\ \mathrm{sgn}(x) + \displaystyle\sum_{j=1}^{5} K_{1j}\left[\mathrm{sgn}\left(x - E_{1j}\right) + \mathrm{sgn}\left(x + E_{1j}\right)\right]\right\}, \\ f_2(y) =\dfrac{|V_{sat}|}{R_y}\displaystyle\sum_{j=1}^{2} K_{2j}\left[\mathrm{sgn}\left(y - E_{2j}\right) + \mathrm{sgn}\left(y + E_{2j}\right)\right], \\ f_3(z) =\dfrac{|V_{sat}|}{R_z}K_{31}\left[\mathrm{sgn}\left(z - E_{31}\right) + \mathrm{sgn}\left(z + E_{31}\right)\right], \\ f_4(w) =\dfrac{|V_{sat}|}{R_{\mathrm{sw}}}K_{41}\left[\mathrm{sgn}\left(w - E_{41}\right) + \mathrm{sgn}\left(w + E_{41}\right)\right]. \end{cases} \tag{8.18}$$

Here, the resistors R_i, $1 \leq i \leq 8$, are precisely adjustable resistors. In practice, one can modify the system parameters $\alpha = \beta = \gamma = \delta$ by adjusting a set of resistors: R_1, R_5, R_2, R_6, R_3, R_7, R_4, and R_8. Additionally, $1/(R_0 C_0)$ is the integrator constant for the integrator N_0 shown in Fig. 8.13, and it acts for the time-scale transformation, satisfying $\tau = t/(R_0 C_0)$. Moreover, the distribution of the chaotic signal's frequency spectrum can be adjusted by varying the integral resistor R_0 or the integral capacitance C_0. Specifically, when R_0 or C_0 decreases, the chaotic signal's frequency spectrum shifts toward the high-frequency end. Conversely, when R_0 or C_0 increases, the spectrum shifts towards the low-frequency end.

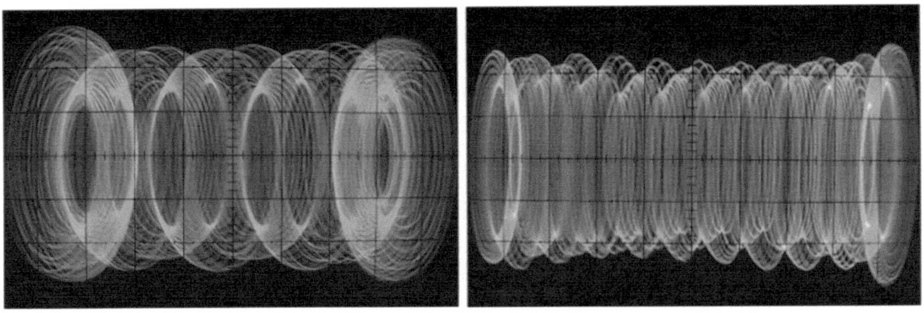

(a) (b)

Fig. 8.14 Experimental observations of 1-D multitorus chaotic attractors on the x-y plane. **a** 8-torus, where $x = 0.25$ V/div, $y = 0.1$ V/div. **b** 20-torus, where $x = 0.55$ V/div, $y = 0.1$ V/div. Reprinted from [3], with permission by the IEEE

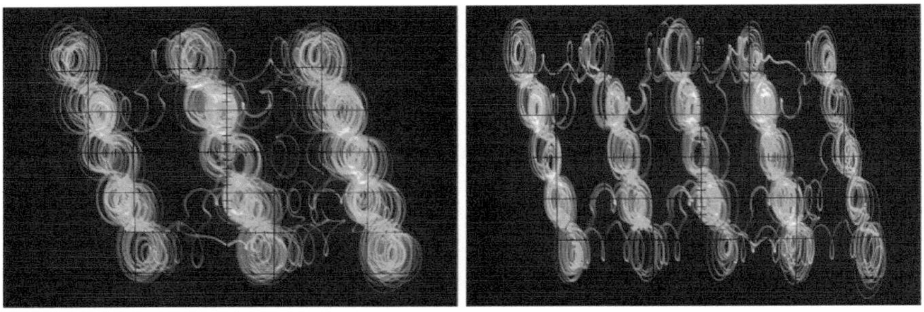

(a) (b)

Fig. 8.15 Experimental observations of 2-D multitorus chaotic attractors on the x-y plane. **a** 3×5-torus, where $x = 0.36$ V/div, $y = 0.2$ V/div. **b** 5×5-torus, where $x = 0.6$ V/div, $y = 0.2$ V/div. Reprinted from [3], with permission by the IEEE

Referring to the circuit diagram in Fig. 8.13 and Tables I and II in [3], one can obtain 1-D, 2-D, 6-D, 8-D, 12-D, 16-D, and 20-D torus chaotic attractors. Figure 8.14 shows the experimental observations of the 8-torus and 20-torus chaotic attractors.

Based on the circuit diagram shown in Fig. 8.13 and the details provided by Tables III and IV in [3], 2-D 3×5- and 5×5-torus chaotic attractors can be generated, as shown in Fig. 8.15.

When $K_1, K_2, K_3, K_{11}, K_{12}$, and K_{22} are switched on, while K_{10}, K_{13}, K_{14}, and K_{15} are switched off, the circuit can generate a 3-D $5 \times 5 \times 3 \times 3$-torus chaotic attractor, as shown in Fig. 8.16. The corresponding circuit parameters are $R_x = 13.5\,k\Omega$, $R_y = R_z = R_w = 54\,k\Omega$, $\pm E_{11} = \pm 1.00\,V$, $\pm E_{12} = \pm 3.00\,V$, $\pm E_{21} = \pm E_{31} = \pm E_{41} = \pm 0.25\,V$, and $\pm E_{22} = \pm 0.75\,V$.

Similarly, 3-D multitorus chaotic attractors can be experimentally validated by adjusting the switch linkages in Fig. 8.13. For simplicity, detailed experimental procedures and results are omitted here. To improve the coupling between the sub-circuits,

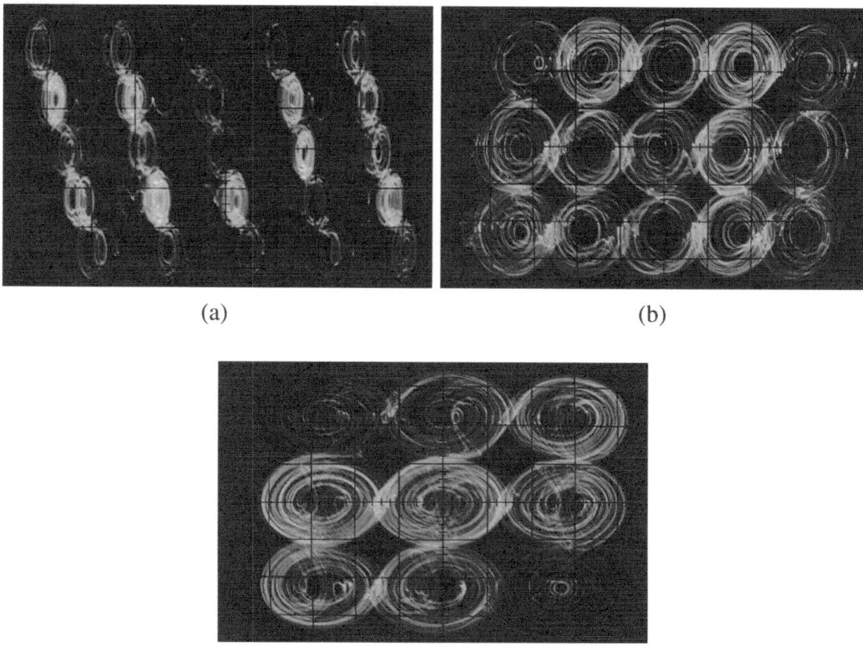

Fig. 8.16 Experimental observations of 4-D $5 \times 5 \times 3 \times 3$-torus chaotic attractors on the x-y plane. **a** the x-y plane, where $x = 0.6$ V/div, $y = 0.2$ V/div. **b** the y-z plane, where $x = 0.16$ V/div, $y = 0.1$ V/div. **c** the z-w plane, where $x = 0.12$ V/div, $y = 0.12$ V/div. Reprinted from [3], with permission by the IEEE

high-capacity capacitances C_1, C_2, C_3, C_5, and C_6, C_7, C_8 are added, with $2\mu F$, as shown in Fig. 8.13. The experiments demonstrate that these capacitances allow the signal trajectory to swiftly and evenly traverse the entire region of all tori. The primary role of these capacitors is to enhance the uniformity of current distribution and prevent the current from staying too long in any localized region. However, it is also possible to observe similar phenomena without these additional capacitances.

8.3 Multifold Torus Chaotic Attractors Generated by a PWL Odd Function

Doungmo Goufo [4] utilizes piecewise functions applied to a modified multifold torus system.

The double-fold torus chaotic system, originally introduced by Matsumoto et al. [2] and subsequently refined by others, e.g., Yu et al. [3], is described by

$$\begin{cases} \dfrac{dx}{dt} = -af(y-x), \\ \dfrac{dy}{dt} = -f(y-x) - z, \\ \dfrac{dz}{dt} = by, \end{cases} \tag{8.19}$$

where $a, b \in \mathbb{R}$, and f is a PWL odd function defined as

$$f(y-x) = \alpha_1 (y-x) + \frac{1}{2}(\alpha_0 - \alpha_1)[|y-x+x_1| - |y-x-x_1|], \tag{8.20}$$

which satisfies $f(x-y) = -f(y-x)$.

Numerical simulations on the model defined by Eqs. (8.19) and (8.20) on the x-y plane demonstrate a two-fold torus chaotic attractor, as shown in Fig. 8.17, with parameters $\alpha_0 = 0.1$, $\alpha_1 = -0.07$, $a = 15$, $b = 1$, and $x_1 = 1$. This attractor has a maximum Lyapunov exponent of 0.0270.

The model (8.19) was further enhanced by Matsumoto et al. [2] and Yu et al. [3] to create multitorus chaotic attractors with more than two folds. Through an N-torus breakdown process, it became feasible to generate an N-fold torus chaotic attractor, for $N \geq 2$. This approach defines the N-torus as a quotient of $\mathbf{R}^N$, obtaining a modified multifold torus system, where the PWL odd function f is modified as follows:

$$\begin{aligned} &f(y-x) \\ &= \alpha_{n-1}(y-x) + \tfrac{1}{2}\sum_{k=1}^{n-1}(\alpha_{k-1} - \alpha_k)[\,|y-x+x_k| - |y-x-x_k|\,]. \end{aligned} \tag{8.21}$$

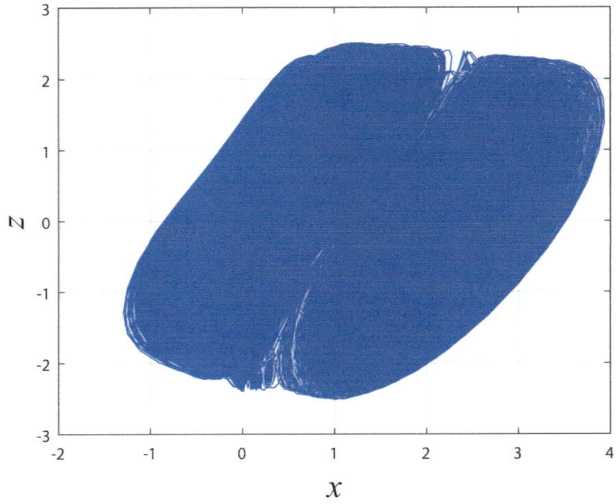

Fig. 8.17 Numerical simulation of the two-fold torus chaotic attractor [4]

The PWL function f can be represented in the following segmented form:

$$f(y-x) = \begin{cases} \alpha_0 (y-x), & \text{if } |y-x| < x_1, \\ \alpha_k (y-x) + \sum_{l=1}^{k} (\alpha_{l-1} - \alpha_l)\, \text{sgn}(y-x)\, x_l, & \text{if } \begin{cases} x_k \le |y-x| \le x_{k+1}, \\ 1 \le k \le n-2, \end{cases} \\ \alpha_{n-1} (y-x) + \sum_{l=1}^{n-1} (\alpha_{l-1} - \alpha_l)\, \text{sgn}(y-x)\, x_l, & \text{if } |y-x| > x_{N-1}. \end{cases} \tag{8.22}$$

Here, α_k represents the slope of each segment across different piecewise-defined regions, with $\pm x_k$, where $x_k \ge 0$, for $0 \le k \le n-1$, denoting the switching points for $1 \le k \le n-1$, as shown in Fig. 8.17.

Let $Q \in \mathbf{R}^3$ and $a, b \in \mathbf{R}$. Suppose that the function $v(q,t)$ is defined on $Q \times (a,b)$ and is t-fractal differentiable of order θ over the interval (a,b). Then, the piecewise derivative of the function v on the interval $[a,b]$, with respect to the derivative order θ, combining both fractal-fractional and classical derivatives in a local manner, is defined by

$$^{\text{pw}}D_t^{\theta} v(q,t) = \begin{cases} ^{\text{fre}}D_t^{\theta} v(q,t), & \text{when} \quad a \le t \le \tau_1, \\ \dfrac{\partial}{\partial t} v(q,t), & \text{when} \quad \tau_1 \le t \le b, \end{cases} \tag{8.23}$$

and

$$^{\text{pw}}D_t^{\theta} v(q,t) = \begin{cases} \dfrac{p(\theta)}{(1-\theta)} \dfrac{\partial}{\partial t^{\theta}} \displaystyle\int_a^t v(q,t) \times \exp\left(\dfrac{-\theta(t-r)}{1-\theta}\right) dr, & \text{if } a \le t \le \tau_1, \\ \dfrac{\partial}{\partial t} v(q,t), & \text{if } \tau_1 \le t \le b, \end{cases} \tag{8.24}$$

where $p(\theta)$ is a regularization function and $p(0) = p(1) = 1$, the generalized version is defined as

$$^{\text{pw}}D_t^{\theta,\varsigma} v(q,t) = \begin{cases} ^{\text{fre}}D_t^{\theta,\varsigma} v(q,t), & \text{when} \quad a \le t \le \tau_1, \\ \dfrac{\partial}{\partial t} v(q,t), & \text{when} \quad \tau_1 \le t \le b, \end{cases}$$

where $^{\text{fre}}D_t^{\theta} v$ and $^{\text{fre}}D_t^{\theta,\varsigma} v$ are defined as follows:

$$^{\text{fre}}D_t^{\theta} v(x,t) = \frac{p(\theta)}{(1-\theta)} \frac{\partial}{\partial t^{\theta}} \int_a^t v(x,r) \times \exp\left(\frac{-\theta(t-r)}{(1-\theta)}\right) dr$$

and

$$^{\text{fre}}D_t^{\theta,\varsigma} v(x,t) = \frac{p(\theta)}{(1-\theta)} \frac{\partial^{\varsigma}}{\partial t^{\theta}} \int_a^t v(x,r) \times \exp\left(\frac{-\theta(t-r)}{(1-\theta)}\right) dr.$$

Doungmo Goufo [4] examines and solves the modified multifold torus system using the above-defined piecewise derivative. The resulting combined model is expressed as follows:

$$\begin{cases} {}^{\mathrm{pw}}D_t^{\theta} x(t) = -af(y(t) - x(t)), \\ {}^{\mathrm{pw}}D_t^{\theta} y(t) = -af(y(t) - x(t)) - z(t), \\ {}^{\mathrm{pw}}D_t^{\theta} z(t) = by(t), \end{cases} \tag{8.25}$$

where f is a PWL odd function defined by

$$\begin{aligned} &f(y(t) - x(t)) \\ &= \alpha_{n-1}(y(t) - x(t)) + \tfrac{1}{2}\sum_{k=1}^{n-1}(\alpha_{k-1} - \alpha_k) \times \begin{bmatrix} |y(t) - x(t) + x_k| \\ -|y(t) - x(t) - x_k| \end{bmatrix}, \end{aligned} \tag{8.26}$$

satisfying $f(x(t) - y(t)) = -f(y(t) - x(t))$.

To facilitate solving the piecewise system (8.3), it is associated with the following initial conditions:

$$(x(0) = \tilde{x}(x), y(0) = \tilde{y}(y), z(0) = \tilde{z}(z)). \tag{8.27}$$

Note that the piecewise system (8.3) can be rewritten as

$$\begin{cases} \begin{cases} {}^{\mathrm{fre}}D_t^{\theta} x(t) = -af(y(t) - x(t)), & \text{when } a \le t \le \tau_1, \\ \dfrac{\partial}{\partial t} x(t) = -af(y(t) - x(t)), & \text{when } \tau_1 \le t \le b, \end{cases} \\ \begin{cases} {}^{\mathrm{fre}}D_t^{\theta} y(t) = -f(y(t) - x(t)) - z(t), & \text{when } a \le t \le \tau_1, \\ \dfrac{\partial}{\partial t} y(t) = -f(y(t) - x(t)) - z(t), & \text{when } \tau_1 \le t \le b, \end{cases} \\ \begin{cases} {}^{\mathrm{fre}}D_t^{\theta} z(t) = by(t), & \text{when } a \le t \le \tau_1, \\ \dfrac{\partial}{\partial t} z(t) = by(t), & \text{when } \tau_1 \le t \le b. \end{cases} \end{cases} \tag{8.28}$$

This approach enables to generate numerical representations of the multifold torus chaotic attractors involved in the piecewise process, as illustrated by Figs. 8.18, 8.19, 8.20 and 8.21. Start with Fig. 8.18a, which presents numerical simulations for the piecewise two-fold torus chaotic attractor related to the system (8.25)–(8.28), where $\theta = 0.9$, $a = 0$, $b = 200$, $\tau_1 = 120$, and the parameters are set as $\alpha_0 = 0.1$, $\alpha_1 = -0.07$, $a = 15$, $b = 1$, with $x_1 = 1$. In this setting, the generated attractor evolves from the initial 7-fold torus chaotic attractor shown in Fig. 8.17 to a stretched one-fold torus chaotic attractor through the piecewise process.

A similar transformation is shown in Fig. 8.18b, where the attractor transits from the initial two-fold torus chaotic attractor, depicted in Fig. 8.17, to a stretched two-fold torus chaotic attractor, through the piecewise process.

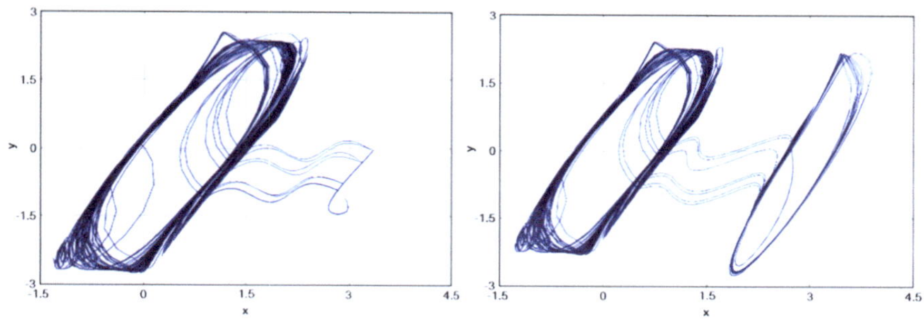

(a) (b)

Fig. 8.18 Numerical simulation results for the piecewise two-fold torus chaotic attractor. **a** It transits from the initial two-fold torus chaotic attractor shown in Fig. 8.17 to a stretched one-fold torus chaotic attractor, through a piecewise process. **b** It transits from the initial two-fold torus chaotic attractor shown in Fig. 8.17 to a stretched two-fold torus chaotic attractor, through a piecewise process. Reprinted from [4], with permission from World Scientific Pub. Co.

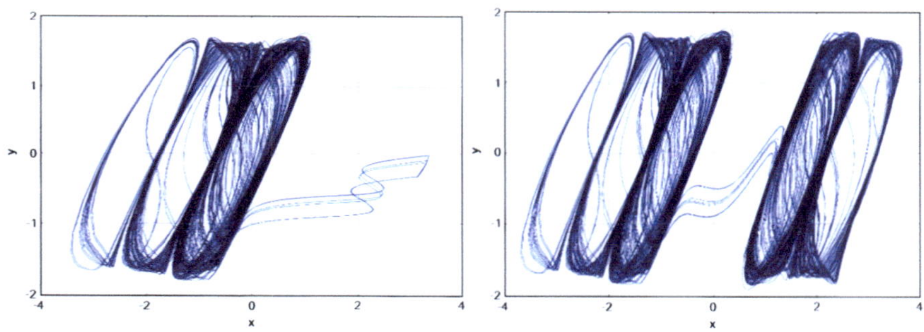

(a) (b)

Fig. 8.19 Numerical simulation results for the piecewise multifold torus chaotic attractor. **a** It transits from the initial three-fold torus chaotic attractor to a stretched three-fold torus chaotic attractor, through a piecewise process. **b** It transits from the initial three-fold torus chaotic attractor to a stretched 5-fold torus chaotic attractor, as a result of the piecewise process. Reprinted from [4], with permission from World Scientific Pub. Co.

Figure 8.19a shows the numerical simulation results for the piecewise three-fold torus chaotic attractor related to the model (8.25)–(8.28), where $\theta = 0.9$, $a = 0$, $b = 300$, $\tau_1 = 160$, and the parameters are $a = 14.5$, $b = 1.25$, $n = 2$, $\alpha_0 = 0.15$, $\alpha_1 = -0.17$, and $x_1 = 0.75$. It can be observed that the attractor transits from the initial three-fold torus chaotic attractor to a stretched three-fold torus chaotic attractor, through a piecewise process.

A similar transformation is shown in Fig. 8.19b, where the attractor transits from the initial three-fold torus chaotic attractor to a stretched 5-fold torus chaotic attractor, through a piecewise process.

Doungmo Goufo [4] demonstrated Fig. 8.20a, which presents numerical simulation results for the piecewise multifold torus chaotic attractor related to the model

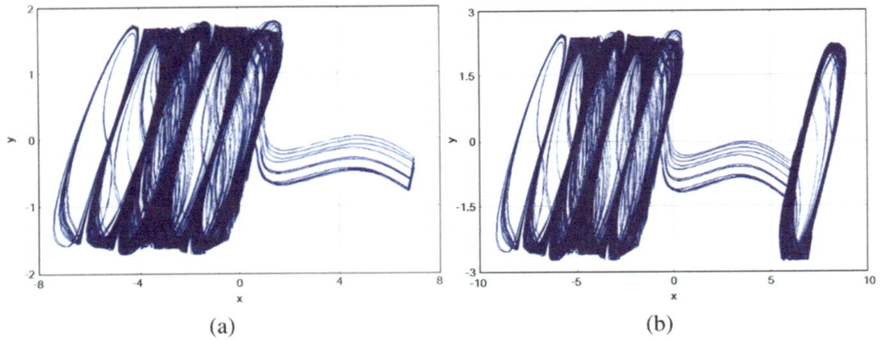

Fig. 8.20 Numerical simulation results for the piecewise multifold torus chaotic attractor. **a** It transits from the initial 7-fold torus chaotic attractor to a stretched three-fold torus chaotic attractor, through a piecewise process. **b** It transits from the initial 7-fold torus chaotic attractor to a stretched 6-fold torus chaotic attractor, as a result of the piecewise process. Reprinted from [4], with permission from World Scientific Pub. Co.

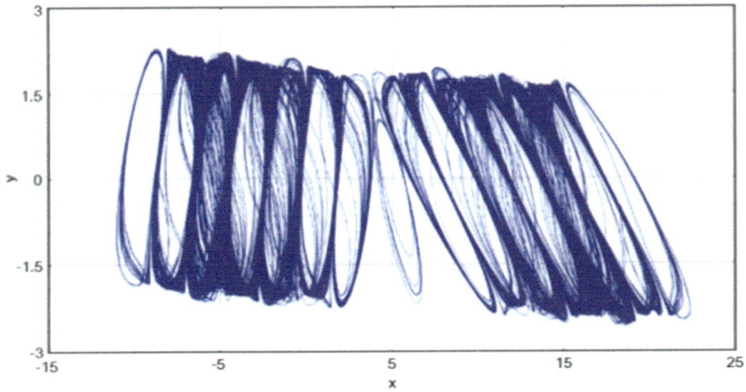

Fig. 8.21 Numerical simulation results for the piecewise 15-fold torus chaotic attractor. Reprinted from [4], with permission from World Scientific Pub. Co.

(8.25)–(8.28), with $\theta = 0.9$, $a = 0$, $b = 700$, $\tau_1 = 580$, and $n = 4$. The parameters are set as $\alpha_0 = 0.15$, $\alpha_1 = -0.17$, $\alpha_2 = 0.15$, $\alpha_3 = -0.17$, with $x_2 = 2.0735$ and $x_3 = 3.5735$. It can be observed that the attractor transits from the initial 7-fold torus chaotic attractor to a stretched 5-fold torus chaotic attractor through a piecewise process.

A similar transition is shown in Fig. 8.20b, where the attractor transits from the initial 7-fold torus chaotic attractor to a stretched 6-fold torus chaotic attractor, again as a result of the piecewise process.

Finally, in Fig. 8.21, Doungmo Goufo [4] presents numerical simulation results for the piecewise multifold torus chaotic attractor related to the model (8.25)–(8.28), where $\theta = 0.9$, $a = 0$, $b = 900$, $\tau_1 = 880$, $n = 5$. The parameters are set as $\alpha_0 = -0.17$, $\alpha_1 = 0.15$, $\alpha_2 = -0.17$, $\alpha_3 = 0.15$, $\alpha_4 = -0.17$, with $x_2 = 2.45$, $x_3 = 3.95$

and $x_4 = 5.65$. In this setting, the attractor transits from the initial 9-fold torus chaotic attractor to a stretched 15-fold torus chaotic attractor, through a piecewise process.

8.4 Conclusion

This chapter studies multifold torus chaotic attractors, discussing their numerical generation, theoretical design and practical implementation. By extending the conventional fold torus circuit using a generalized PWL function, the chapter illustrates how to generate multifold torus chaotic attractors that possess greater complexity and higher dimensionality. Additionally, the introduction of multidirectional multitorus (MDMT) chaotic attractors provides a systematic approach to constructing higher-dimensional complex chaotic systems. Circuit-based implementations demonstrate the practicality of these theoretical designs, laying a basis for future real-world applications. These findings significantly advance chaos theory and create new opportunities for potential applications in secure communication, signal processing and nonlinear systems control.

References

1. Yu, S., Lü, J., Chen, G.: Multifolded torus chaotic attractors: theoretical design and circuit implementation. Chaos **17**, 013118 (2007)
2. Matsumoto, T., Chua, L., Tokunaga, R.: Chaos via torus breakdown. IEEE Trans. Circuits Syst. **34**, 240–253 (1987)
3. Yu, X., Lü, J., Chen, G.: Theoretical design and circuit implementation of multidirectional multitorus chaotic attractors. IEEE Trans. Circuits Syst.-I **54**, 2087–2098 (2007)
4. Doungmo Goufo, E.F.: Implementation of multi-folded torus attractors via a piecewise system with a piecewise linear odd function. Fractals **30**, 2240232 (2022)

Chapter 9
Generating Hyperchaotic Multiscroll Attractors

Abstract This chapter introduces some methodologies for generating hyperchaotic multiscroll attractors from various systems. The techniques include modifying a 4-D Chua's circuit, the MCK circuit, CNNs, and coupled Chua's circuits with sine functions. The chapter also highlights the extension of PWL functions to create n-scroll chaotic attractors and investigates higher-dimensional grid-scroll hyperchaotic attractors. Theoretical models presented in the chapter are supported by experimental simulations, demonstrating their feasibility and potential applications to engineering and nonlinear dynamics research.

9.1 Hyperchaotic Multiscroll Attractors from the 4-D Chua's Circuit

A simple 4-D modified Chua's circuit can generate complex n-scroll chaotic attractors [1–3].

Yin [1] proposed a modified Chua's circuit, depicted in Fig. 9.1a, which differs from the classic Chua's circuit by incorporating an R_C parallel circuit composed of R_3 and C_3 into the L-branch of the original circuit. This modified circuit is described by

$$\begin{cases} \dot{x} = \alpha\,(y - x - g\,(x)), \\ \dot{y} = x - y + z, \\ \dot{z} = -\beta\,(y - w), \\ \dot{w} = -\gamma_2\,(z + \gamma_1 w), \end{cases} \tag{9.1}$$

where

$$g\,(x) = m_1 x + \frac{1}{2}\,(m_0 - m_1) \times (|x + b_1| - |x - b_1|). \tag{9.2}$$

J. Lü et al., *Chaotic Systems with Multiscroll Attractors*,
https://doi.org/10.1007/978-3-032-09854-2_9

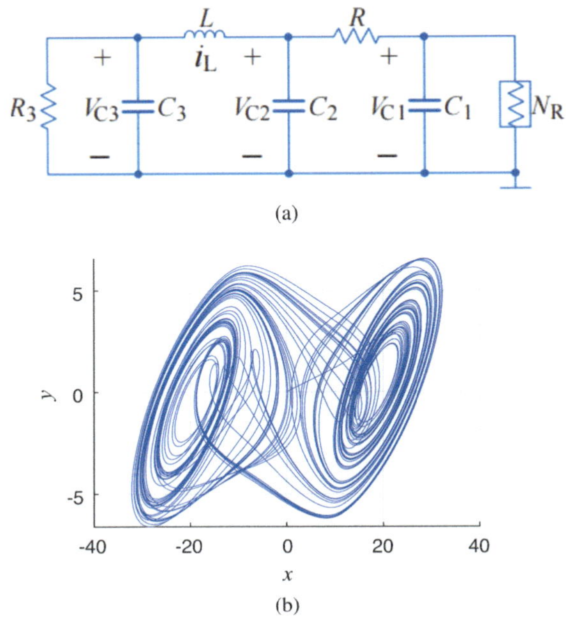

Fig. 9.1 4-D modified Chua's circuit. **a** The modified Chua's circuit. **b** Double-scroll chaotic attractor. Reprinted from [4], with permission by World Scientific Pub. Co.

With parameters $\alpha = 9.934$, $\beta = 14.334$, $\gamma_1 = 27.333$, $\gamma_2 = 0.0497$, $m_0 = -1.246$, $m_1 = -0.6724$, and $b_1 = 10$, system (9.1) produces a double-scroll chaotic attractor, as shown in Fig. 9.1b.

Yu et al. [2, 5] further extended the PWL function in Eq. (9.2) to generate n-scroll chaotic attractors, with

$$g(x) = m_{N-1}x + \frac{1}{2}\sum_{i=1}^{N-1}(m_{i-1} - m_i) \times (|x + b_i| - |x - b_i|). \tag{9.3}$$

When the parameters are set as follows: $\alpha = 9.934$, $\beta = 14.334$, $\gamma_1 = 27.333$, $\gamma_2 = 0.0497$, $m_0 = m_2 = m_4 = -1.246$, $m_1 = m_3 = m_5 = -0.6724$, $b_1 = 10$, $b_2 = 29.2466$, $b_3 = 49.2466$, $b_4 = 68.4931$, and $b_5 = 88.4931$, system (9.1) with the PWL function in Eq. (9.2) generates a 6-scroll chaotic attractor, as shown in Fig. 9.2a.

When $\alpha = 9.934$, $\beta = 14.334$, $\gamma_1 = 27.333$, $\gamma_2 = 0.0497$, $m_0 = m_2 = m_4 = m_6 = m_8 = m_{10} = -1.246$, $m_1 = m_3 = m_5 = m_7 = m_9 = m_{11} = -0.6724$, $b_1 = 12.4$, $b_2 = 36.2657$, $b_3 = 61.0657$, $b_4 = 84.9315$, $b_5 = 109.7315$, $b_6 = 133.5972$, $b_7 = 158.3972$, $b_8 = 182.2629$, $b_9 = 207.0629$, $b_{10} = 230.9287$, and $b_{11} = 279.5944$, system (9.1) with (9.2) generates a 10-scroll chaotic attractor, as shown in Fig. 9.2b.

To experimentally verify the n-scroll chaotic attractors, Yu et al. [2] reformulated the modified Chua's circuit (9.1) with Eq. (9.2), as follows:

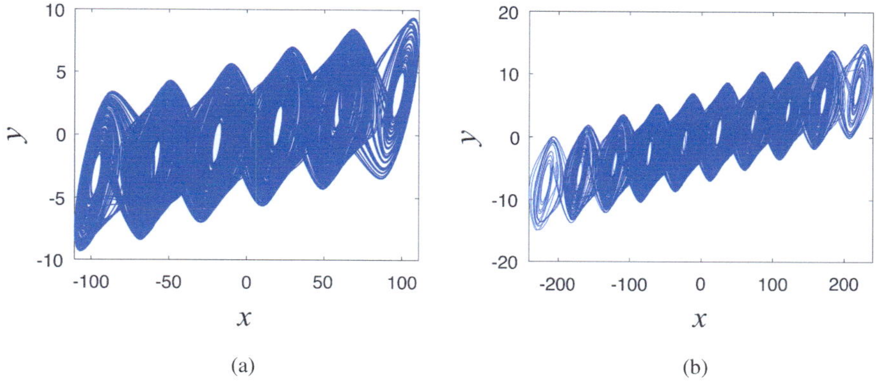

Fig. 9.2 n-Scroll chaotic attractors from the 4-D modified Chua's circuit [4]. **a** 6-Scroll chaotic attractor. **b** 10-Scroll chaotic attractor.

$$\begin{cases} \dfrac{dV_{C1}}{dt} = \dfrac{1}{RC_1}\left((V_{C2} - V_{C1}) - \dfrac{1}{C_1} f\left(V_{C1}\right)\right), \\ \dfrac{dV_{C2}}{dt} = \dfrac{1}{RC_2}\left((V_{C1} - V_{C2}) + \dfrac{1}{C_2} i_L\right), \\ \dfrac{di_L}{dt} = -\dfrac{1}{L}\left(V_{C3} - V_{C2}\right), \\ \dfrac{dV_{C3}}{dt} = -\dfrac{1}{C_3} i_L - \dfrac{1}{R_0 C_3} V_{C3}, \end{cases} \tag{9.4}$$

where

$$\begin{aligned} f\left(V_{C1}\right) &= \frac{m_{N-1}}{R} V_{C1} + \frac{1}{2}\sum_{i=1}^{N-1}\left(\frac{m_{i-1}}{R} - \frac{m_i}{R}\right) \times \left(|V_{C1} + E_i| - |V_{C1} - E_i|\right) \\ &= G_{N-1} V_{C1} + \frac{1}{2}\sum_{i=1}^{N-1}\left(G_{i-1} - G_i\right) \times \left(|V_{C1} + E_i| - |V_{C1} - E_i|\right), \end{aligned} \tag{9.5}$$

with $C_2/C_1 = \alpha$, $R_2C_2/L = \beta$, $R/R_0 = \gamma_1$, $C_2/C_3 = \gamma_2$, $V_{C1}/B_P = x$, $V_{C2}/B_P = y$, $i_L R/B_P = z$, $V_{C3}/B_P = w$, $t/RC_2 = \tau$, $B_P = 1V$, $f(V_{C1}) = g(V_{C1})/R$, $E_i = x_i B_P$, and $G_i = m_i/R$.

Thereafter, another circuit configuration was developed in [2] to implement 4- and 6-state chaotic attractors. The circuit diagram is shown in Fig. 9.3, where all operational amplifiers are TL082. The function block within the dotted line represents the diode NR in Chua's circuit. Each operational amplifier, along with its associated resistors, generates a subcurrent of N_R with switching voltages $E_i, i = 1, 2, 3, 4, 5$. When switch K is turned on, the circuit generates a 6-scroll chaotic attractor, while turning K off generates a 4-scroll chaotic attractor. The electronic source voltages are ± 15V, with the corresponding saturating voltage of ± 13.5V. The circuit parameters

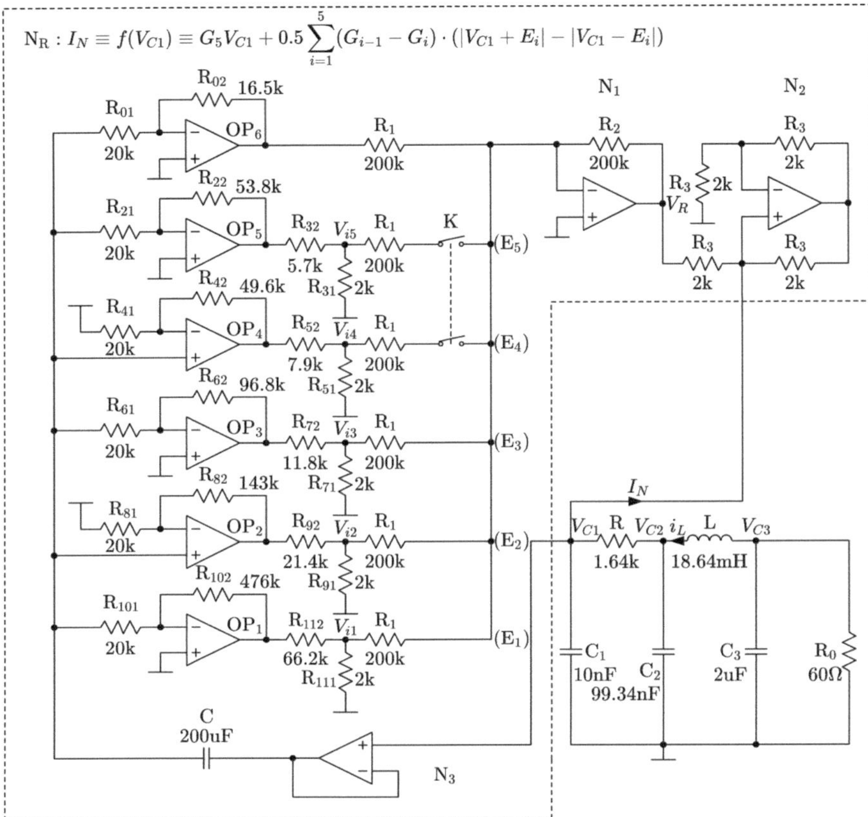

Fig. 9.3 Circuit diagram of the 4-D modified Chua's circuit. Reprinted from [6], with permission by World Scientific Pub. Co.

are detailed in Fig. 9.3. Figure 9.4 show the experimentally observed 4- and 6-scroll chaotic attractors.

9.2 Hyperchaotic Multiscroll Attractors from the Modified MCK Circuit

Matsumoto et al. [7] proposed the first hyperchaotic circuit, known as the MCK circuit, which is described by

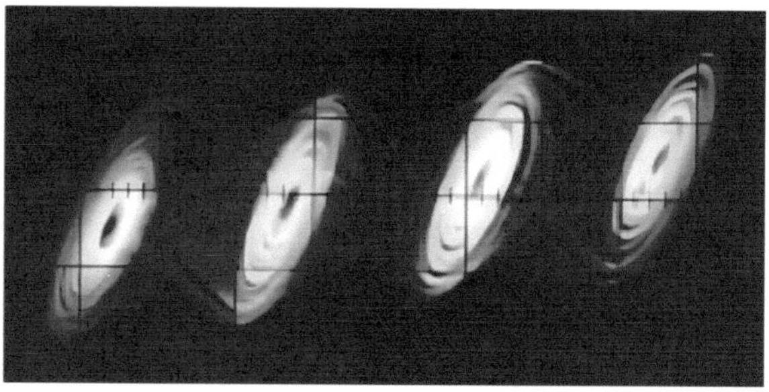

(a)

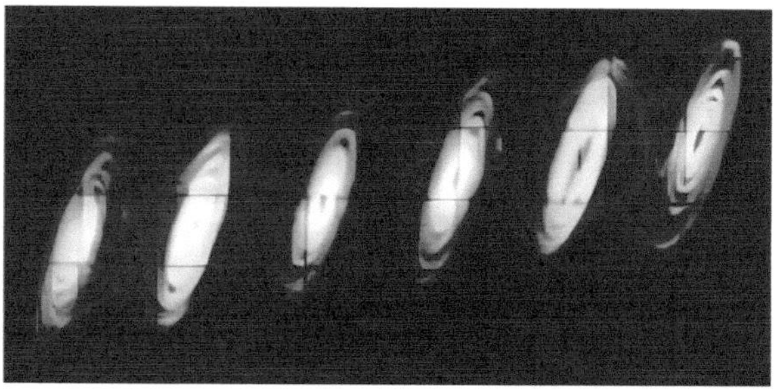

(b)

Fig. 9.4 Experimental observations of 4- and 6-scroll chaotic attractors. **a** 4-Scroll chaotic attractor. **b** 6-Scroll chaotic attractor. Reprinted from [6], with permission by World Scientific Pub. Co.

$$\begin{cases} C_1 \dfrac{dv_{C1}}{dt} = g\,(v_{C2} - v_{C1}) - i_{L1}, \\ C_2 \dfrac{dv_{C2}}{dt} = -g\,(v_{C2} - v_{C1}) - i_{L2}, \\ L_1 \dfrac{di_{L1}}{dt} = v_{C1} + Ri_{L1}, \\ L_2 \dfrac{di_{L2}}{dt} = v_{C2}, \end{cases} \tag{9.6}$$

where

$$\begin{aligned} g\,(v_{C2} - v_{C1}) = m_0\,(v_{C2} - v_{C1}) + 0.5\,(m_1 - m_0) \\ \times\,(|v_{C2} - v_{C1} - 1| - |v_{C2} - v_{C1} + 1|). \end{aligned} \tag{9.7}$$

Figure 9.5a illustrates the MCK hyperchaotic circuit, where the nonlinear resistor is depicted in Fig. 9.5b, while all other components are linear and passive, except for a negative resistance $-R$. The implementation of the MCK hyperchaotic circuit is shown in Fig. 9.6a, where N_1 represents the nonlinear resistor and N_2 corresponds to the negative resistance. With $C_1 = 0.5$, $C_2 = 0.05$, $L_1 = 1$, $L_2 = 2/3$, $R = 1$, $m_0 = 3$, and $m_1 = -0.2$, a hyperchaotic attractor is generated, as shown in Fig. 9.6b.

The dimensionless state equation of the MCK circuit shown in Eq. (9.6) is expressed as

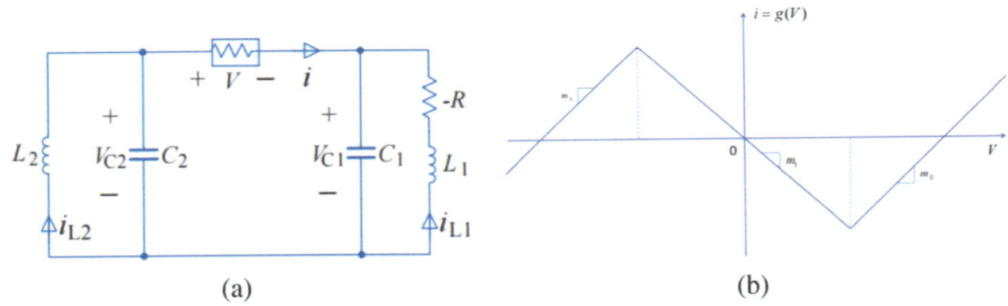

Fig. 9.5 4-D MCK hyperchaotic circuit. **a** MCK circuit. **b** v-i characteristic of N_1. Reprinted from [7], with permission by the IEEE

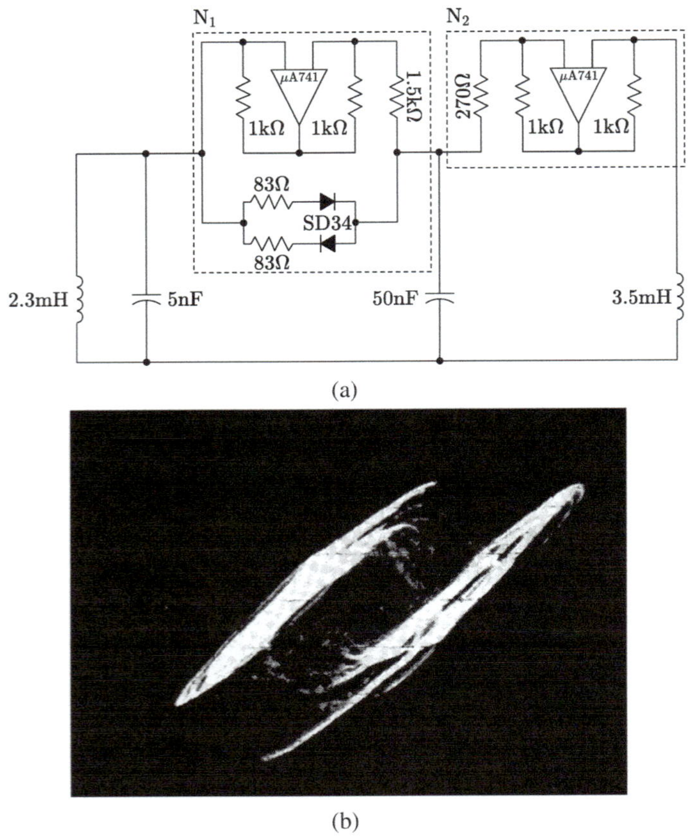

Fig. 9.6 Realization of the MCK hyperchaotic circuit. **a** Circuit diagram. **b** Experimental observation. Reprinted from [7], with permission by the IEEE

$$\begin{cases} \dot{x} = \alpha\,(g\,(y - x) - z), \\ \dot{y} = \beta\,(-g\,(y - x) - w), \\ \dot{z} = x + z, \\ \dot{w} = \gamma y, \end{cases} \tag{9.8}$$

where

$$g\,(y - x) = m_0 x + \frac{1}{2}\,(m_1 - m_0) \times (|y - x + 1| - |y - x - 1|). \tag{9.9}$$

To generate n-double scroll hyperchaotic attractors from Eq. (9.8), Yalcin et al. [8] altered the PWL function in Eq. (9.9), as follows:

$$\begin{aligned} g\,(y - x) = m_0\,(y - x) + \tfrac{1}{2} \sum_{i=1}^{2n-1} (m_i - m_{i-1}) \\ \times (|y - x + c_i| - |y - x - c_i|). \end{aligned} \tag{9.10}$$

Furthermore, Yu et al. [2] expanded Eq. (9.10) to generate n-scroll hyperchaotic attractors, as follows:

$$\begin{aligned} g\,(y - x) = m_{N-1}\,(y - x) + \tfrac{1}{2} \sum_{i=1}^{N-1} (m_{i-1} - m_i) \\ \times (|y - x + c_i| - |y - x - c_i|). \end{aligned} \tag{9.11}$$

When $\alpha = 2$, $\beta = 20$, $\gamma = 1.5$, $m_0 = m_2 = m_4 = m_6 = m_8 = 3$, $m_1 = m_3 = m_5 = -0.9$, $m_7 = -0.8$, $c_1 = 1$, $c_2 = 3.2105$, $c_3 = 5.5205$, $c_4 = 8.1953$, $c_5 = 10.9904$, $c_6 = 14.2268$, $c_7 = 17.7699$, and $c_8 = 22.9435$, system (9.8) with Eq. (9.10) generates a 9-scroll hyperchaotic attractor, as shown in Fig. 9.7a. When

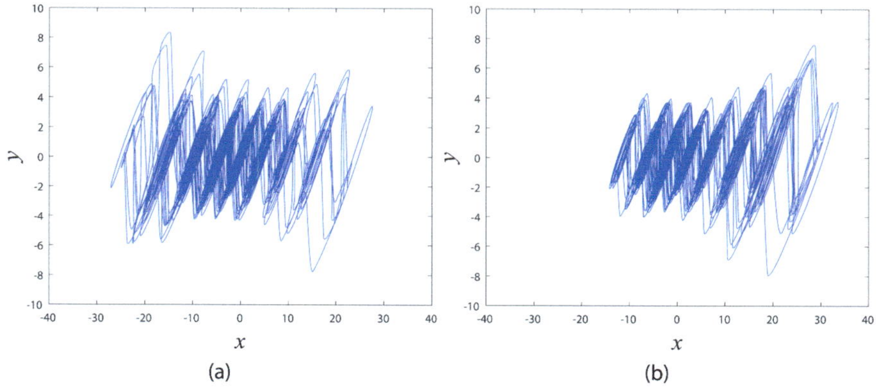

Fig. 9.7 n-Scroll hyperchaotic attractors from the modified MCK circuit. **a** 9-Scroll hyperchaotic attractor. **b** 10-Scroll hyperchaotic attractor. Reprinted from [6], with permission by World Scientific Pub. Co.

$\alpha = 2$, $\beta = 20$, $\gamma = 1.5$, $m_0 = m_2 = m_4 = m_6 = -0.9$, $m_8 = -0.8$, $m_1 = m_3 = m_5 = m_7 = m_9 = 2.9, c_1 = 1, c_2 = 3.1, c_3 = 5.41, c_4 = 7.951, c_5 = 10.7461, c_6 = 13.8207, c_7 = 17.2028, c_8 = 21.2774$, and $c_9 = 26.9686$, the system generates a 10-scroll hyperchaotic attractor, as shown in Fig. 9.7b.

9.3 Hyperchaotic Multiscroll Attractors from CNNs

Suykens and Chua [9] proposed a 1-D CNN for generating n-scroll hypercube attractors.

Design a CNN by composing m identical generalized Chua's circuits, with unidirectional coupling between the second equations of the cells. The model is described by

$$\begin{cases} \dot{x}_i = \alpha\left(y_i - h\left(x_i\right)\right), \\ \dot{y}_i = x_i - y_i + z_i + k_{i-1}\left(y_i - y_{i-i}\right), \quad i{=}1, 2, \ldots, m, \\ \dot{z}_i = -\beta y_i, \end{cases} \tag{9.12}$$

or, with diffusive couplings between each pair of the first equations of the cells, as described by

$$\begin{cases} \dot{x}_i = \alpha\left(y_i - h\left(x_i\right)\right) + d_x\left(x_{i-1} - x_i + x_{i+1}\right), \\ \dot{y}_i = x_i - y_i + z_i, \qquad i{=}1, 2, \ldots, m, \\ \dot{z}_i = -\beta y_i, \end{cases} \tag{9.13}$$

or, with diffusive couplings between each pair of the first and the second equations of the cells, as described by

$$\begin{cases} \dot{x}_i = \alpha\left(y_i - h\left(x_i\right)\right), \\ \dot{y}_i = x_i - y_i + z_i + d_y\left(x_{i-1} - x_i + x_{i+1}\right), \quad i{=}1, 2, \ldots, m. \\ \dot{z}_i = -\beta y_i, \end{cases} \tag{9.14}$$

Set $y_0 = y_m$, $k_0 = 0$, and $k_i = k$, $i = 1, \ldots, m-1$, in Eq. (9.12), set $x_0 = x_m$ and $x_{m+1} = x_1$ in Eqs. (9.13) and (9.14).

Then, by using n-scroll attractors as cells, Suykens and Chua [9] generated n-scroll hypercube attractors, $m \geq 4$, with weak unidirectional and diffusive coupling in the common state subspace of the cells; that is, (x_1, x_2) for $m = 2$, (x_1, x_2, x_3) for $m = 3$, and so on. The dimension of the hypercube equals the number of cells, m. Hyperchaotic attracors are observed for the cases with $m > 2$. The coupling coefficients k, d_x and d_y are set to be some small values, such as 0.01, and the initial values are chosen to be close to zero.

Figure 9.8 displays the 2-double-scroll square and cube from a 1-D CNN of 2 and 3 cells, respectively, each consisting of 2-double-scroll circuits with unidirectional coupling $k = 0.002$ between the cells.

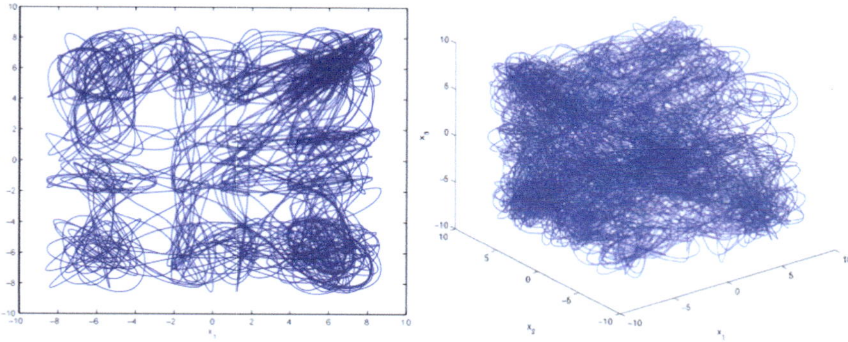

(a) (b)

Fig. 9.8 1-D CNN of 2, 3 cells in the 2-double-scroll circuit with unidirectional coupling $k = 0.002$ between the cells. **a** 2-double-scroll square. **b** 2-double-scroll cube. Reprinted from [6], with permission by World Scientific Pub. Co.

9.4 Hyperchaotic Coupled Chua's Circuits

Cafagna and Grassi [10, 11] developed a method for generating 2-D and 3-D hyperchaotic attractors using coupled Chua's circuits with sine functions. Two Chua's circuits [10] with sine functions coupled in the following manner:

$$\begin{cases} \dot{x}_1 = \alpha\,(x_2 - f_1\,(x_1)), \\ \dot{x}_2 = x_1 - x_2 + x_3 + h\,(x_5 - x_2), \\ \dot{x}_3 = -\beta x_2, \\ \dot{x}_4 = \alpha\,(x_5 - f_2\,(x_4)), \\ \dot{x}_5 = x_4 - x_5 + x_6 + h\,(x_2 - x_5), \\ \dot{x}_6 = -\beta x_5, \end{cases} \tag{9.15}$$

where

$$f_1\,(x_1) = \begin{cases} \dfrac{b\pi}{2a}\,(x_1 - 2ac_1), & \text{if} \quad x_1 \geq 2ac_1, \\ -b\sin\left(\dfrac{\pi x_1}{2a} + d_1\right), & \text{if} \quad -2ac_1 < x_1 < 2ac_1, \\ \dfrac{b\pi}{2a}\,(x_1 + 2ac_1), & \text{if} \quad x_1 \leq -2ac_1, \end{cases}$$

and

$$f_2\,(x_4) = \begin{cases} \dfrac{b\pi}{2a}\,(x_4 - 2ac_2), & \text{if} \quad x_4 \geq 2ac_2, \\ -b\sin\left(\dfrac{\pi x_4}{2a} + d_2\right), & \text{if} \quad -2ac_2 < x_4 < 2ac_2, \\ \dfrac{b\pi}{2a}\,(x_4 + 2ac_2), & \text{if} \quad x_4 \leq -2ac_2\,. \end{cases}$$

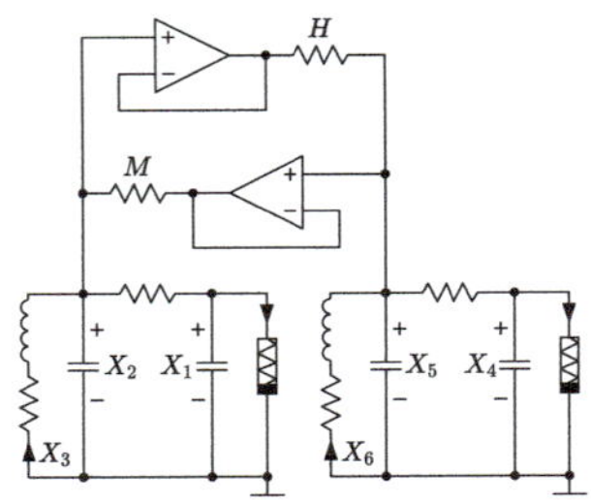

Fig. 9.9 Two-coupled Chua's circuits with sine resistors. Reprinted from [10], with permission by World Scientific Pub. Co.

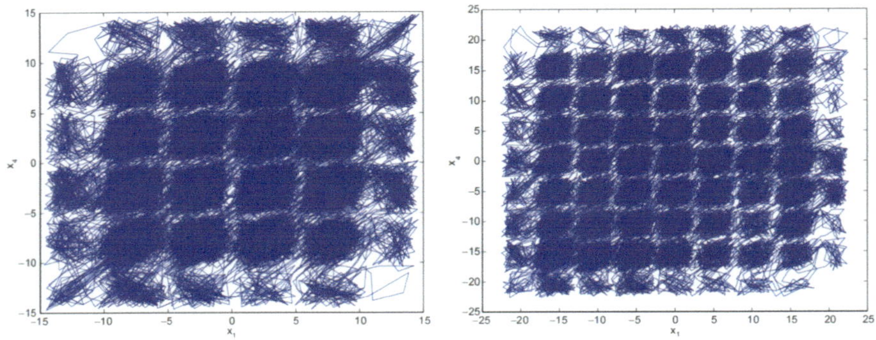

Fig. 9.10 Numerical simulations of 2-D $m \times n$-grid scroll hyperchaotic attractors. **a** 2-D 6 × 6-grid scroll attractor. **b** 2-D 9 × 9-grid scroll attractor. Reprinted from [6], with permission by World Scientific Pub. Co.

Here, $\alpha, \beta, a, b, c_1, c_2, d_1, d_2$ are real constants, with

$$d_i = \begin{cases} \pi, & \text{for } c_i \text{ is even,} \\ 0, & \text{for } c_i \text{ is odd,} \end{cases}$$

where $i = 1, 2$.

Figure 9.9 displays the coupling arrangement of two Chua's circuits with sine functions. The parameters are: $\alpha = 10.814$, $\beta = 14.0$, $a = 1.3$, $b = 0.11$, and $h = 0.25$, with c_1, c_2, d_1, d_2 as design parameters. For suitable values of these design parameters, system (9.15) can generate 2-D $m \times n$-grid scroll hyperchaotic attractors. Specifically, when $c_1 = c_2 = 5$ and $d_1 = d_2 = 0$, system (9.15) generates a 2-D 6 × 6-grid scroll hyperchaotic attractor, as shown in Fig. 9.10a. With $c_1 = c_2 = 8$ and $d_1 = d_2 = \pi$, the system generates a 2-D 9 × 9-grid scroll hyperchaotic attractor, as shown in Fig. 9.10b. The Lyapunov exponent spectrum for the 9 × 9-grid scroll attractor is $LE1 = 0.334$, $LE2 = 0.172$, $LE3 = 0$, $LE4 = -0.169$, $LE5 = -1.347$, and $LE6 = -1.749$.

In a similar fashion, three Chua's circuits [11] with sine functions are coupled as follows:

$$\begin{cases} \dot{x}_1 = \alpha\left(x_2 - f_1\left(x_1\right)\right), \\ \dot{x}_2 = x_1 - x_2 + x_3 + h\left(x_8 - x_2\right), \\ \dot{x}_3 = -\beta x_2, \\ \dot{x}_4 = \alpha\left(x_5 - f_2\left(x_4\right)\right), \\ \dot{x}_5 = x_4 - x_5 + x_6 + h\left(x_2 - x_5\right), \\ \dot{x}_6 = -\beta x_5, \\ \dot{x}_7 = \alpha\left(x_8 - f_3\left(x_7\right)\right), \\ \dot{x}_8 = x_7 - x_8 + x_9 + h\left(x_5 - x_8\right), \\ \dot{x}_9 = -\beta x_8, \end{cases} \tag{9.16}$$

where $f_1(x_1)$ and $f_2(x_4)$ are as above, with

$$f_3\left(x_7\right) = \begin{cases} \dfrac{b\pi}{2a}\left(x_7 - 2ac_3\right), & \text{if} \quad x_7 \geq 2ac_3, \\ -b\sin\left(\dfrac{\pi x_7}{2a} + d_3\right), & \text{if} \quad -2ac_3 < x_7 < 2ac_1, \\ \dfrac{b\pi}{2a}\left(x_7 + 2ac_1\right), & \text{if} \quad x_7 \leq -2ac_3, \end{cases}$$

where $\alpha, \beta, a, b, c_1, c_2, c_3, d_1, d_2, d_3$ are real constants, and

$$d_i = \begin{cases} \pi, & \text{for} \quad c_i \quad \text{is} \quad \text{even}, \\ 0, & \text{for} \quad c_i \quad \text{is} \quad \text{odd}, \end{cases}$$

where $i = 1, 2, 3$.

Figure 9.11 shows the coupling arrangement of three Chua's circuits in a ring configuration, each incorporating sine functions. Given parameters $\alpha = 10.814$, $\beta = 14.0$, $a = 1.3$, $b = 0.11$, $h = 0.18$, with $c_1, c_2, c_3, d_1, d_2, d_3$ as design parameters, system (9.16) can generate 3-D $m \times n \times l$-grid scroll hyperchaotic attractors.

With parameters $c_1 = 5$, $c_2 = 6$, $c_3 = 7$, $d_1 = d_3 = 0$, and $d_2 = \pi$, system (9.16) generates a 3-D $6 \times 7 \times 8$-grid scroll hyperchaotic attractor, as shown in Fig. 9.12. The Lyapunov exponent spectrum for this $6 \times 7 \times 8$-grid scroll attractor is given by $LE1 = 0.675$, $LE2 = 0.297$, $LE3 = 0.180$, $LE4 = 0$, $LE5 = -0.127$, $LE6 = -0.718$, $LE7 = -1.078$, $LE8 = -1.302$, and $LE9 = -1.518$.

9.5 Conclusion

This chapter explores the design and generation of hyperchaotic multiscroll attractors using a variety of advanced techniques and PWL circuits. By extending some classical models such as the 4-D Chua's circuit and the MCK circuit, effective methods can be developed for generating n-scroll and grid-scroll hyperchaotic attractors.

Additionally, CNNs and coupled Chua's circuits with sine functions further enhance the scope of hyperchaotic attractor design, allowing for constructing more

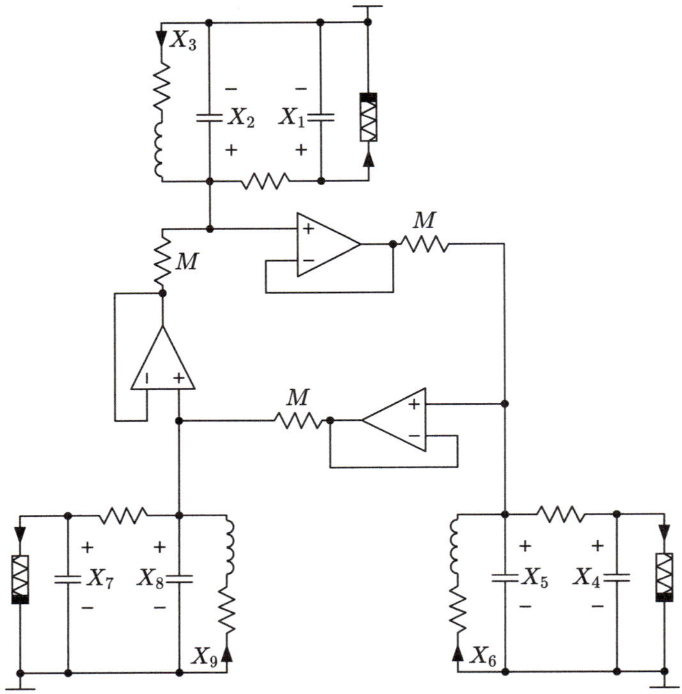

Fig. 9.11 Two-coupled Chua's circuits with sine resistors. Reprinted from [11], with permission by World Scientific Pub. Co.

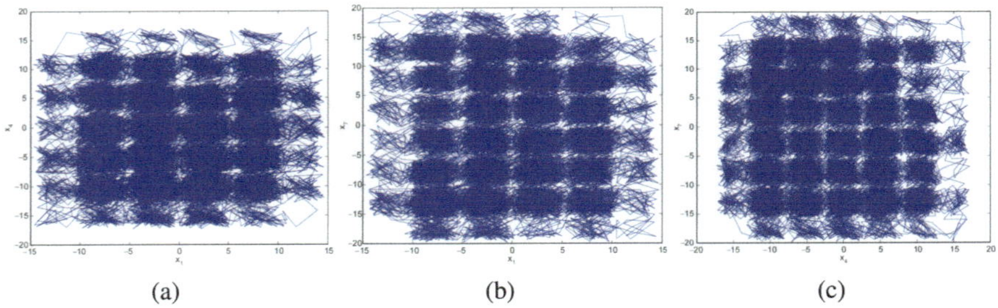

Fig. 9.12 3-D 6 × 7 × 8-grid scroll hyperchaotic attractor. **a** x_1-x_4 plane. **b** x_1-x_7 plane. **c** x_4-x_7 plane. Reprinted from [6], with permission by World Scientific Pub. Co.

complex but also more powerful multi-dimensional circuit structures. Theoretical analyses are supported by experimental simulations, showcasing the practical feasibility of these designs and methods. These advanced methodologies significantly advance chaos theory, providing robust tools for applications to secure communications, signal processing and nonlinear dynamic systems.

References

1. Yin, Y.: Synchronization of chaos in a modified Chua's circuit using continuous control. Int. J. Bifur. Chaos **6**, 2101–2117 (1996)
2. Yu, S., Lin, Q.H., Qiu, S.: Simulation investigation on multi-scroll chaotic and hyperchaotic attractors for four-dimensional systems. Acta Phy. Sinica **52**, 25–33 (2003)
3. Yu, S., Lü, J., Leung, H., Chen, G.: N-scroll chaotic attractors from a general jerk circuit. In: IEEE International of Symposium Circuits, Systems, May 23–26, Kobe, Japan, pp. 1473–1476 (2005)
4. Lü, J., Chen, G.: A new chaotic attractor coined. Int. J. Bifur. Chaos **12**, 659–661 (2002)
5. Yu, S., Qiu, S., Lin, Q.: New results of study on generating multiple-scroll chaotic attractors. Sci. in China, Ser. F: Inf. Sci. **46**, 104–115 (2003)
6. Lü, J., Chen, G.: Generating multiscroll chaotic attractors: theories, methods and applications. Int. J. Bifur. Chaos **16**, 775–858 (2006)
7. Matsumoto, T., Chua, L., Kobayashi, K.: Hyper chaos: laboratory experiment and numerical confirmation. IEEE Trans. Circuits Syst. **33**, 1143–1147 (1986)
8. Yalcin, M. E., Suykens, J., Vandewalle, J.: Hyperchaotic n-scroll attractors. Nonlinear Dyn. Electronic Syst., 25–28 (2000)
9. Suykens, J.A.K., Chua, L.O.: n-Double scroll hypercubes in 1-D CNNs. Int. J. Bifur. Chaos **7**, 1873–1885 (1997)
10. Cafagna, D., Grassi, G.: Hyperchaotic coupled Chua circuits: an approach for generating new $n \times m$-scroll attractors. Int. J. Bifur. Chaos **13**, 2537–2550 (2003)
11. Cafagna, D., Grassi, G.: New 3-D scroll attractors in hyperchaotic Chua's circuit forming a ring. Int. J. Bifur. Chaos **13**, 2889–2903 (2003)

Chapter 10
Generating Multiscroll Chaotic Attractors from Nonautonomous Systems

Abstract This chapter examines methods for generating multiscroll chaotic attractors from nonautonomous systems. Basic techniques include using a pulse-driven oscillator based on Chua's circuit and a multi-level logic pulse excitation scheme, which generate multi-double-scroll attractors without altering the original nonlinear functions. The task is performed using Chua's circuit, a jerk circuit and a modified Lorenz system, highlighting the scalability and versatility of the methods. PSpice simulations verify and demonstrate these methods, confirming their effectiveness in circuit implementation. Additionally, a 4-D autonomous chaotic system is introduced, which can generate coexisting butterfly-like attractors. These findings enhance the understanding and design of nonautonomous chaotic systems towards engineering and scientific applications.

10.1 Multiscroll Chaotic Attractors from Nonautonomous Systems

This section presents a simple nonautonomous chaotic system to generate n-scroll chaotic attractors.

Aziz-Alaoui [1] examined a nonautonomous circuit defined by

$$\begin{cases} C\dfrac{dv_C}{d\tau} = i_L - g\left(v_C\right) , \\ L\dfrac{di_L}{d\tau} = -Ri_L - i_C + V_S \sin\left(\omega\tau\right) , \end{cases} \tag{10.1}$$

where

$$g\left(v_C\right) = G_A v_C + \frac{1}{2}\left(G_B - G_A\right)\left(\left|v_C + B_p\right| - \left|v_C - B_p\right|\right) ,$$

in which $g\left(v_C\right)$ is a PWL function. The circuit diagram of the circuit is depicted in Fig. 10.1.

J. Lü et al., *Chaotic Systems with Multiscroll Attractors*,
https://doi.org/10.1007/978-3-032-09854-2_10

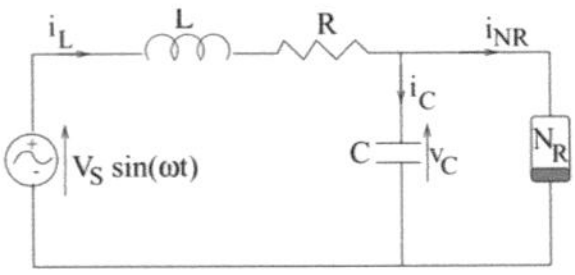

Fig. 10.1 Circuit diagram of the n-scroll nonautonomous circuit. Reprinted from [1], with permission from World Scientific Pub. Co

For circuit (10.1), define $x = v_C/b_1$, $y = i_L R/b_1$, $t = \tau/RC$, $\Omega = \omega/2\pi RC$, $\alpha = CR^2/L$, and $K = V_S/b_1$. The resulting dimensionless state equations are expressed as

$$\begin{cases} \dfrac{dx}{dt} = y - f_2(x) \ , \\ \dfrac{dy}{dt} = -\alpha(x+y) + \alpha K \sin(2\pi\Omega t) \ , \end{cases} \tag{10.2}$$

where

$$f_2(x) = m_1 x + \frac{1}{2}(m_0 - m_1)\left[|x+1| - |x-1|\right].$$

To generate n-scroll chaotic attractors from system (10.2), Aziz-Alaoui [1] revised system (10.2) to be in the following form:

$$\begin{cases} \dfrac{dx}{dt} = y - f_N(x) \ , \\ \dfrac{dy}{dt} = -\alpha(x+y) + \alpha K \sin(2\pi\Omega t) \ , \end{cases} \tag{10.3}$$

where

$$f_N(x) = \begin{cases} m_k x + \operatorname{sgn}(x)\,\xi_k \ , & \text{if } s_{k-1} \le |x| \le s_{k-1},\ k \in I_{N-2} \ , \\ m_{N-1} x + \operatorname{sgn}(x)\,\xi_{N-1} \ , & \text{if } |x| \ge s_{N-2} \ . \end{cases}$$

Aziz-Alaoui [1] analyzed the equilibrium points and eigenspaces of system (10.3) with parameters $N = 4$, $\alpha = 0.81$, $K = 1.19$, $\Omega = 0.22$, $m_0 = -1.45$, $m_1 = 0.66$, $s_1 = 1.45$, and $s_2 = 2.49$. Under these conditions, system (10.3) generates a 4-scroll chaotic attractor, as shown in Fig. 10.2a. With $N = 6$, $\alpha = 0.81$, $K = 1.19$, $\Omega = 0.22$, $m_0 = -1.45$, $m_1 = 0.66$, $s_1 = 1.45$, $s_2 = 2.49$, $s_3 = 2.94$, and $s_4 = 4.1$, system (10.3) generates a 6-scroll chaotic attractor, as shown in Fig. 10.2b.

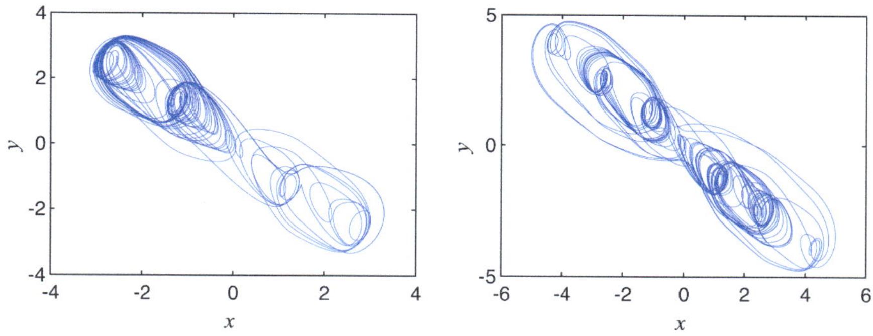

Fig. 10.2 n-Scroll chaotic attractors from the nonautonomous system (10.3) [1]. **a** 4-scroll chaotic attractor. **b** 6-scroll chaotic attractor

10.2 Multiscroll Chaotic Attractors from Pulse-Driven Nonautonomous Systems

This section introduces a pulse-driven nonautonomous chaotic oscillator, based on the passive structure of Chua's circuit, proposed by Elwakil [2].

While most nonautonomous chaotic circuits are controlled by sinusoidal input signals, which cause the equilibrium points to vary over time, Elwakil [2] introduced a pulse-driven nonautonomous chaotic oscillator that can fix the positions of the equilibrium points in the phase space.

For the circuit shown in Fig. 10.3a, the system equations are given by

$$\begin{cases} L\dot{I}_L = V_{C2}\,, \\ C_2\dot{V}_{C2} = \dfrac{V_{C1}}{R} - \left(\dfrac{1}{R} + \dfrac{1}{R_S} + \dfrac{1}{R_F}\right) V_{C2} + \dfrac{V_P}{R_S} + \dfrac{V_N}{R_F} - I_L\,, \\ C_1\dot{V}_{C1} = \dfrac{V_{C2} - V_{C1}}{R}\,, \end{cases} \tag{10.4}$$

where

$$V_N = V_{CC}\mathrm{sgn}\,(V_{C1}) = \begin{cases} V_{CC}, & \text{if } V_{C1} \geq 0\,, \\ -V_{CC}, & \text{if } V_{C1} < 0\,, \end{cases}$$

with $V_P = V_{CC}\,\mathrm{sgn}(\sin(\omega pt))$ and ω_p representing the oscillation frequency of the pulse train.

Let $X = V_{C1}/V_{CC}, Y = V_{C2}/V_{CC}, Z = RI_L/V_{CC}, \tau = \omega Pt, \varepsilon_c = C_1/C_2, \alpha_1 = L\omega P/R, \alpha_2 = RC_2\omega P, \beta_F = R/R_F$, and $\beta_S = R/R_S$. System (10.4) then takes the following form:

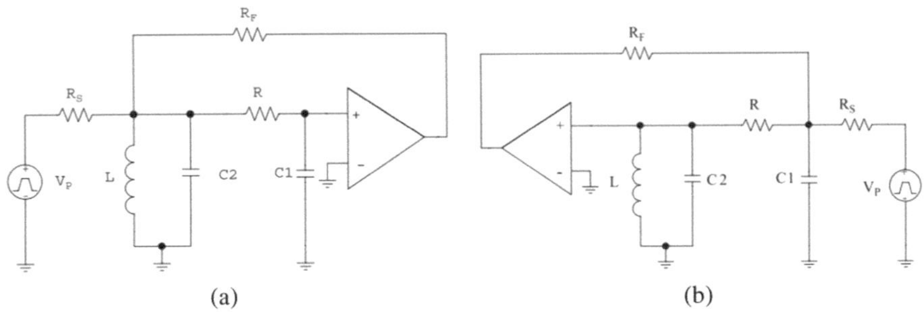

Fig. 10.3 Pulse-driven nonautonomous chaotic oscillator. **a** Circuit structure with excitation across C_2. **b** Circuit structure with excitation across C_2. Reprinted from [2], with permission by Elsevier

$$\begin{pmatrix} \varepsilon_c \dot{X} \\ \dot{Y} \\ \dot{Z} \end{pmatrix} = \begin{pmatrix} -1/\alpha_2 & 1/\alpha_2 & 0 \\ 1/\alpha_2 & (-1+\beta_F+\beta_S)/\alpha_2 & -1/\alpha_2 \\ 0 & 1/\alpha_1 & 0 \end{pmatrix} \begin{pmatrix} X \\ Y \\ Z \end{pmatrix} + \begin{pmatrix} 0 \\ a/\alpha_2 \\ 0 \end{pmatrix} \tag{10.5}$$

where $\alpha_1, \alpha_2, \beta_F, \beta_S$ are real constants, and the single parameter a is defined as

$$a = \begin{cases} \beta_F + \beta_S P(\tau), & \text{if } X \geq 0, \\ -\beta_F + \beta_S P(\tau), & \text{if } X < 0, \end{cases} \tag{10.6}$$

in which

$$P(\tau) = \operatorname{sgn}(\sin\tau) = \begin{cases} 1, & \text{if } \sin\tau \geq 0, \\ -1, & \text{if } \sin\tau < 0. \end{cases}$$

Figure 10.4a shows the numerical simulation result of a 3-scroll chaotic attractor, with parameters $\alpha_1 = 0.05$, $\alpha_2 = 50$, $\beta_F = \beta_S = 3$, $\tau = 0.1t$, and $\varepsilon_c = 0.2$. It is evident that the three equilibrium points are located at the origin and at $(0, 0, \pm 6)$, respectively.

Elwakil [2] designed a circuit, with its diagram shown in Fig. 10.3a, to observe the chaotic behavior. The circuit components include $L = 1\,\text{mH}$, $C_1 = 10\,\text{nF}$, $C_2 = 100\,\text{nF}$, $R = 5\,\text{k}\Omega$, and $F_F = 1\,\text{k}\Omega$. Additionally, R_S is a variable resistor of $5\,\text{k}\Omega$, with the resonant frequency of the L_{C2} around 16 kHz. The frequency of the pulse-driven generator varies between 1 and 20 kHz. The operational amplifier is biased with ± 5 V supply, while the pulse generator output is fixed at ± 5 V. Figure 10.4b shows the observed chaotic attractor on the V_{C1}-V_{C2} (X-Y) plane, with $R_S = 1.636\,\text{k}\Omega$. It is important to note that measuring the current in the inductor $I_L(Z)$ is quite difficult.

The passive structure of Chua's circuit is composed of a two-node configuration. In Fig. 10.3a, one of these nodes, specifically the one across C_2, is excited. Subsequently, the other node, across C_1, is alternately excited, as shown in Fig. 10.3b. The corresponding circuit equations are described by

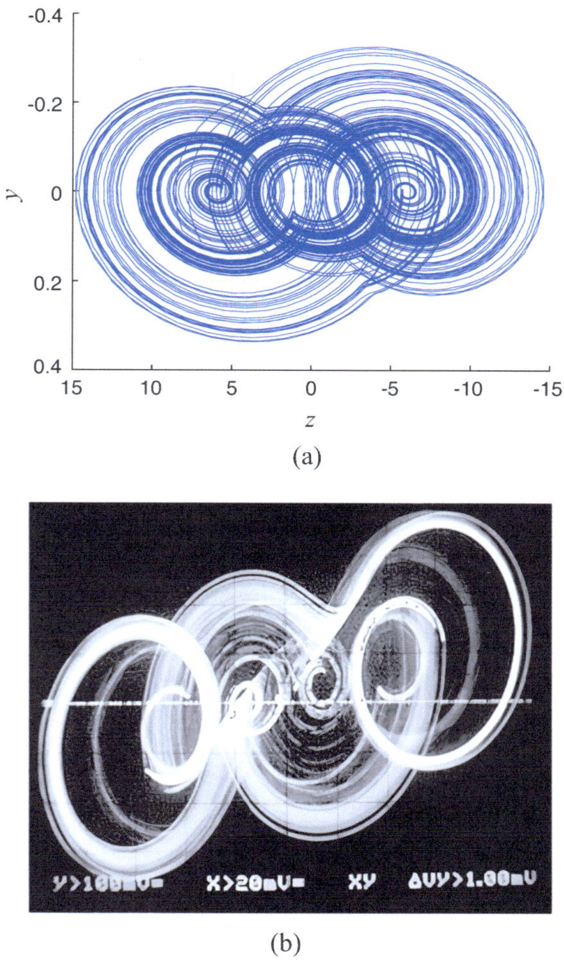

Fig. 10.4 Pulse-driven nonautonomous chaotic attractor. **a** 3-scroll attractor generated from numerical simulation with $(\beta_F, \beta_S) = (2, 3)$. **b** Experimental observation of 3-Scroll attractor on the V_{C1}-V_{C2} plane (X-axis: 20 mV/div, Y-axis: 100 mV/div). Reprinted from [2], with permission from Elsevier

$$\begin{pmatrix} \varepsilon_c \dot{X} \\ \dot{Y} \\ \dot{Z} \end{pmatrix} = \begin{pmatrix} 1 & -(1+\beta_F+\beta_S) & 0 \\ 1/\alpha_2 & -1/\alpha_2 & -1/\alpha_2 \\ 0 & 1/\alpha_1 & 0 \end{pmatrix} \begin{pmatrix} X \\ Y \\ Z \end{pmatrix} + \begin{pmatrix} a \\ 0 \\ 0 \end{pmatrix}, \qquad (10.7)$$

where a is defined in Eq. (10.6), with the switching condition depending on y rather than x.

Figure 10.5 displays the observed 4-scroll chaotic attractor, where $c = 0.1$, $\alpha_1 = 0.07$, $\alpha_2 = 0.5$, and $\beta_F = \beta_S = 3$. A single positive Lyapunov exponent is computed from the y-time series, obtaining a value of 0.0089.

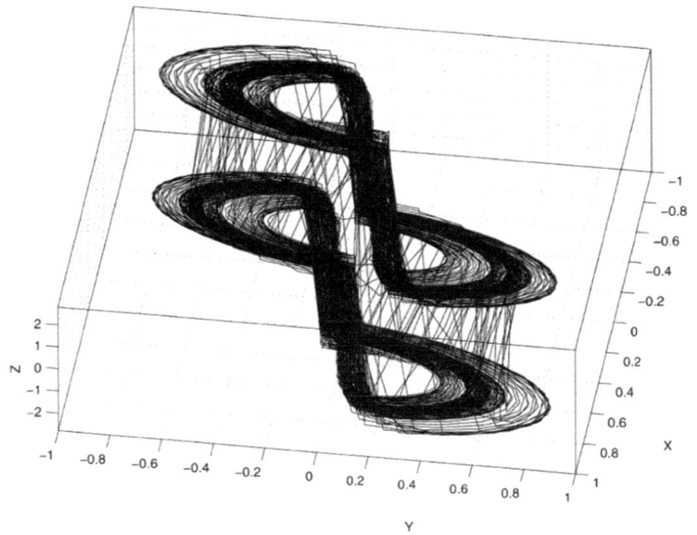

Fig. 10.5 4-Scroll chaotic attractor observed from the model of alternatively excited circuit. Reprinted from [2], with permission by Elsevier

10.3 Multiscroll Chaotic Attractors from a Multilevel-Logic-Pulse Driven System

Hong et al. [3] constructed a nonautonomous system to generate multi-double-scroll attractors without modifying the original nonlinear functions. Multi-double-scroll attractors were created by applying a multilevel-logic-pulse excitation technique to double-scroll chaotic systems.

10.3.1 Chua's Circuit with Multiple-Pulse Excitation

The generation of multi-double-scroll attractors from Chua's circuit is demonstrated in Fig. 10.6, utilizing a multiple-pulse excitation source. An external pulse current source (i_p) is connected in parallel with C_2. The circuit shown in Fig. 10.6 is described by

$$\begin{cases} \dfrac{dv_1}{dt} = \dfrac{1}{RC_1}(v_2 - v_1) - \dfrac{1}{C_1} f(v_1), \\ \dfrac{dv_2}{dt} = \dfrac{1}{RC_2}(v_1 - v_2) - \dfrac{1}{C_2} i_L - \dfrac{1}{C_2} i_p(t), \\ \dfrac{di_L}{dt} = \dfrac{1}{L} v_2, \end{cases} \tag{10.8}$$

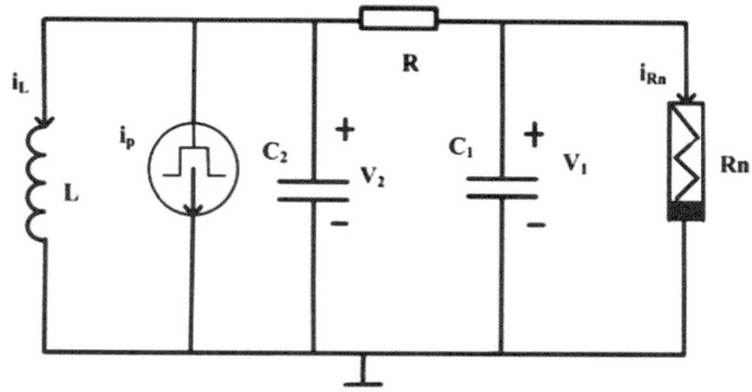

Fig. 10.6 Chua's circuit with current pulse-excitation. Reprinted from [3], with permission from the AIP

where $f(v_1)$ represents the i-v characteristic equation of the Chua's diode, which is

$$f(v_1) = \begin{cases} G_b v_1 + E(G_a - G_b), & \text{if } \; v_1 > E, \\ G_a v_1, & \text{if } \; |v_1| \le E, \\ G_b v_1 - E(G_a - G_b), & \text{if } \; v_1 < -E, \end{cases}$$

in which the current source $i_p(t)$ is an external pulse, defined by

$$i_p(t) = A\,\mathrm{sgn}(\sin(\omega t)) = \begin{cases} A, & \text{if } \; \sin(\omega t) \ge 0, \\ -A, & \text{if } \; \sin(\omega t) < 0. \end{cases} \tag{10.9}$$

Set the right-hand side of system (10.8) to zero. It results in the equations $v_1 = -Rf(v_1)$, $v_2 = 0$, and $i_L = v_1/R - i_p(t)$.

By combining the above equations, one obtains six equilibrium points as follows:

$$(v_1, v_2, i_L) = [(0, 0, 0 \pm A),\ (2.19, 0, 0.0012 \pm A),\ (-2.19, 0, -0.0012 \pm A)],$$

where the parameters are $R = 1850\,\Omega$, $C_1 = 10.38\,\text{nF}$, $C_2 = 102.5\,\text{nF}$, $L = 18.81\,\text{mH}$, $G_a = 0.72\,\text{ms}$, $G_b = 0.39\,\text{ms}$, $E = 1\,\text{V}$, and the initial conditions are $(0.1, 0, 0)$. The system's equilibrium points are clearly affected by the external excitation i_p in the i_L-direction, where i_p provides two fixed time-invariant values ($\pm A$) with an alternating frequency ω. When $A = 0.001$ and $\omega = 1000$, a 2-double-scroll attractor emerges, as shown in Fig. 10.7a. It can be observed that the equilibrium points in the i_L-direction are located at (±0.0022, ±0.0002), consistent with the theoretical analysis.

By analogy, a three-level-logic source can be built using two excitation sources, expressed as

$$i_p(t) = 0.5\left[A\,\mathrm{sgn}(\sin(\omega_1 t)) + A\,\mathrm{sgn}(\sin(\omega_2 t))\right],$$

resulting in three values, $(0, A, -A)$. This configuration produces a 3-double-scroll attractor, as shown in Fig. 10.7b, with the parameter set $(A, \omega_1, \omega_2) = (0.001, 1000, 200)$. Similarly, Fig. 10.7c illustrates the generation of a 4-double-scroll attractor.

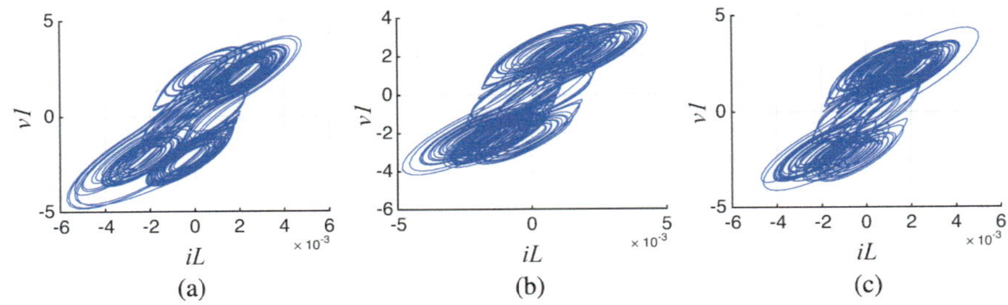

Fig. 10.7 n-double-scroll attractors from the current pulse-excited Chua's circuit [3]. **a** 2-double-scroll attractor. **b** 3-double-scroll attractor. **c** 4-double-scroll attractor

10.3.2 Jerk Circuit via Multiple-Pulse Excitation

Another example of a 3-D chaotic system capable of generating a double-scroll attractor is the jerk circuit. To generate multiple double-scroll attractors, a composite excitation is applied to the jerk circuit modeled by

$$\begin{cases} \dot{x} = y - f(y) , \\ \dot{y} = z , \\ \dot{z} = -a(x + y + z) + f(t) . \end{cases} \tag{10.10}$$

In this setting, the system parameter is $a = 0.8$ and the nonlinear function is $f(y) = \text{sgn}(y)$. Here, $f(t)$ is considered as a composite pulse excitation, given by

$$f(t) = 0.5 \sum_{n=1}^{N} A_n \,\text{sgn}\left(\sin(\omega_n t)\right) ,$$

where A_n and ω_n represent the amplitude and frequency of the pulse excitation source, respectively.

The equilibrium points of the circuit are obtained from

$$\begin{cases} y = f(y) , \\ z = 0 , \\ x = -y + f(t)/a . \end{cases} \tag{10.11}$$

When $n = 1$, the system has six equilibrium points:

$$(x, y, z) = \left(\left(0 \pm \frac{A_1}{a}, 0, 0\right) , \left(1 \pm \frac{A_1}{a}, -1, 0\right) , \left(-1 \pm \frac{A_1}{a}, 1, 0\right) \right) .$$

The generated 2-double-scroll attractor is shown in Fig. 10.8a, with parameters $A_1 = 1$, $\omega_1 = 0.02$, and initial condition $(0.2, 0, 0)$. When $n = 2$, 3 and 4 double-

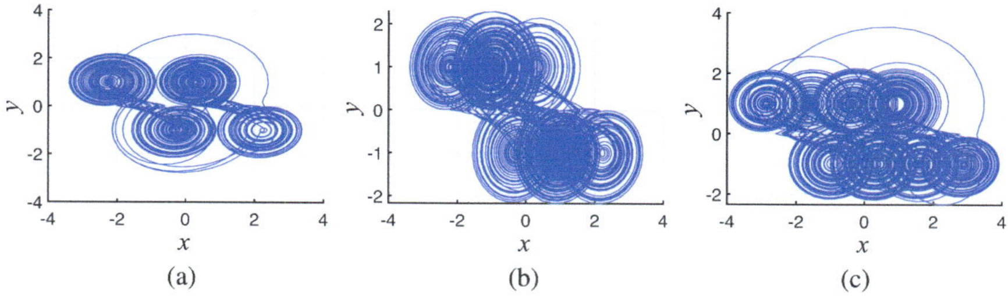

Fig. 10.8 n-double-scroll attractors from the jerk circuit using various combinations of pulse-excitations [3]. **a** 2-double-scroll attractor. **b** 3-double-scroll attractor. **c** 4-double-scroll attractor

scroll attractors can be generated, as shown in Fig. 10.8b and c, where the frequencies are $\omega_1 = 0.02$ and $\omega_2 = 0.005$, and the amplitudes (A_1, A_2) are $(1, 1)$ and $(1, 2)$, respectively.

10.3.3 *Modified Lorenz System with Multiple-Pulse Excitation*

The modified Lorenz system with multi-level pulse excitation is modeled by

$$\begin{cases} \dot{x} = a\,(y - x) + f\,(t)\ , \\ \dot{y} = K\,(b - z)\ , \\ \dot{z} = |x| - cz\ , \end{cases} \tag{10.12}$$

where $a = 0.9, b = 2, c = 0.2$, and $K = \mathrm{sgn}(x)$. The multipulse excitation function $f(t)$ is

$$f(t) = 0.5\left[A_1\,\mathrm{sgn}\left(\sin(\omega_1 t)\right) + A_2\,\mathrm{sgn}\left(\sin(\omega_2 t)\right)\right]\ ,$$

where $n = 2$.

With frequencies $\omega_1 = 0.05$ and $\omega_2 = 0.03$, 3- and 4-double-scroll attractors are generated, as shown in Fig. 10.9a and b, for different amplitude combinations: $A_1 = A_2 = 1.2$, $A_1 = 0.5$, and $A_2 = 1.2$. Figure 10.9c displays the 5-double-scroll attractor for $n = 3$.

10.3.4 *Pspice Simulation Verification*

In this subsection, a Pspice simulation is performed to verify the above-discussed method for implementation in a circuit.

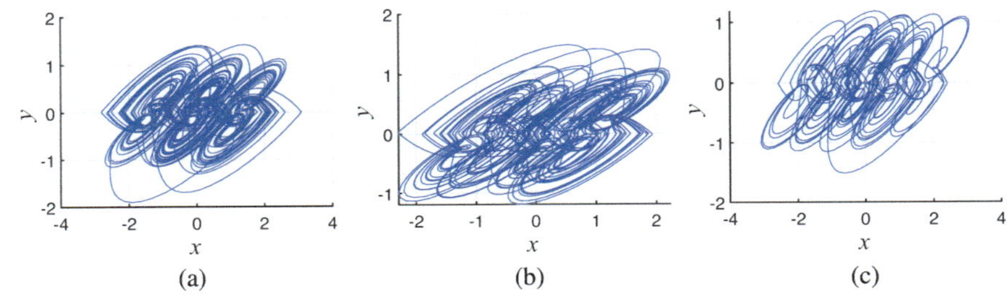

Fig. 10.9 n-double-scroll attractors generated from the modified Lorenz system with various combined pulse-excitations [3]. **a** 3-double-scroll attractor. **b** 4-double-scroll attractor. **c** 5-double-scroll attractor

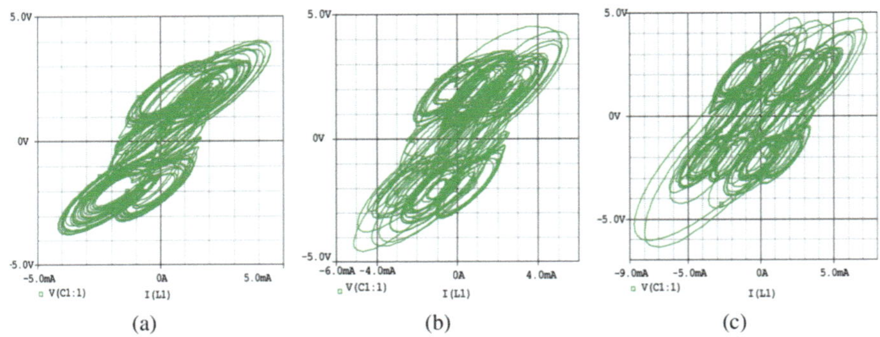

Fig. 10.10 Pspice simulation results of the excited Chua's circuit. **a** 2-double-scroll attractor. **b** 3-double-scroll attractor. **c** 4-double-scroll attractor. Reprinted from [3], with permission by AIP

To begin, Chua's multi-scroll is demonstrated using the circuit shown in Fig. 10.6, where the Chua's diode is simulated by a circuit with two operational amplifiers. When an excitation source i_{p1} is applied, with parameters $i_{p1} = 0.9\,\text{mA}$ and $\omega_1 = 1.4\,\text{kHz}$, 2-double-scrolls are obtained, as shown in Fig. 10.10a. By adding an additional source i_{p2} in parallel, 3- and 4-double-scrolls can also be generated, as shown in Fig. 10.10b and c, with parameters $(i_{p2}, \omega_2) = (0.9\,\text{mA}, 0.1\,\text{kHz})$ and $(i_{p2}, \omega_2) = (1.8\,\text{mA}, 0.1\,\text{kHz})$, respectively.

The pulse-excited jerk circuit described by

$$i_p(t) = 0.5\left[A\,\text{sgn}\,(\sin(\omega_1 t)) + A\,\text{sgn}\,(\sin(\omega_2 t))\right]$$

is implemented with CFOA AD844. The circuit diagram, which includes the realization of the nonlinear function $f(x)$, is depicted in Fig. 10.11.

Applying Kirchhoff's circuit laws, the corresponding circuit equations can be obtained, as

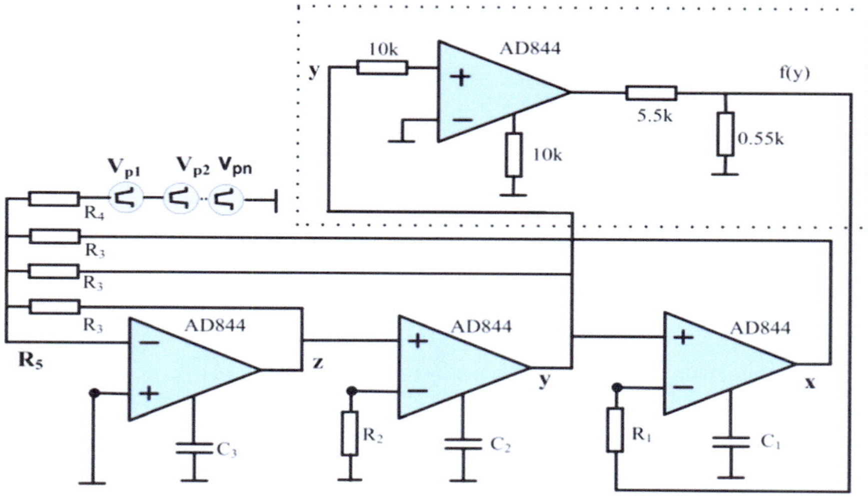

Fig. 10.11 Circuit realization of the pulse-excited jerk system. Reprinted from [3], with permission by AIP

$$\begin{cases} \dot{x} = \dfrac{y - f(y)}{R_1 C_1}, \\ \dot{y} = \dfrac{z}{R_2 C_2}, \\ \dot{z} = -\dfrac{x + y + z}{R_3 C_3} - \dfrac{x + y + z}{R_4 C_3}. \end{cases} \tag{10.13}$$

To ensure that Eq. (10.13) is equivalent to Eq. (10.10), the following circuit parameters are chosen: $R_1 = R_2 = R_4 = 10\,\text{k}\Omega$, $R_3 = 12.5\,\text{k}\Omega$, and $C_1 = C_2 = C_3 = 10\,\text{nF}$.

Figure 10.12 demonstrates various multi-double-scroll chaotic attractors obtained by different power combinations. In Fig. 10.12a, a 2-double-scroll attractor is observed when the voltage source v_{p1} is set to $v_{p1} = 0.75\,\text{V}$ and $\omega_1 = 16.6\,\text{Hz}$. By introducing a second exciting current source with frequency $\omega_2 = 6.6\,\text{Hz}$ and amplitudes $v_{p2} = 0.75V$ and $1.5V$, 3- and 4-double-scroll attractors are observed as shown in Fig. 10.12b and c, respectively.

Finally, the circuit designed to implement Eq. (10.12) is shown in Fig. 10.13. It includes summator, integrator and inverter components, all using AD711 operational amplifiers, described by

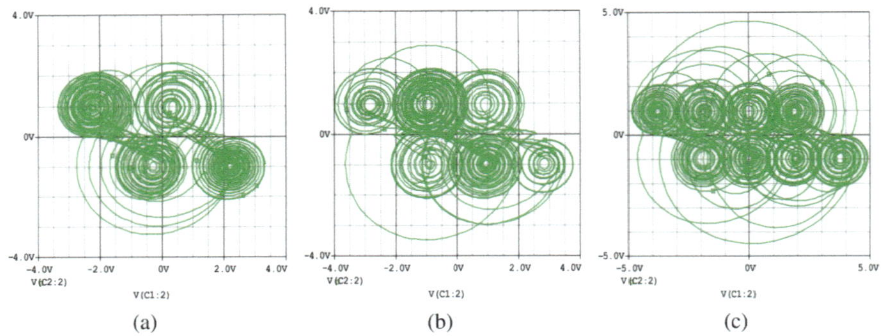

Fig. 10.12 Generation of multi-scroll attractors by Pspice from the excited jerk circuit. **a** 2-double-scroll attractor. **b** 3-double-scroll attractor. **c** 4-double-scroll attractor. Reprinted from [3], with permission by AIP

$$\begin{cases} \dot{x} = \dfrac{1}{RC}\left[\dfrac{R_1}{R_2}y - \dfrac{R_1}{R_3}x + \dfrac{R_1}{R_4}\left(v_{p1} + v_{p2}\right)\right], \\ \dot{y} = \dfrac{1}{RC}\left[K\left(2\dfrac{R_5}{R_7} - \dfrac{R_1}{R_2}z\right)\right], \\ \dot{z} = \dfrac{1}{RC}\left(\dfrac{R_8}{R_9}Kx - \dfrac{R_8}{R_{10}}z\right). \end{cases} \tag{10.14}$$

The modified Lorenz system can generate varying numbers of multi-double-scroll attractors by different parameter combinations. The Pspice simulation results for the excited modified Lorenz system are shown in Fig. 10.14. Figure 10.14a and b display 3- and 4-double-scroll attractors, respectively, with parameters $(\omega_1, \omega_2, A_1, A_2) =$ (25 Hz, 6.3 Hz, 1 V, 1 V), and (25 Hz, 6.3 Hz, 1 V, 2 V), respectively. Additionally, Fig. 10.14c shows a five-double-scroll attractor.

10.4 Butterfly Chaotic Attractors in 4-D Autonomous Systems

Lai et al. [4] introduced a 4-D autonomous system with a variety of coexisting attractors.

This 4-D chaotic system is described by

$$\begin{cases} \dot{x} = a\left(y - x\right), \\ \dot{y} = xz + w, \\ \dot{z} = b - xy, \\ \dot{w} = cyz, \end{cases} \tag{10.15}$$

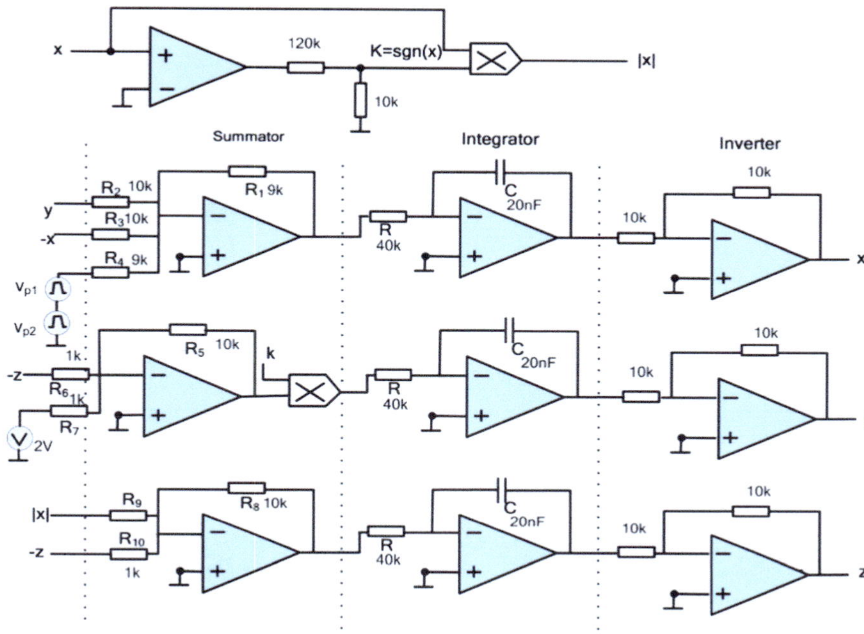

Fig. 10.13 Circuit for realizing the modified Lorenz system (10.12). Reprinted from [3], with permission by AIP

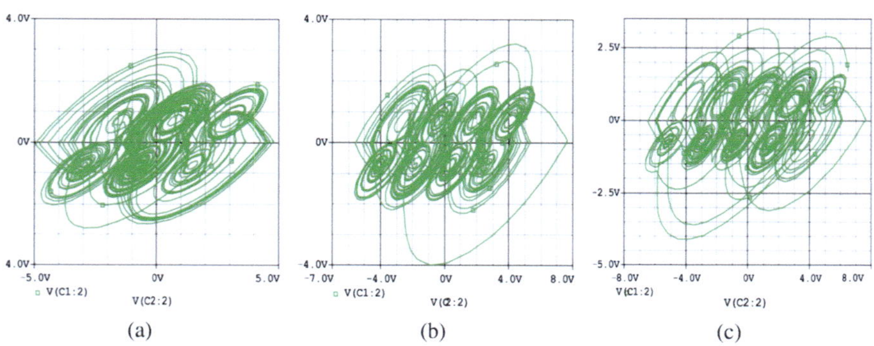

Fig. 10.14 Pspice simulation results of the excited modified Lorenz system. **a** 3-double-scroll attractor. **b** 4-double-scroll attractor. **c** 5-double-scroll attractor. Reprinted from [3], with permission by AIP

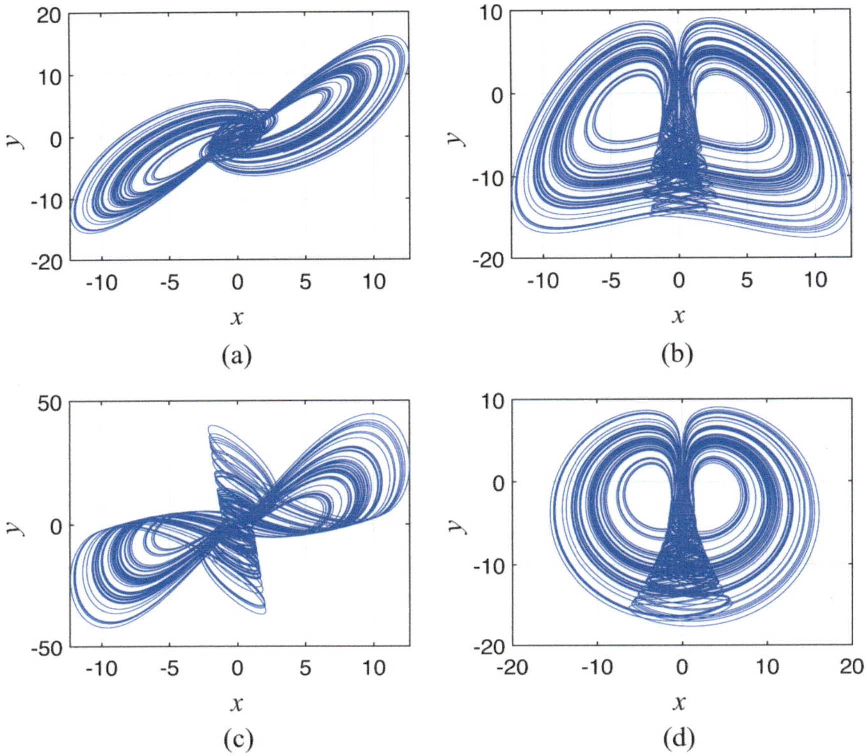

Fig. 10.15 A butterfly attractor of system (10.15), with $a = 10$, $b = 10$, $c = 3$ [4]. **a** x-y plane. **b** x-z plane. **c** x-w plane. **d** y-z plane

where x, y, z, w are state variables, and a, b, c are positive real numbers.

For the parameter values $a = 10$, $b = 10$, $c = 3$, and the initial condition $(1, 1, 0, 0)$, a butterfly attractor is generated from system (10.15). Its projections onto different planes are presented in Fig. 10.15. By adjusting the parameter to $c = 1$ while keeping a and b fixed, two independent attractors arise from system (10.15) with the initial conditions $(\pm 1, \pm 1, 0, 0)$. These attractors are shown in Fig. 10.16, revealing distinct and separated chaotic structures.

10.5 Conclusion

This chapter discusses the generation of multiscroll chaotic attractors from nonautonomous systems, presenting some approaches to enhancing the design and complexity of various chaotic systems. By utilizing pulse-driven oscillators and multi-level logic pulse excitation, multi-double-scroll attractors can be generated from

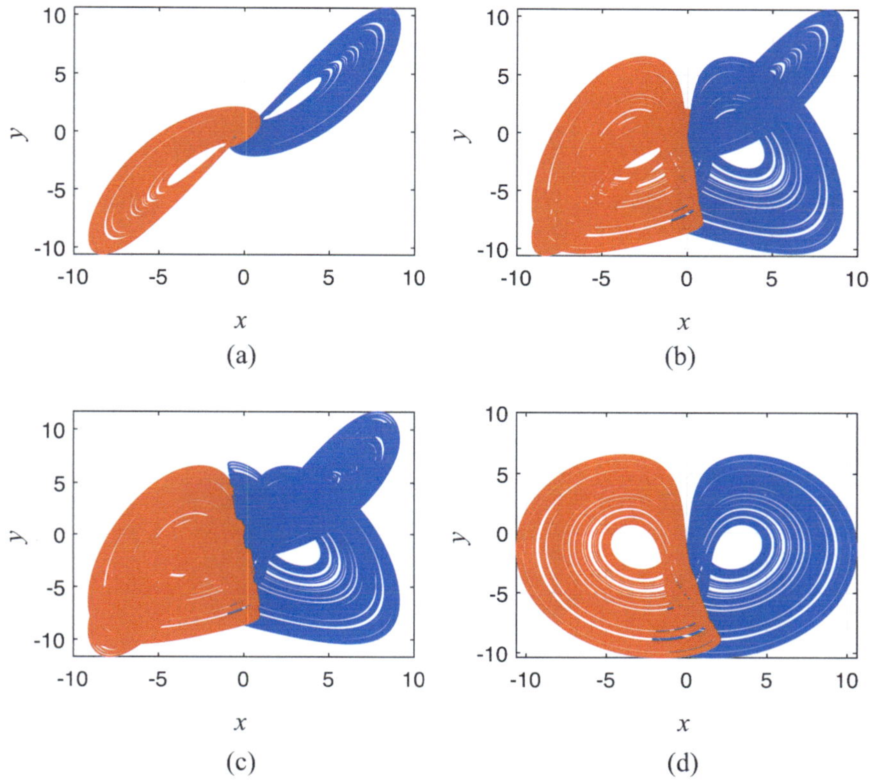

Fig. 10.16 One pair of chaotic attractors from system (10.15), with $a = 10, b = 10, c = 1$ [4]. **a** x-y plane. **b** x-z plane. **c** x-w plane. **d** y-z plane

chaotic systems such as Chua's circuit, jerk circuit, and a modified Lorenz system, without altering their nonlinear functions.

Experimental verification and PSpice simulations confirm the practicality and scalability of these methods. Additionally, the introduction of a 4-D autonomous chaotic system demonstrates the generation of coexisting butterfly-like attractors, which expands the spectrum of chaotic phenomena. These advancements contribute to the development of chaos-based applications in secure communications, nonlinear control and signal processing.

References

1. Aziz-Alaoui, M.A.: Differential equations with multispiral attractors. Int. J. Bifur. Chaos **9**, 1009–1039 (1999)
2. Elwakil, A.S.: Nonautonomous pulse-driven chaotic oscillator based on Chua's circuit. Microelectron. J. **33**, 479–486 (2002)
3. Hong, Q., Xie, Q., Shen, Y., Wang, X.: Generating multi-double-scroll attractors via nonautonomous approach. Chaos **26**, 083110 (2016)
4. Lai, Q., Nestor, T., Kengne, J., Zhao, X.: Coexisting attractors and circuit implementation of a new 4-D chaotic system with two equilibria. Chaos Solit. Fract. **107**, 92–102 (2018)

Chapter 11
Generating Multiscroll Chaotic Attractors by Other Techniques

Abstract This chapter explores advanced techniques for generating multiscroll chaotic attractors from some nonlinear dynamical systems. Various effective methods are introduced, including modifications to the modified Lorenz system, a chaotic system bridging the Lorenz and Chen systems, and a non-autonomous forced systems. Additionally, other generating methods are presented based on PWL van der Pol and Duffing systems, oscillatory potential wells and a thresholding mechanism. The chapter furthermore discusses innovative techniques for generating $n \times m$-grid chaotic attractors using arctangent function series. These methodologies for generating complex multiscroll attractors expand the application domain of chaos theory across engineering and technological fields.

11.1 Multiscroll Chaotic Attractors from the Modified Lorenz System

This section explores a modified Lorenz system [1–3] to generate a four-wing butterfly chaotic attractor.

Recall the classical Lorenz system (1.31), where a, b, and c are real parameters, while the nonlinear terms xz and xy are key drivers of producing chaotic behaviors.

Elwakil et al. [2] proposed a modified Lorenz system by relaxing the constraint on the positive z-axis, to extend the range of potential chaotic behaviors. The system is described by

$$\begin{cases} \dot{x} = a\,(y - x)\,, \\ \dot{y}_{\mp} = \mp K z, \\ \dot{z}_{\pm} = \pm\,|x| \mp 1, \end{cases} \tag{11.1}$$

where

$$K = \begin{cases} 1, & x \geq 0, \\ -1, & x < 0, \end{cases}$$

and $0 < a < 1$.

J. Lü et al., *Chaotic Systems with Multiscroll Attractors*,
https://doi.org/10.1007/978-3-032-09854-2_11

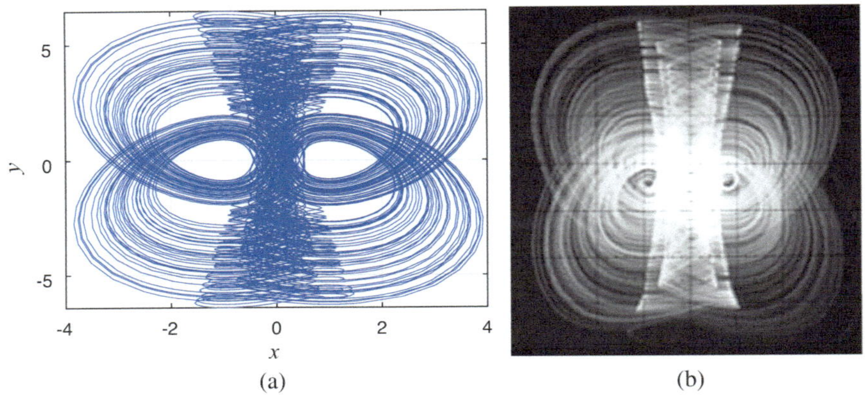

Fig. 11.1 4-wing butterfly chaotic attractor from the modified Lorenz system. **a** Numerical simulation. **b** Experimental observation. Reprinted from [2], with permission by the IEEE

From the original Lorenz system (1.31), the nonlinear terms xy and xz are replaced by the absolute value function $|x|$ and the sign function Kz, respectively. Additionally, the threshold constant b and the damping constant c are removed from the original system, by which the system is simplified while maintaining its chaotic nature.

System (11.1) functions as a dual system with two complementary modes. The mode $S(-,+)$ represents the case where $\dot{y} = \dot{y}_-$ and $\dot{z} = \dot{z}_+$, while $S(+,-)$ represents the case where $\dot{y} = \dot{y}_+$ and $\dot{z} = \dot{z}_-$. In this new configuration, the threshold effect originally imposed by b in Eq. (1.31) is realized through the absolute value function $|x|$ in (11.1). Additionally, in Eq. (11.1), it is the sign of z, rather than $(b - z)$, that changes at the threshold, removing the corresponding constraint from the original Lorenz system.

When $a = 0.55$, system (11.1) can generate two-wing butterfly attractors in both modes $S(-,+)$ and $S(+,-)$, similarly to the chaotic behavior observed from the original Lorenz system. Importantly, the eigenvalue pattern remains unaffected by the mode in which the system operates, enabling the use of an external source to trigger the switching between $S(-,+)$ and $S(+,-)$.

Figure 11.1a shows the 4-wing butterfly chaotic attractor generated under these conditions, with $a = 0.55$ and a pulse train of period $T_F = 250T_S$ used to facilitate the switching.

Elwakil et al. [4] also designed a circuit diagram to experimentally verify the 4-wing butterfly chaotic attractor. The circuit diagram, shown in Fig. 11.2, includes three capacitors, (C_X, C_Y, C_Z), twelve bilateral MOS analog switches, six CFOAs, one external voltage source V_P and one reference voltage V_R.

By choosing $C_X = C_Y = C_Z = C$ and $R_1 = R_2 = R_3 = R$ with $R_a = R/a$, and defining $x = V_X/V_R$, $y = V_Y/V_R$, and $z = V_Z/V_R$, while normalizing the time based on RC, it can be verified that the circuit diagram in Fig. 11.2 correctly implements system (11.1). Figure 11.1b shows the experimental observation of the 4-wing but-

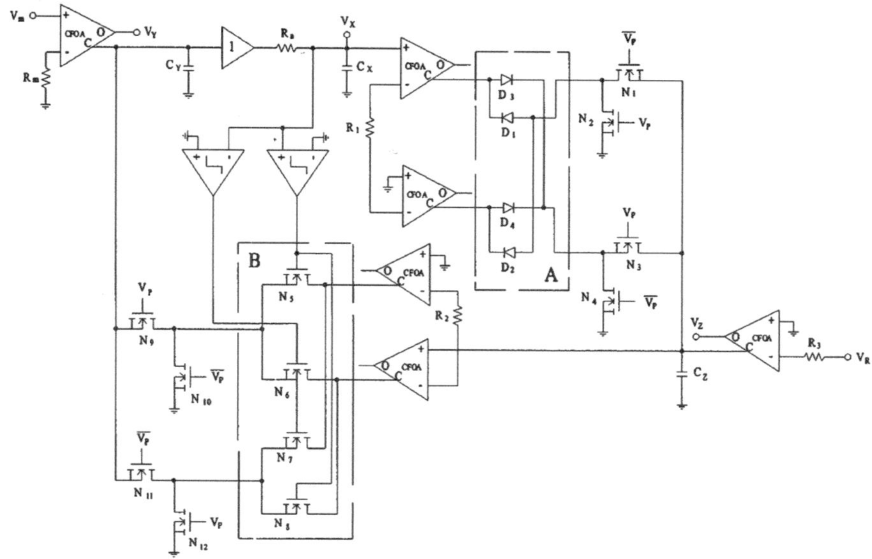

Fig. 11.2 Circuit diagram for the realization of the modified Lorenz system. Reprinted from [2], with permission by the IEEE

terfly chaotic attractor, with a source frequency of 150 Hz and the circuit's center frequency around 31 kHz.

11.2 Multiscroll Chaotic Attractors from the Modified Proto-Lorenz System

This section introduces the proto-Lorenz system [5], which is a quotient of the Lorenz system. The proto-Lorenz system can be utilized to generate n-scroll chaotic attractors.

Recall the Lorenz system (1.31), where $(x, y, z) \in X = \mathbf{R}^3$. It is clear that the Lorenz system (1.31) is invariant under the linear transformation $L_1 : (-x, -y, z) \to (x, y, z)$.

Define a map, $\pi : X \to Y$ by

$$u = \pi\,(x) = x^2 - y^2, \quad v = \pi\,(y) = 2xy, \quad z = \pi\,(z) = z. \tag{11.2}$$

Under this transformation L_1, the Lorenz system (1.31) remains invariant in X, resulting in a new vector field F in the space Y. Miranda and Stone [5] derived the equations for this new vector field F in the space Y as follows:

$$\begin{cases} \dot{u} = -(a+1)\,u + (a-c+z)\,v + (1-c)\sqrt{u^2+v^2}, \\ \dot{v} = (c-a)\,u - (a+1)\,v + (a+c-z)\sqrt{u^2+v^2} - uz, \\ \dot{z} = \dfrac{1}{2}v - bz. \end{cases} \tag{11.3}$$

This system is referred to as the proto-Lorenz system.

It is important to note that the proto-Lorenz system F represents a continuous vector field in the entire space Y. However, it is not differentiable because of the norm $\sqrt{u^2+v^2}$, though it remains differentiable except on the z-axis.

Based on the construction of the proto-Lorenz system mentioned above, the proto-Lorenz system has the Lorenz system as a double covering, and the double covering explains the two-eared nature of the attractor for the Lorenz system. The double cover F_2 corresponds to the original Lorenz system. Additionally, the nth cover F_n of the proto-Lorenz system can generate n-scrolls. The triple cover F_3 is described by [5]

$$\begin{cases} \dot{p} = \begin{pmatrix} \frac{1}{3\sqrt{p^2+q^2}}\left((1-a)\left(p^2+q^2\right) + 2\,(a+c-z)\,pq\right) \\ +\frac{1}{3}\left(-(a+1)\,p + (a-c+z)\,q\right) \end{pmatrix}, \\ \dot{q} = \begin{pmatrix} \frac{1}{3\sqrt{p^2+q^2}}\left(2\,(a-1)\,pq + (a+c-z)\left(p^2+q^2\right)\right) \\ +\frac{1}{3}\left(-(a+1)\,p + (a-c+z)\,q\right) \end{pmatrix}, \\ \dot{z} = \frac{1}{2}\left(3p^2q - q^3\right) - bz, \end{cases} \tag{11.4}$$

where $a = 10$, $b = 8/3$, and $c = 28$, the triple-cover system (11.4) generates a 3-scroll chaotic attractor, as shown in Fig. 11.3a. This attractor showcases the complex behavior of the system, demonstrating its capacity to produce various chaotic patterns through the interplay of its parameters.

This structure not only highlights the chaotic characteristics of the system but also serves for illustrating the sophisticated dynamics emerging from the proto-Lorenz framework.

Similarly, the quartic cover F_4 of the proto-Lorenz system, constructed by Miranda and Stone [5], is represented by

$$\begin{cases} \dot{s} = \dfrac{1}{2\left(s^2+t^2\right)}\begin{bmatrix} -as^3 + (2a+c-z)\,s^2t \\ +(a-2)\,st^2 - (c-z)\,t^3 \end{bmatrix}, \\ \dot{t} = \dfrac{1}{2\left(s^2+t^2\right)}\begin{bmatrix} (c-z)\,s^3 + (a-2)\,s^2t \\ +(-2a-c+z)\,st^2 - at^3 \end{bmatrix}, \\ \dot{z} = 2s^3 - 2\,st^3 - bz, \end{cases} \tag{11.5}$$

where $a = 10$, $b = 8/3$, and $c = 28$. The quartic-cover system (11.5) generates a 4-scroll chaotic attractor, as shown in Fig. 11.3b.

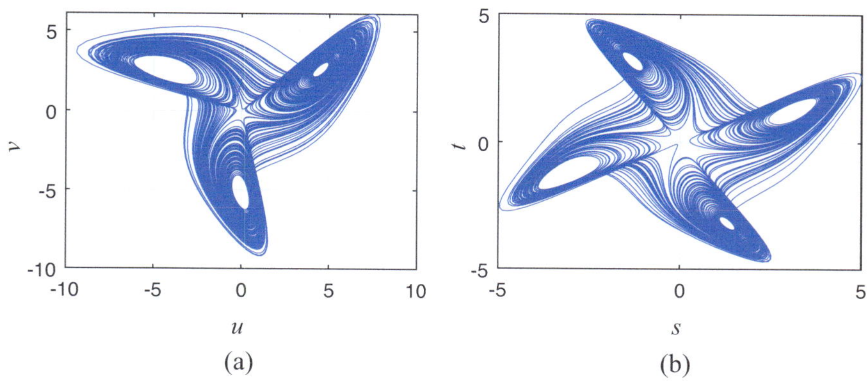

Fig. 11.3 Multiscroll chaotic attractors from the modified proto-Lorenz system [5]. **a** 3-scroll chaotic attractor. **b** 4-scroll chaotic attractor

11.3 Multiscroll Chaotic Attractors from a Bridging Chaotic System

This section presents a 3-D quadratic autonomous chaotic system, which bridges the Lorenz system and the Chen system, demonstrating various complex dynamic behaviors. Specifically, the system is capable of exhibiting:

(i) Two 1-scroll chaotic attractors simultaneously, even with only three equilibria, highlighting its ability to produce complex chaotic dynamics with a minimal necessary number of equilibrium points.
(ii) Two 2-scroll chaotic attractors simultaneously, when the system has five equilibria, it allows the system to transit among more intricate multi-scroll chaotic states.

These behaviors, as observed by Lü and Chen [6] and further investigated by Lü et al. [7, 8], illustrate the rich dynamical potential of this type of simple quadratic chaotic systems.

Lü et al. [7, 8] introduced the following simple quadratic chaotic system, which bridges the Lorenz system and the Chen system, with distinctive dynamic characteristics:

$$\begin{cases} \dot{x} = -\dfrac{ab}{a+b}x - yz + c, \\ \dot{y} = ay + xz, \\ \dot{z} = bz + xy, \end{cases} \tag{11.6}$$

where a, b, c are real parameters.

System (11.6) exhibits chaotic behavior over a broad range of parameter values and showcases several captivating dynamic features [7]. Specifically, with parame-

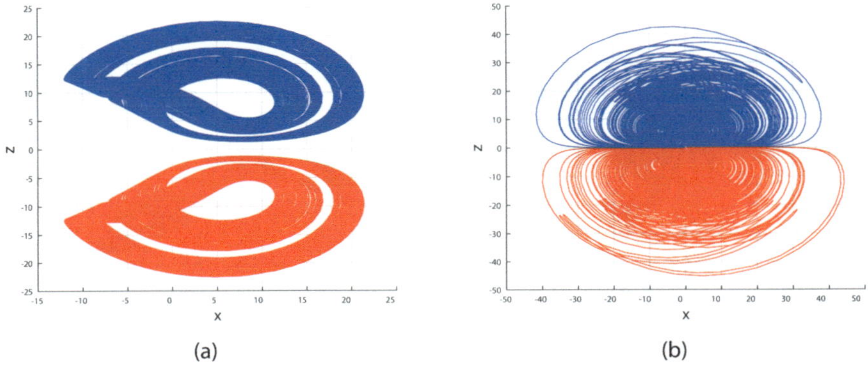

Fig. 11.4 Multiscroll chaotic attractors from the bridging chaotic system [7, 8]. **a** Two coexisting 1-scroll chaotic attractors. **b** Two coexisting 2-scroll chaotic attractors

ters $a = -10, b = -4, c = 18.1$, or $a = -10, b = -4, c = 0$, the system generates two coexisting 1-scroll chaotic attractors, as shown in Fig. 11.4a. The corresponding Lyapunov exponent spectrum is $\lambda_1 = 0.253223$, $\lambda_2 = 0$, and $\lambda_3 = -11.3944$, resulting in a Lyapunov dimension of $d_L = 2.0221$ for the initial condition $(1, 1, 1)$.

Additionally, when parameters are adjusted to $a = -10$, $b = -4$, and $c = 0$, system (11.6) transits to having two coexisting 2-scroll chaotic attractors, as shown in Fig. 11.4b. The corresponding Lyapunov exponent spectrum is $\lambda_1 = 1.1662, \lambda_2 = 0$, and $\lambda_3 = -12.3090$, leading to a Lyapunov dimension of $d_L = 2.0947$ for the same initial conditions $(1, 1, 1)$ and $(1, 1, -1)$.

Notably, Liu and Chen [9] introduced a similar but distinct chaotic system capable of generating two closely spaced 2-scroll chaotic attractors, separating by a single point, which together form an apparently single 4-scroll attractor, as shown by Fig. 11.5.

11.4 Multiscroll Chaotic and Nonchaotic Aattractors from a Forced System

This section presents a general quasi-periodically forced system capable of generating both chaotic and non-chaotic multiscroll attractors. The system utilizes external quasi-periodic forcing to induce complex dynamic behaviors, forming multiscroll attractors that exhibit both chaotic and non-chaotic features.

Aziz-Alaoui [10, 11] studied the dynamics of a nonautonomous circuit, specifically a classical forced negative-resistance oscillator, which can produce strange nonchaotic two-frequency torus attractors. The circuit equations are

$$\begin{cases} \dot{x} = y - f_2(x), \\ \dot{y} = -\beta\left[x + (v + 1)\,y + A\left(\sin(\omega_1 t) + \sin(\omega_2 t)\right)\right], \end{cases} \tag{11.7}$$

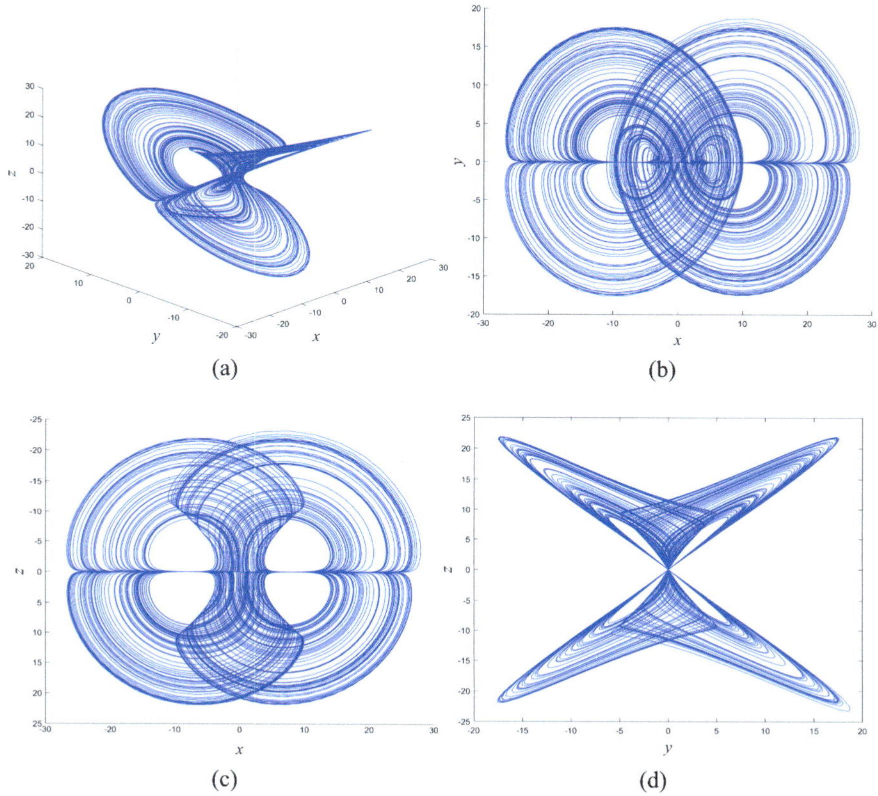

Fig. 11.5 The single 4-scroll chaotic attractor [9]. **a** 3-D space. **b** x-y plane. **c** x-z plane. **d** y-z plane

where

$$f_2(x) = m_1 x + \frac{1}{2}(m_0 - m_1)\left[|x+1| - |x-1|\right],$$

and β, v, ω_1, ω_2, A are real parameters.

To generate multiscroll chaotic or nonchaotic attractors, Aziz-Alaoui [10] altered the function f_2 in Eq. (11.7) to construct the following modified system:

$$\begin{cases} \dot{x} = y - f_N(x), \\ \dot{y} = -\beta\left[x + (v+1)y + A(\sin(\omega_1 t) + \sin(\omega_2 t))\right], \end{cases} \tag{11.8}$$

with

$$f_N(x) = \begin{cases} m_k x + \operatorname{sgn}(x)\xi_k, & \text{if } s_{k-1} \le |x| \le s_k, k \in I_{N-2}, \\ m_{N-1} x + \operatorname{sgn}(x)\xi_{N-1}, & \text{if } |x| \ge s_{N-2}, \end{cases} \tag{11.9}$$

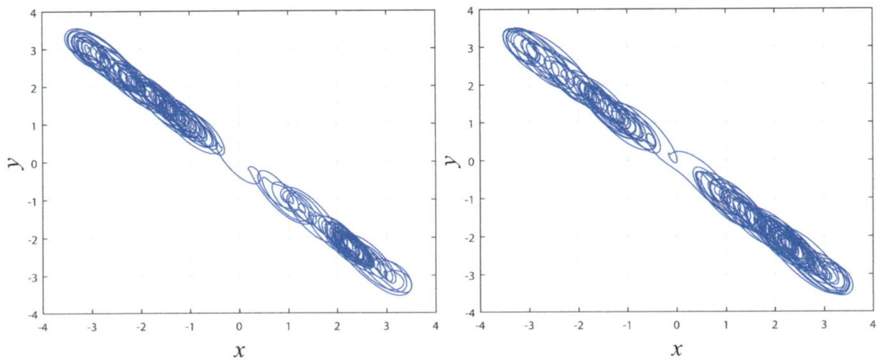

Fig. 11.6 Multiscroll chaotic attractors from the quasi-periodically forced system (11.7) [10, 11]. **a** Two coexisting 3-scroll attractors with $A = 0.173$. **b** A double 3-scroll attractor with $A = 0.19$

where the parameter set is $B_N^K = \{\beta, v, \omega_1, \omega_2, A\} \cup B_N \subset \mathbf{R}^{2N+4}$, with $B_N = \{(s_k)_{k \in I_{N-2}}, (m_k)_{k \in I_{N-1}}\} \subset \mathbf{R}^{2N-1}$. Set $m_0 = m_{2i} = \cdots = -1.02$ and $m_1 = m_{2i+1} = \cdots = -0.55$, $i = 1, 2, \ldots$.

Figure 11.6a shows two coexisting 3-scroll chaotic attractors, with the parameters $N = 6$, $A = 0.173$, and B_6^K. Figure 11.6b shows the combination of these two chaotic attractors, resulting in a double 3-scroll chaotic attractor, with $N = 6$, $A = 0.18$ and B_6^K, where B_6^K is defined by the parameters $\beta = 1.0$, $v = 0.015$, $\omega_1 = 0.75$, $\omega_2 = \sqrt{2}$, $m_0 = -1.02$, $m_2 = -0.55$, $s_0 = 1.0$, $s_1 = 1.11$, $s_2 = 2.06$, $s_3 = 2.14$, $s_4 = 3.06 \subset A$.

Concerning nonchaotic strange attractors, see, e.g. [12, 13].

11.5 Multiscroll Chaotic Attractors in the PWL Van Der Pol System

This section presents a generalized van der Pol system with periodic forcing [10], which is capable of generating multiscroll chaotic attractors.

The equations of this generalized van der Pol system with periodic forcing are given by

$$\begin{cases} \varepsilon \dfrac{dx}{dt} = y - \left(\dfrac{x^2}{3} - x\right), \\ \dfrac{dy}{dt} = -\alpha x + b\cos(t). \end{cases} \tag{11.10}$$

The associated PWL system of Eq. (11.10), where α and ε are sufficiently small and $b > 0$, is expressed as [10]

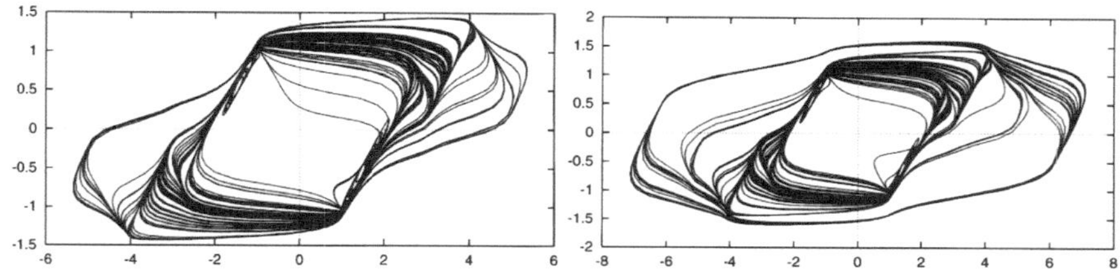

Fig. 11.7 Multiscroll chaotic attractors from the generalized van der Pol system (11.11). **a** Chaotic attractor, $N = 6$. **b** Chaotic attractor, $N = 8$. Reprinted from [10], with permission by World Scientific Pub. Co.

$$\begin{cases} \varepsilon \dfrac{dx}{dt} = y - f_2(x), \\ \dfrac{dy}{dt} = -\alpha x + b\cos(t), \end{cases} \tag{11.11}$$

where $f_2(x)$ is the same as defined as in the last section, with $m_0 = -1$ and $m_1 = 1$ here. When the parameters are set to $\alpha = \varepsilon = 0.167$ and $b = 0.59983$, system (11.10) has a positive Lyapunov exponent, indicating its chaotic nature. However, the parameter range corresponding to this chaotic attractor is quite narrow, and the attractor is highly sensitive to small variations, which can easily destabilize its dynamical structure.

To generate n-scroll chaotic attractors, Aziz-Alaoui [10] further modified system (11.11) as

$$\begin{cases} \varepsilon \dfrac{dx}{dt} = y - f_N(x), \\ \dfrac{dy}{dt} = -\alpha x + b\cos(t), \end{cases} \tag{11.12}$$

where $f_N(x)$ is defined by Eq. (11.9), and the parameter set $B_N^V = \{\alpha, \epsilon, b\} \cup B_N \subset \mathbf{R}^{2N+2}$, with $B_N = \left\{(s_k)_{k\in I_{N-2}}, (m_k)_{k\in I_{N-1}}\right\} \subset \mathbf{R}^{2N-1}$. Here, set $m_0 = m_{2i} = \cdots = -1.0$ and $m_1 = m_{2i+1} = \cdots = 1.0$, $i = 1, 2, \ldots$.

Figure 11.7a shows a chaotic attractor for $N = 6$ with the parameters $s_1 = 2.7$, $s_2 = 3.0482$, $s_3 = 4.028$, and $s_4 = 4.975$ [10]. In contrast, Fig. 11.7b shows a chaotic attractor for $N = 8$ with the parameters $s_1 = 2.7$, $s_2 = 3.0482$, $s_3 = 4.028$, $s_4 = 4.975$, $s_5 = 5.2$, and $s_6 = 6.1$ [10].

11.6 Multiscroll Chaotic Attractors from the PWL-Duffing System

This section presents a generalized PWL-Duffing system, constructed to generate multiscroll chaotic attractors. By combining the nonlinear function in the classical Duffing oscillator with a PWL function, the PWL-Duffing system exhibits intricate dynamical behaviors, including multiscroll attractors.

The classical Duffing system is represented by

$$\begin{cases} \dot{x} = y, \\ \dot{y} = x - x^3 - \varepsilon y + \gamma \cos(\omega t), \end{cases} \tag{11.13}$$

where x and y are functions of time t. With $\varepsilon = 0.25$, $\gamma = 0.3$, and $\omega = 1.0$, system (11.13) generates a chaotic attractor.

The simplest PWL-Duffing system, introduced by Aziz-Alaoui [10], is constructed by substituting the nonlinear term $x^3 - x$ with $f_2(x)$. This results in the following system:

$$\begin{cases} \dot{x} = y, \\ \dot{y} = -f_2(x) - \varepsilon y + \gamma \cos(\omega t), \end{cases} \tag{11.14}$$

where $f_2(x)$ is the same as defined above. For the parameter values $\varepsilon = 0.25$, $\gamma = 0.3$, $\omega = 1.0$, $m_0 = -0.0845$, and $m_1 = 0.66$, system (11.14) can generate chaotic attractors.

To generate n-scroll chaotic attractors, Aziz-Alaoui [10] further modified system (11.14) as follows:

$$\begin{cases} \dot{x} = y, \\ \dot{y} = -f_N(x) - \varepsilon y + \gamma \cos(\omega t), \end{cases} \tag{11.15}$$

where $f_N(x)$ is defined by Eq. (11.9).

Figure 11.8a shows a 4-scroll chaotic attractor with parameters $\varepsilon = 0.25$, $\gamma = 0.3$, $\omega = 1.0$, $m_0 = m_2 = -0.0845$, $m_1 = m_3 = 0.66$, and segment values $s_1 = 1.4$, $s_2 = 2.7$. In contrast, Fig. 11.8b displays a 6-scroll chaotic attractor, with the same values of ε, γ, and ω, but now $m_0 = m_2 = m_4 = -0.0845$ and $m_1 = m_3 = m_5 = 0.66$, along with segment values $s_1 = 1.4$, $s_2 = 2.7$, $s_3 = 3.67$, and $s_4 = 4.5$.

11.7 Multiscroll Multiwing Chaotic Attractors from Oscillatory Potential Wells

The oscillatory potential well can drive chaotic motion in mass particles confined within it, offering a physical mechanism for chaos generation. Building on this concept, Cheng et al. [14] used an oscillatory multi-well potential with several

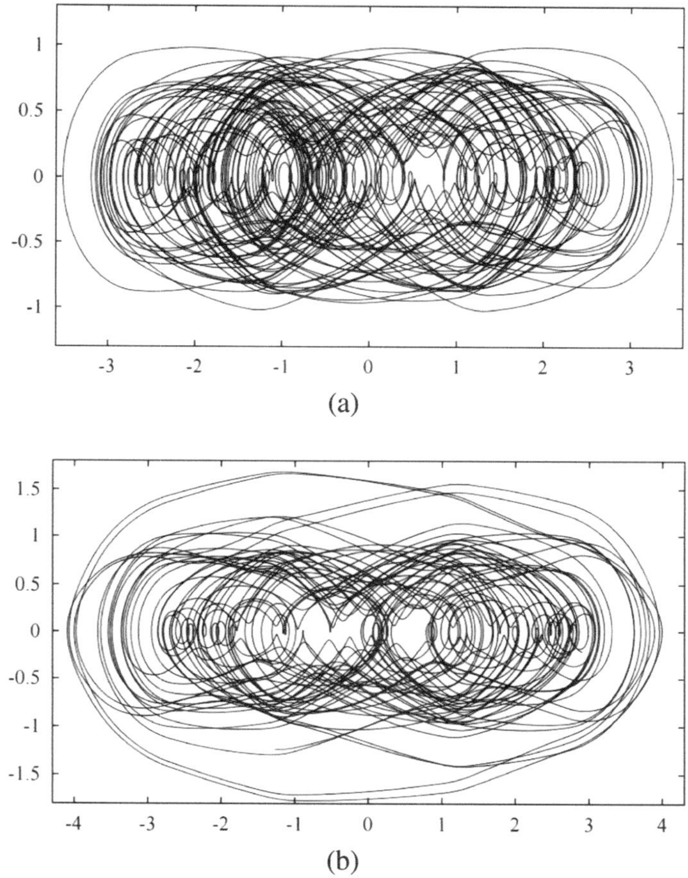

Fig. 11.8 Multiscroll chaotic attractors from the PWL-Duffing system. **a** 4-scroll chaotic attractor. **b** 6-scroll chaotic attractor. Reprinted from [11], with permission by the IEEE

concave regions to generate controllable multiscroll multiwing chaotic attractors. These chaotic attractors consist of two distinct topological components: the scroll and the wing. The interaction of these components creates the intricate geometry of the chaotic attractors, enabling precise control and manipulation of complex dynamics.

This system is rdescribed by

$$
\begin{aligned}
\dot{x} &= y, \\
\dot{y} &= -\gamma\, y + (-a\, x + b\, \sin x) \cdot \left(A \cos^2 z + d\right), \\
\dot{z} &= \omega,
\end{aligned}
\tag{11.16}
$$

where a, b, A, d, γ, and ω are real parameters.

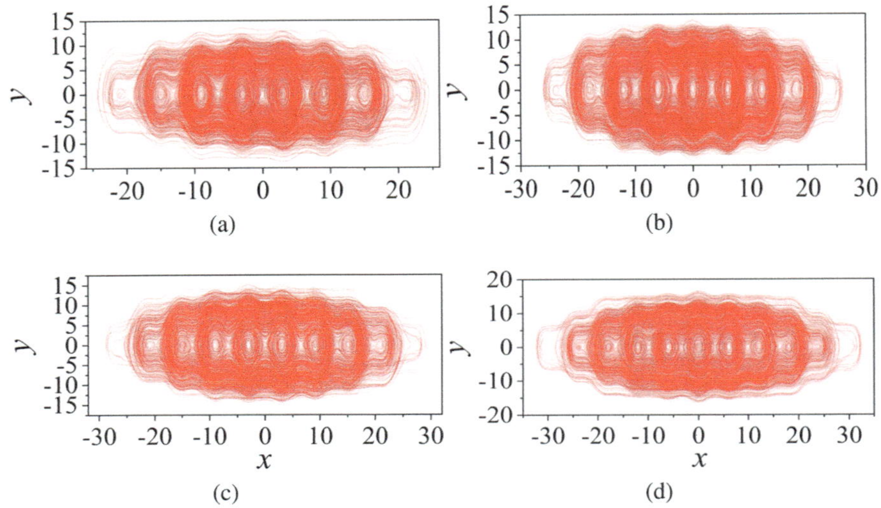

Fig. 11.9 Chaotic attractors. **a** 8-scroll chaotic attractor with $A = 12.5$, $\gamma = 0.04$. **b** 9-scroll chaotic attractor with $A = 11$, $\gamma = 0.0014$. **c** 8-scroll chaotic attractor with $A = 12.5$, $\gamma = 0.01$. **d** 6-scroll chaotic attractor with $A = 12.5$, $\gamma = 0.004$. Reprinted from [14], with permission by Elsevier

A single-scroll attractor corresponds to the potential with one well. However, the addition of two wings arises from the concave curves on both sides of the well. This suggests that wings can form without requiring equilibrium points, allowing for complex dynamics to emerge even in the absence of equilibrium states of such systems.

As the number of potential wells increases, the structure of the attractor becomes more intricate, leading to the formation of multiscroll chaotic attractors. The amplitude A and the damping coefficient γ play a significant role in determining the number of scrolls in the chaotic attractor. Figure 11.9 presents the phase diagrams of chaotic attractors with varying numbers of scrolls. As the number of scrolls increases, the complexity of the attractor also increases, reflecting a more intricate dynamic behavior.

To generate additional wings, the system with an oscillatory potential well is reformulated as

$$\begin{aligned} \dot{x} &= y, \\ \dot{y} &= -\gamma\, y + (-1.1\ \tanh(0.25x) + \sin x) \cdot \left(A \cos^4 z + d\right), \\ \dot{z} &= \omega, \end{aligned} \tag{11.17}$$

where A, d, γ, and ω are real parameters.

Figure 11.10 displays double-scroll chaotic attractors with 4-wings, 6-wings and 22-wings, respectively.

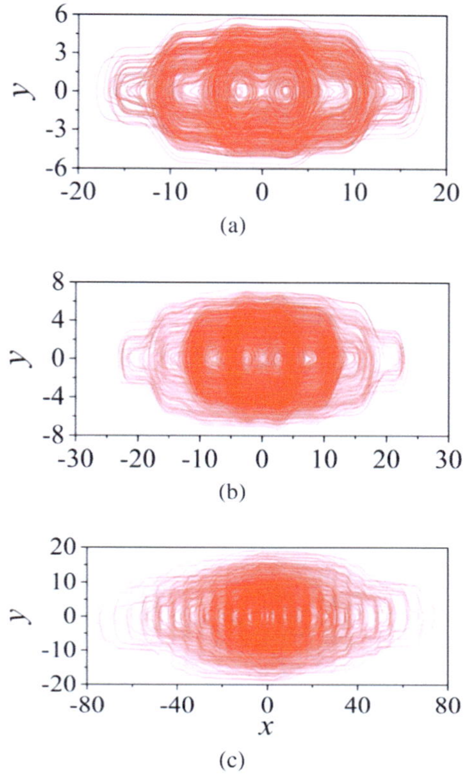

Fig. 11.10 Chaotic attractors with $\gamma = 0.01$, $\omega = 1$, and $d = 0.1$. **a** Double-scroll chaotic attractor with 4-wing, $A = 2$. **b** Double-scroll chaotic attractor with 6-wing, $A = 2.72$. (c) Double-scroll chaotic attractor with 8-wing, $A = 6$. Reprinted from [14], with permission by Elsevier

To facilitate the circuit implementation of the multi-scroll chaotic attractor with multiple wings, redefine $f_1(x)$ and $f_2(x)$ using a saturation function and a sawtooth wave function, respectively, as follows:

$$\begin{aligned} f_1(x) &= \tfrac{a}{2b}\left(|x+b| - |x-b|\right), \\ f_2(x) &= -x + 0.5 \sum_{i=-n}^{n} \operatorname{sgn}(x+i). \end{aligned} \tag{11.18}$$

These functions simplify the circuit design and implementation.

The Duffing equation with an oscillatory potential well is expressed as

$$\begin{aligned} \dot{x} &= y, \\ \dot{y} &= -\gamma y + (-f_1(x) + f_2(x)) \cdot A\ (\cos 2\pi f t + d). \end{aligned} \tag{11.19}$$

In this formulation, the oscillation factor is given by $A\,(\cos(2\pi f t) + d)$, which simplifies the circuit design by reducing the number of analog multipliers needed. Here, d is a constant greater than 1.

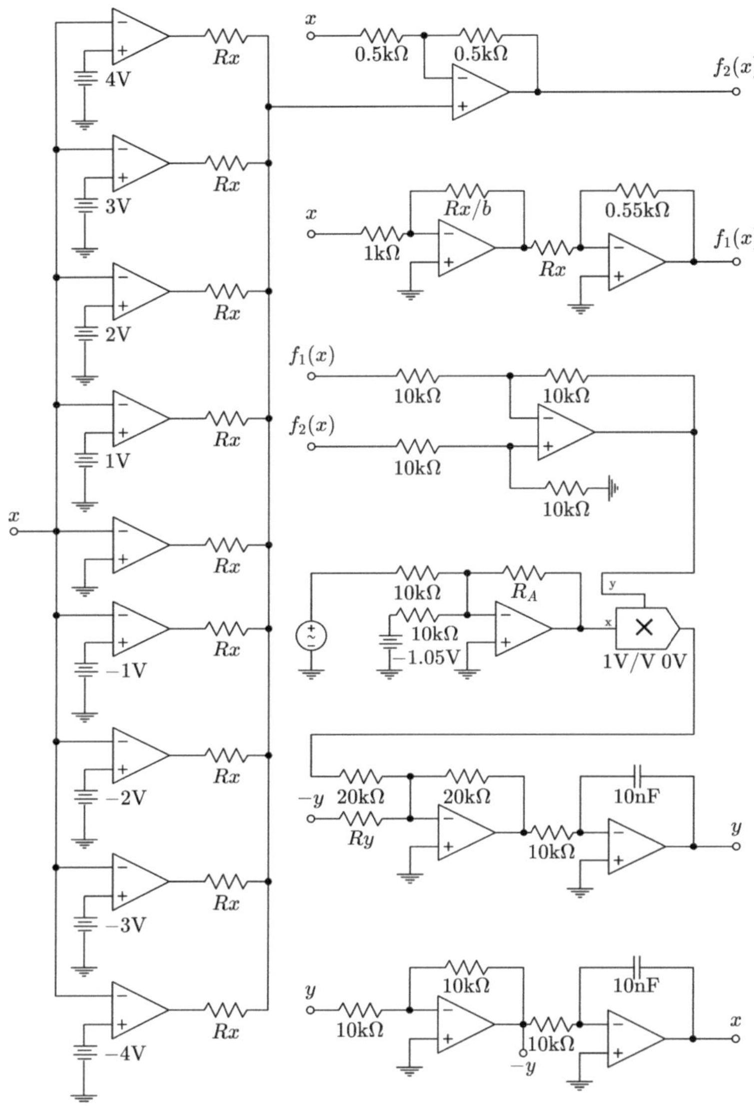

Fig. 11.11 Multisim simulation circuit for Eq. (11.19). Reprinted from [14], with permission by Elsevier

Figure 11.11 presents the Multisim simulation circuit for Eq. (11.19), with parameters $a = 0.55$ and $b = 0.9$ for a double-well potential, and $b = 1.9$ for a 4-well potential. The system exhibits a scroll count of $n = 4$, with $d = 1.05$. The resistance is $R_x = E_s$ (kΩ), where E_s denotes the saturation voltage of the operational amplifier, which is used to control the oscillation amplitude.

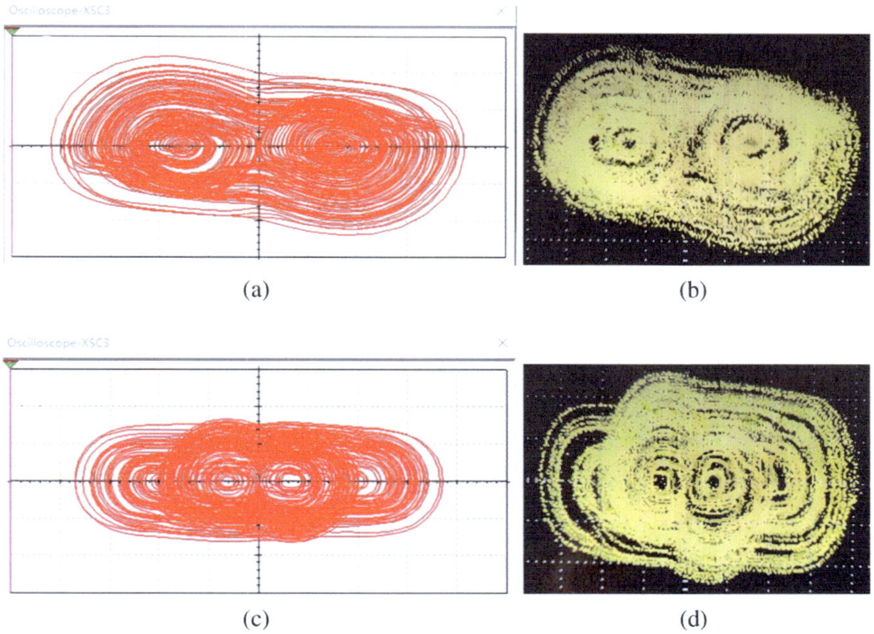

Fig. 11.12 Multisim simulation circuit and analog circuit results. **a** Double-scroll chaotic attractor from the Multisim simulation circuit with $A = 2$V, $f = 5$ kHz, $\gamma = 0.3$. **b** Double-scroll chaotic attractor with double-wings from the analog circuit with $A = 2$V, $f = 5$ kHz, $\gamma = 0.3$. **c** Double-scroll chaotic attractor from the Multisim simulation circuit with $A = 2$V, $f = 5$ kHz, $\gamma = 0.1$. (d) Double-scroll chaotic attractor with double-wings from the analog circuit with $A = 2$V, $f = 5$ kHz, $\gamma = 0.1$. Reprinted from [14], with permission by Elsevier

The oscillation amplitude A of the potential well and the damping coefficient γ can be controlled by the resistors R_A and R_γ, respectively. This provides the ability to fine-tune the chaotic attractor's characteristics.

An analog circuit is constructed using operational amplifiers OP275, analog multipliers AD633JN, and other standard electronic components. This hardware implementation verify and demonstrate the feasibility of generating multiscroll chaotic attractors from physical circuits, showcasing the effectiveness of the methodology.

For the double-well system, Fig. 11.12 presents the double-scroll chaotic attractor and the double-scroll double-wing chaotic attractor.

11.8 Multiscroll Chaotic Attractors Generated via Thresholding

Lü et al. [15] introduced a thresholding technique for generating multi-scroll chaotic attractors. A general jerk circuit and Chua's circuit with a sine function are used as examples to demonstrate the effectiveness of the methodology. By incorporating thresholding into the system, multi-scroll chaotic attractors can be effectively generated.

Consider a general n-D controlled system, described by

$$\dot{X} = F(X, t) + H(X), \tag{11.20}$$

where $X = (x_1, x_2, \ldots, x_n)^T \in \mathbf{R}^n$ represents the state vector, $\dot{X} = F(X, t)$ defines the dynamics of the system, and the function $H(X)$ acts as a feedback threshold controller. When the input function $\varphi(X)$ exceeds a specified threshold h_0, the controller is activated and the system output is reset to a new state X^*.

The threshold controller $H(X)$ depends on the dynamics of the original system $\dot{X} = F(X, t)$, and it operates based on these dynamics. While the algebraic forms of $F(X, t)$ and $H(X)$ can vary, making it difficult to define a general framework for analyzing the controller's behavior. There are certain types of nonlinear controllers, such as PWL or triangular functions, which allow for specific design for different purposes.

The key idea behind thresholding is to "clip" both linear and nonlinear functions, including input and output variables, to reach the desired system states through thresholding. The distinguishing features of this method include the following:

(i) The saturation function series increases linearly between adjacent saturation levels.
(ii) The function has a staircase-like structure.
(iii) The series is based on independent variables, or their linear combinations, with the threshold controller designed using either independent or dependent variables as well as their certain functions.

This multivariate approach facilitates the manipulation and control of the system to achieve the intended chaotic or controlled system states.

Now, consider a general jerk circuit, described by

$$\dddot{x} + c\ddot{x} + b\dot{x} + ax = G(x), \tag{11.21}$$

where the jerk dynamics are described using real parameters a, b, and c, along with a nonlinear threshold function $G(x)$.

The system's derivatives with respect to time τ include the velocity $\dot{x} = dx/d\tau$, the acceleration $\ddot{x} = d^2x/d\tau^2$, and the jerk $\dddot{x} = d^3x/d\tau^3$, which characterizes the rate of change of acceleration. The time scale τ is related to the time variable t with a transformation factor $1/R_0C_0$, where R_0 and C_0 represent the resistance and capacitance in the circuit, respectively. This factor serves as the integration constant for the circuit's integrator, controlling the system's dynamics and its temporal evolution.

To generate multi-scroll chaotic attractors from Eq. (11.21), additional thresholds are set. Specifically, the threshold function is divided into multiple sections with high precision. In a more generalized formulation, the function $G_1(x; k_1, k_2, k_3, k_4)$ is introduced, as

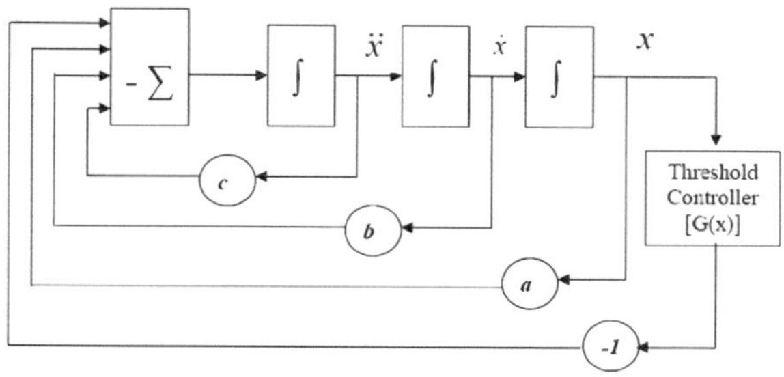

Fig. 11.13 Diagrammatic sketch of the threshold-control system (11.23). Reprinted from [15], with permission by Elsevier

$$G_1(x; k_1, k_2, k_3, k_4) = \begin{cases} k_4 - k_3, & \text{if} \quad x > k_4, \\ x - k_3, & \text{if} \quad k_3 \le x \le k_4, \\ 0, & \text{if} \quad k_2 < x < k_3, \\ x - k_2, & \text{if} \quad k_1 \le x \le k_2, \\ k_1 - k_2, & \text{if} \quad x < k_1, \end{cases} \tag{11.22}$$

where $k_1 < k_2 < k_3 < k_4$ represent the lower, middle-left, middle-right, and upper threshold values, respectively.

Putting together, the system described by Eq. (11.21) with the threshold controller Eq. (11.22) is as follows:

$$\dddot{x} + c\ddot{x} + b\dot{x} + ax = G_1(x). \tag{11.23}$$

System (11.23) can be separated into five distinct regions or subspaces: $V_1 = \{X \mid x \ge k_4\}$, $V_2 = \{X \mid k_3 \le x \le k_4\}$, $V_3 = \{X \mid k_2 \le x \le k_3\}$, $V_4 = \{X \mid k_1 \le x \le k_2\}$, $V_5 = \{X \mid x \le k_1\}$, where $X = (x, \dot{x}, \ddot{x})^T \in \mathbf{R}^3$.

The system demonstrates inherent symmetry when $k_1 = -k_4$ and $k_2 = -k_3$, under the transformation of coordinates $(x, \dot{x}, \ddot{x}) \rightarrow (-x, -\dot{x}, -\ddot{x})$. Moreover, system (11.23) remains dissipative in each subspace for $c > 0$, ensuring stability within these regions.

Figure 11.13 shows the schematic structure of the proposed threshold control system (11.23), illustrating the fundamental process of its implementation. In this setting, three lossless integrators are connected in series, and a summing amplifier is used to establish a feedback loop. The threshold controller is then incorporated to realize the function $G(x)$.

Figure 11.14a shows the a 1-scroll chaotic attractor on the $(x, \dot{x})$ plane, with the threshold control voltages set to $B_1 = 0.3$ V and $B_2 = -0.7$ V. A slight modification of the threshold control voltage B_2 to -0.3 V results in a 2-scroll chaotic attractor, as shown in Fig. 11.14b.

For voltage levels of $B_1 = +1.3$ V, $B_2 = -1.3$ V, $B_3 = -0.3$ V, and $B_4 = +0.3$ V, a 3-scroll chaotic attractor is generated, as shown in Fig. 11.14c. Experimental results confirm that the system is highly resistant to small perturbations or noise, demonstrating its robustness.

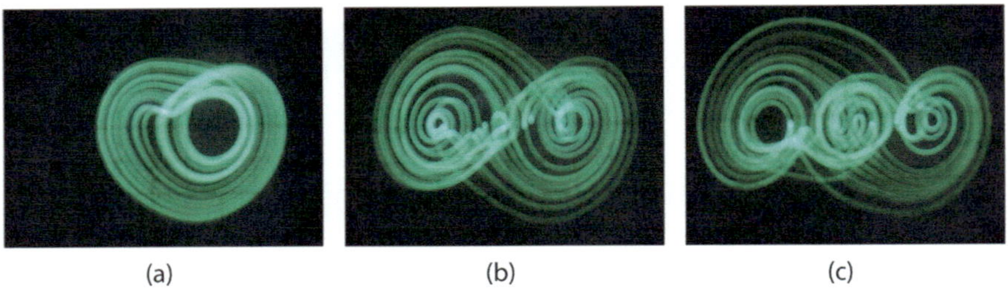

Fig. 11.14 Experimental observations of n-scroll chaotic attractors on the x-$\dot{x}$ plane. **a** $n = 1$. **b** $n = 2$. **c** $n = 3$. Reprinted from [15], with permission by Elsevier

11.9 Multiscroll Chaotic Attractors from SAF-TLSN Model

Bao et al. [16] presented a 2-D nonautonomous tabu learning single neuron (TLSN) model incorporating a sinusoidal activation function (SAF). This model is capable of generating various multiscroll chaotic attractors, with the number of scrolls being controllable through some parameters. The SAF-based TLSN (SAF-TLSN) model includes a periodically varying equilibrium point, influenced by the model parameters. As a result, the equilibrium trajectories can exhibit different types of stability, distributing across the phase plane and eventually forming multiscroll chaotic attractors.

Let x represent the neuron firing state and y denote the tabu learning state. The mathematical model of the TLSN [17] is described by

$$\begin{aligned} C\dot{x} &= -x/R + bf(x) + y + I(t), \\ \dot{y} &= -cy - df(x), \end{aligned} \tag{11.24}$$

where $C = 1$, and $a = 1/R$, b, c and d are four control parameters. The function $f(x)$ represents an activation function, bounded from above and below, and $I(t)$ denotes an external stimulus current.

To demonstrate the complex range of neuron firing patterns, referred to as attractors, a sinusoidal activation function is used:

$$f(x) = \sin(x).$$

The external stimulus current is mathematically expressed as $I(t) = H\sin(2\pi Ft)$, where H represents the amplitude and F is the frequency. Consequently, the TLSN model is reformulated as

$$\begin{aligned} \dot{x} &= -x/R + b\sin(x) + y + H\sin(2\pi Ft), \\ \dot{y} &= -cy - d\sin(x). \end{aligned} \tag{11.25}$$

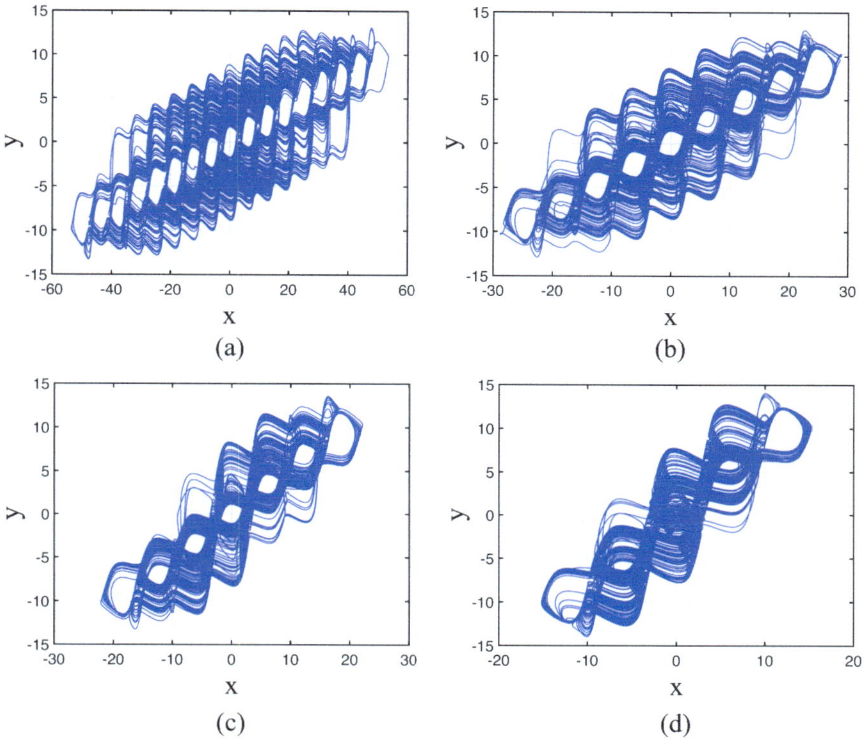

Fig. 11.15 Multiscroll chaotic attractors. **a** 15-scroll chaotic attractor. **b** 9-scroll chaotic attractor. **c** 7-scroll chaotic attractor. **d** 5-scroll chaotic attractor

This model uses an activation function that significantly differs from traditional TLSN models. By incorporating a sinusoidal activation function, an SAF-TLSN model is designed to generate a variety of multiscroll chaotic attractors, with the ability to control the number of scrolls produced. This SAF-TLSN model enhances the dynamical behavior and extends its potential applications in chaotic systems and neural network dynamics.

With $b = 3.7$, $c = 0.1$, $d = 2.3$, $H = 4$, $F = 0.09$, and initial conditions $(x(0), y(0)) = (0, 0)$, the control parameter a is adjusted to values such as $a = 0.2, 0.4, 0.6$, and 1, the SAF-TLSN model continues to generate multiscroll chaotic attractors with the varying control parameter a [16]. The results are demonstrated in Fig. 11.15. Specifically, Fig. 11.15a shows a 15-scroll chaotic attractor with $a = 0.2$, Fig. 11.15b shows a 9-scroll chaotic attractor with $a = 0.4$, Fig. 11.15c shows a 7-scroll chaotic attractor with $a = 0.6$, and Fig. 11.15d shows a 5-scroll chaotic attractor with $a = 1$. It is noted that, as the control parameter a increases, the scrolls at both ends of the chaotic attractors in Fig. 11.15c and d become significantly degraded.

When the system parameters and initial values are set to $a = 0.5$, $b = 3.7$, $d = 2.3$, $H = 4$, $F = 0.09$, and $(x(0), y(0)) = (0, 0)$, the SAF-TLSN model produces

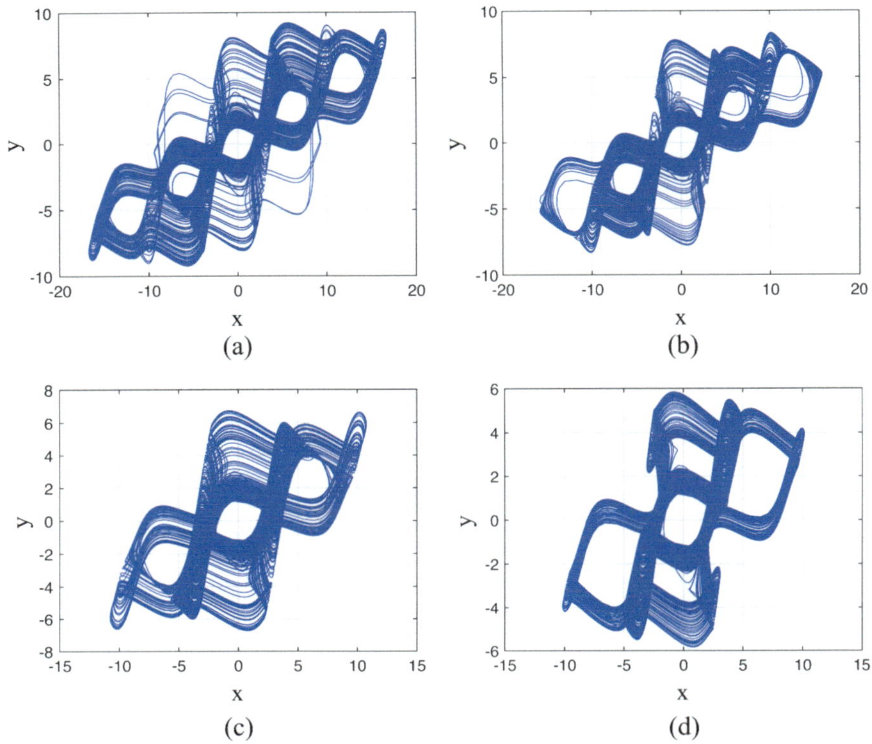

Fig. 11.16 Multiscroll chaotic attractors. **a** 5-scroll chaotic attractor. **b** Degenerated 5-scroll chaotic attractor. **c** 3-scroll chaotic attractor. **d** A variant of 3-scroll chaotic attractor

four distinct multi-scroll chaotic attractors by selecting four different values for another control parameter c: 0.15, 0.18, 0.24, and 0.28 [16]. Figure 11.16a and b display 5-scroll chaotic attractors at $c = 0.15$ and $c = 0.18$, respectively. Notably, the scrolls at both ends of the chaotic attractor shown in Fig. 11.16b appear degraded and seemingly incomplete. Moreover, Fig. 11.16c and d show two 3-scroll chaotic attractors with $c = 0.24$ and $c = 0.28$, respectively, whereas the scrolls in Fig. 11.16d appear with considerable variations.

The SAF-TLSN model is a 2-D nonautonomous system, for which the plane defined by $I(t) = 0$ can be used as the cross-sectional view to observe the positions of the chaotic scrolls and the dynamic state of the attractor. Numerical simulations are conducted on the time interval [100, 20000], representing the multiscroll chaotic attractors shown on the x-z plane in Fig. 11.15 to be viewed on the x-y plane in Fig. 11.17. In Fig. 11.17, (a) $a = 0.2$; (b) $a = 0.4$; (c) $a = 0.6$; (d) $a = 1$. These cross-sections offer valuable insights into the spatial arrangement and interaction of the multiscrol chaotic attractors.

The discrete-time version of the SAF-TLSN model is formulated as follows:

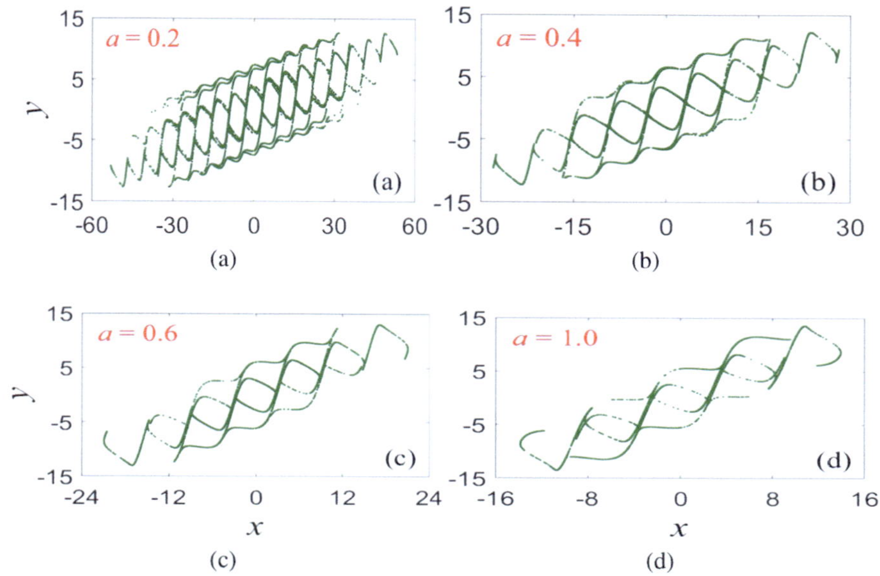

Fig. 11.17 Multiscroll chaotic attractors on the x-z plane in Fig. (11.15) is displayed on the x-y plane here. **a** 15-scroll chaotic attractor. **b** 9-scroll chaotic attractor. **c** 7-scroll chaotic attractor. **d** 5-scroll chaotic attractor. Reprinted from [16], with permission by Elsevier

$$
\begin{aligned}
x_{k+1} &= x_k\left(\bar{x}_1 + 2\bar{x}_2 + 2\bar{x}_3 + \bar{x}_4\right)/6, \\
y_{k+1} &= y_k\left(\bar{y}_1 + 2\bar{y}_2 + 2\bar{y}_3 + \bar{y}_4\right)/6,
\end{aligned}
\tag{11.26}
$$

where x_j and y_j, $j = 1, 2, 3, 4$, represent intermediate variables, defined by

$$
\begin{aligned}
\bar{x}_1 &= -ax_k + b\sin\left(x_k\right) + y_k + I\left(k\right), \\
\bar{x}_2 &= -a\left(x_k + \tfrac{1}{2}h\bar{x}_1\right) + b\sin\left(x_k + \tfrac{1}{2}h\bar{x}_1\right) + y_k + \tfrac{1}{2}h\bar{y}_1 + I\left(k\right), \\
\bar{x}_3 &= -a\left(x_k + \tfrac{1}{2}h\bar{x}_2\right) + b\sin\left(x_k + \tfrac{1}{2}h\bar{x}_2\right) + y_k + \tfrac{1}{2}h\bar{y}_2 + I\left(k\right), \\
\bar{x}_4 &= -a\left(x_k + \tfrac{1}{2}h\bar{x}_3\right) + b\sin\left(x_k + h\bar{x}_3\right) + y_k + h\bar{y}_3 + I\left(k\right), \\
\bar{y}_1 &= -cy_k - d\sin\left(x_k\right), \\
\bar{y}_2 &= -c\left(y_k + \tfrac{1}{2}h\bar{y}_1\right) - d\sin\left(x_k + \tfrac{1}{2}h\bar{x}_1\right), \\
\bar{y}_3 &= -c\left(y_k + \tfrac{1}{2}h\bar{y}_2\right) - d\sin\left(x_k + \tfrac{1}{2}h\bar{x}_2\right), \\
\bar{y}_4 &= -c\left(y_k + h\bar{y}_3\right) - d\sin\left(x_k + h\bar{x}_3\right).
\end{aligned}
\tag{11.27}
$$

For the discrete-time SAF-TLSN model (11.26), the sampling step is chosen as $h = 0.01$. The model parameters and initial conditions are set as $b = 3.7$, $c = 0.1$, $d = 2.3$, $H = 4$, $F = 0.09$, with the initial values $(x(0), y(0)) = (0, 0)$. In the experimentation, four distinct values for a are selected: $a = 1.2, 0.8, 0.5$, and 0.3.

The experimental results, including phase plane portraits obtained from an MCU-based hardware platform and captured on a digital oscilloscope, are shown in Fig. 11.18. To ensure the correct operation of the DAC8563 converter, the volt-

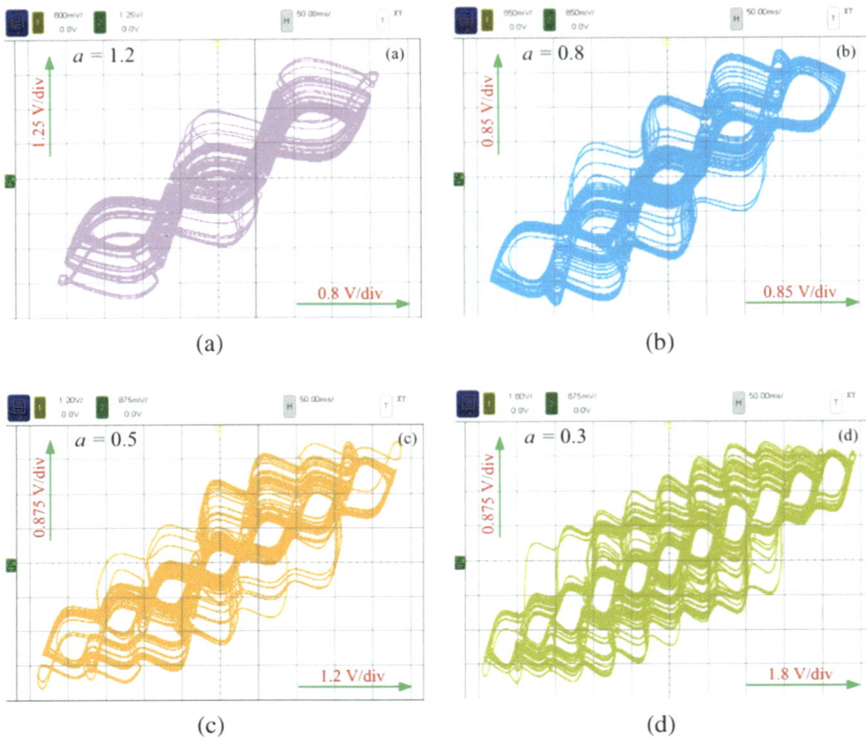

Fig. 11.18 Experimentally measured results of the multiscroll chaotic attractors from the discrete-time SAF-TLSN model. **a** $a = 1.2, b = 3.7, c = 0.1, d = 2.3, H = 4, F = 0.09$. **b** $a = 0.8, b = 3.7, c = 0.1, d = 2.3, H = 4, F = 0.09$. **c** $a = 0.5, b = 3.7, c = 0.1, d = 2.3, H = 4, F = 0.09$. **d** $a = 0.3, b = 3.7, c = 0.1, d = 2.3, H = 4, F = 0.09$. Reprinted from [16], with permission by Elsevier

age sequence dynamic ranges generated by the discrete-time SAF-TLSN model are scaled. Specifically, the dynamic ranges in Fig. 11.18a are scaled by a factor of 3, while those in Fig. 11.18b–d are scaled by a factor of 4.

The experimental results demonstrate the feasibility and effectiveness of the MCU-based hardware platform in generating and analyzing multiscroll chaotic attractors from the discrete-time SAF-TLSN model.

11.10 Generating Multi-grid Double-Scroll Chaotic Attractors from the Rucklidge System

Chen et al. [18] introduced a method for generating $n \times m$-grid double-scroll chaotic attractors from the Rucklidge system. This method utilizes an arctangent function

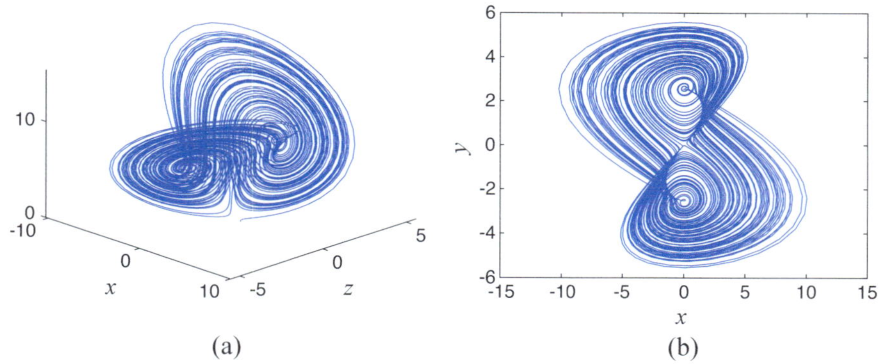

Fig. 11.19 Chaotic attractor from system (11.28). **a** x-y-z space. **b** x-y plane

series to create a nonlinear function to replace the state variables x and y in the original Rucklidge system. This modification leads to the development of a new $n \times m$-grid double-scroll chaotic system, facilitating the study of more intricate dynamical behaviors and attractor patterns.

The Rucklidge chaotic system [18] is described by

$$\begin{cases} \dot{x} = -ax + by - cyz, \\ \dot{y} = dx, \\ \dot{z} = y^2 - z, \end{cases} \tag{11.28}$$

where a, b, c, d are positive constants. With the values of $a = 2, b = 6.7, c = 1$, and $d = 1$, system (11.28) generates a chaotic attractor as shown in Fig. 11.19.

Now, introduce the following arctangent function series:

$$\begin{aligned} g(x, p, n, M, N) &= \frac{2}{\pi} P \begin{bmatrix} \sum_{I=1}^{M} \operatorname{atan}(n(x + 2ip)) - M\frac{\pi}{2} \\ + \sum_{I=1}^{N} \operatorname{atan}(n(x + 2ip)) - N\frac{\pi}{2} + \operatorname{atan}(nx) \end{bmatrix}, \\ &= \frac{2}{\pi} P \sum_{I=1}^{N} \operatorname{atan}(n(x + 2ip)) - (N - M) P, \end{aligned} \tag{11.29}$$

where $\operatorname{atan}(x)$ is the arctangent function, $n > 0$, M, and N are nonnegative integers, and $p > 0$ represents the delay time.

To generate $n \times m$-grid double-scroll chaotic attractors, nonlinear control is applied to the system. Specifically, the arctangent function series are used as the continuous nonlinear controllers. The controlled state variables x and y are thus replaced by the nonlinear functions $f(x, p_1, n_1, M_1, N_1)$ and $f(y, p_2, n_2, M_2, N_2)$, respectively, where

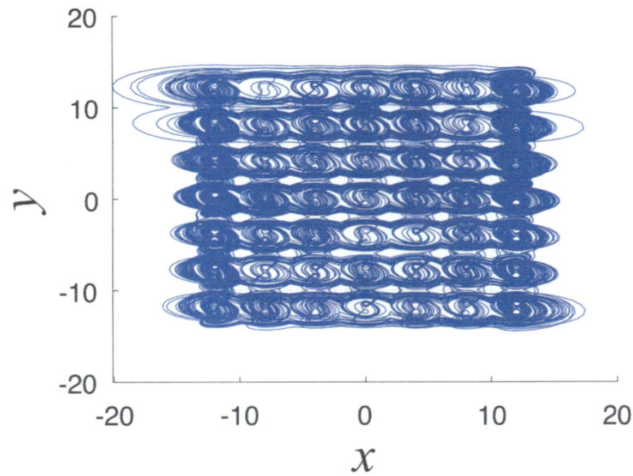

Fig. 11.20 7 × 7-grid double-scroll chaotic attractor generated from system (11.31) with arctangent function series

$$f(x, p, n, M, N) = x - g(x, p, n, M, N). \tag{11.30}$$

For simplicity of discussion, consider the following dynamical system:

$$\begin{cases} \dot{x} = -af(x, p_1, n_1, M_1, N_1) + (b - cz) f(y, p_2, n_2, M_2, N_2), \\ \dot{y} = df(x, p_1, n_1, M_1, N_1), \\ \dot{z} = f(y, p_2, n_2, M_2, N_2)^2 - z, \end{cases} \tag{11.31}$$

where a, b, c, d are positive constants, and $f(x, p_1, n_1, M_1, N_1)$, $f(y, p_2, n_2, M_2, N_2)$ are defined by Eq. (11.30); that is, $f(x, p_1, n_1, M_1, N_1) = x - g(x, p_1, n_1, M_1, N_1)$ and $f(y, p_2, n_2, M_2, N_2) = y - g(y, p_2, n_2, M_2, N_2)$.

With parameters $a = 2, b = 10, c = 20, d = 4, p_1 = p_2 = 2, n_1 = n_2 = 100, M_1 = N_1 = M_2 = N_2 = 2$, a 7 × 7-grid double-scroll chaotic attractor generated from system (11.31) with $M = N$, as shown in Fig. 11.20.

With parameters $a = 2, b = 10, c = 20, d = 4, p_1 = p_2 = 2, n_1 = n_2 = 100$ and $M_1 = N_1 = 4$, a 10 × 10-grid double-scroll chaotic attractor is generated from system (11.31), as shown in Fig. 11.21.

11.11 Generating Multiscroll Chaotic Attractors from Hopfield Neural Networks

Lin et al. [19] introduced a method for generating n-scroll chaotic attractors from a magnetized Hopfield Neural Network (HNN) with three neurons. This model integrates a multi-piecewise memristor to account for the electromagnetic induction effects.

The HNN is described by

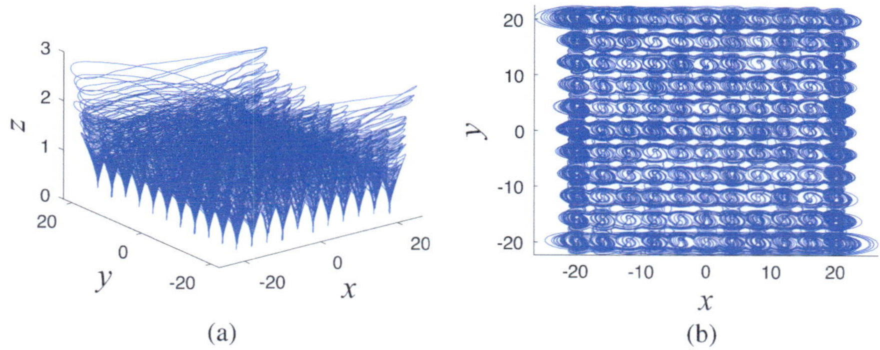

Fig. 11.21 10×10-grid double-scroll chaotic attractor generated from system (11.31) with arctangent function series. **a** x-y-z space. **b** x-y plane

Fig. 11.22 Structure of the three-neuron HNN. Reprinted from [19], with permission by the IEEE

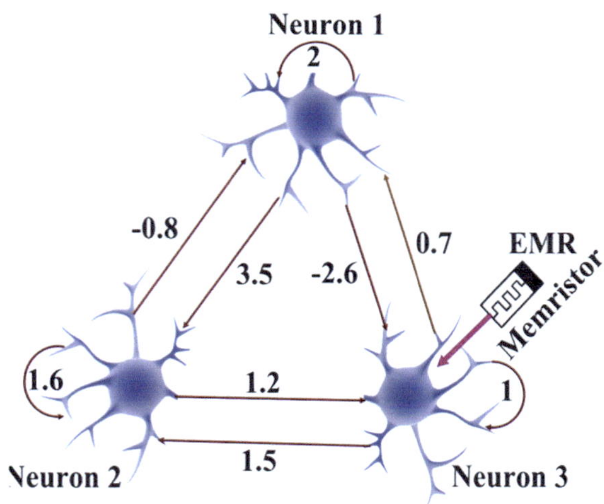

$$C_i \dot{x}_i = -\frac{x_i}{R_i} + \sum_{i}^{n} w_{ij} \tanh\left(x_j\right) + I_i, \quad i, j \in N^*, \tag{11.32}$$

where x_i represents the membrane potential of neuron i, C_i, and R_i denote the corresponding membrane capacitance and resistance, respectively; w_{ij} are the synaptic weight coefficients, which control the coupling strength from neuron j to neuron i; $\tanh(\cdot)$ is the activation function of the neuron, and I_i represents the external input current. For simplicity, the values of C_i, R_i, and I_i are set to 1, 1, and 0, respectively.

With appropriate values of w_{ij}, a three-neuron HNN is constructed, as illustrated in Fig. 11.22.

The mathematical representation of this system is given by

$$\begin{cases} \dot{x}_1 = -x_1 + 2\tanh(x_1) - 0.8\tanh(x_2) + 0.7\tanh(x_3), \\ \dot{x}_2 = -x_1 + 3.5\tanh(x_1) + 1.6\tanh(x_2) + 1.5\tanh(x_3), \\ \dot{x}_3 = -x_1 - 2.6\tanh(x_1) + 1.2\tanh(x_2) + \tanh(x_3), \end{cases} \tag{11.33}$$

where x_i represent the membrane potential of neuron i, $i = 1, 2, 3$.

As illustrated in Fig. 11.22, neuron 3 is exposed to electromagnetic radiation (EMR). To better capture this influence, an improved multi-piecewise flux-controlled memristor is introduced, as follows:

$$\begin{cases} i = W(\varphi)\, v, \\ \dot{\varphi} = cv - dh(\varphi), \\ W(\varphi) = a + b\,|\varphi|, \end{cases} \tag{11.34}$$

where a, b, c, and d are real parameters, $W(\phi)$ denotes the memductance, and $h(\phi)$ is the internal state variable function of the memristor model, given by

$$h(\varphi) = \begin{cases} h_1(\varphi) = \begin{cases} \varphi, & N = 0, \\ \varphi - \sum\limits_{i=1}^{N} \begin{pmatrix} \operatorname{sgn}(\varphi + (2i+1)) \\ +\operatorname{sgn}(\varphi - (2i+1)) \end{pmatrix}, & N = 1, 2, 3, \ldots, \end{cases} \\ h_2(\varphi) = \begin{cases} \varphi - \operatorname{sgn}(\varphi), & M = 0, \\ \varphi - \operatorname{sgn}(\varphi) - \sum\limits_{i=1}^{M} \begin{pmatrix} \operatorname{sgn}(\varphi + 2j) \\ +\operatorname{sgn}(\varphi - 2j) \end{pmatrix}, & M = 1, 2, 3, \ldots, \end{cases} \end{cases} \tag{11.35}$$

where N and M are control parameters of the memristor.

Unlike other memristor models, the memristor used here offers enhanced adjustability and diversity, based on the controllable parameters N and M. This feature is essential for generating multiscroll attractors. Following this, the magnetized HNN is modeled as follows:

$$\begin{cases} \dot{x}_1 = -x_1 + 2\tanh(x_1) - 0.8\tanh(x_2) + 0.7\tanh(x_3), \\ \dot{x}_2 = -x_1 + 3.5\tanh(x_1) + 1.6\tanh(x_2) + 1.5\tanh(x_3), \\ \dot{x}_3 = -x_1 - 2.6\tanh(x_1) + 1.2\tanh(x_2) + \tanh(x_3) - \mu W(\varphi)\, x_3, \\ \dot{\varphi} = cx_3 - dh(\varphi), \end{cases} \tag{11.36}$$

where $\mu W(\varphi)x_3$ represents the current generated by electromagnetic radiation, μ denotes the feedback strength of the electromagnetic induction current, φ is the magnetic flux passing through the membrane of neuron 3, the function $W(\varphi)$ characterizes the relationship between the magnetic flux and the membrane potential, the parameter c scales the influence of magnetic flux on the membrane potential, and $-dh(\varphi)$ accounts for the leakage of the magnetic flux.

For simplicity, consider the case where $h(\varphi)$ is evaluated with $N = 2$, while other parameters are fixed as $a = 1, b = 0.01, c = 1$, and $d = 1.2$. The equilibrium points of the system can be found by solving the following equations:

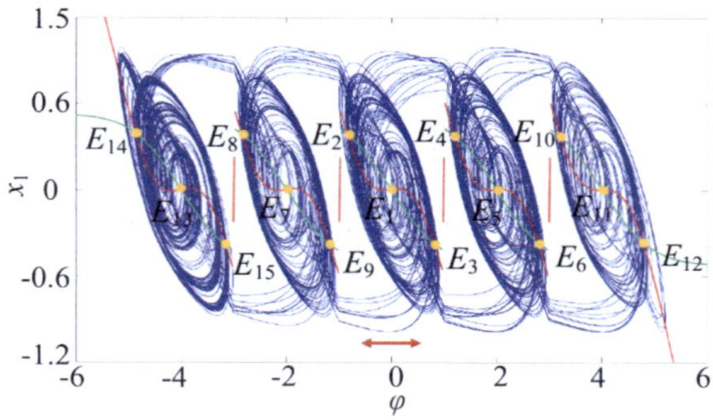

Fig. 11.23 Structure of the HNN with three neurons. Reprinted from [19], with permission by the IEEE

$$\begin{cases} \dot{x}_1 = dh(\varphi)/c, \\ \dot{x}_2 = \operatorname{atanh}(-x_1 + 2\tanh(x_1) + 0.7\tanh(x_3)), \\ f_1(\varphi, x_1) = -x_2 + 3.5\tanh(x_1) + 1.6\tanh(x_2) + 1.5\tanh(x_3) = 0, \\ f_2(\varphi, x_1) = -x_3 - 2.6\tanh(x_1) + 1.2\tanh(x_2) + \tanh(x_3) - \mu W(\varphi)x_3 = 0. \end{cases} \tag{11.37}$$

The graphical analytical method is used to calculate and plot the curves of the functions $f_1(\phi, x_1)$ (in green) and $f_2(\phi, x_1)$ (in red), as shown in Fig. 11.23.

Numerical simulations demonstrate that the magnetized HNN has the capability to generate an arbitrary number of multiscroll attractors. By setting the parameters $a = 1, b = 0.01, c = 1, d = 1.2, \mu = 1$, and initializing the system with state values $(0.1, 0.1, 0.1, 0.1)$, various n-scroll chaotic attractors can be generated by modifying the values of N or M in Eq. (11.35), as illustrated in Fig. 11.24. The number of scrolls created by the magnetized HNN is determined by $2N + 1$ and $2M + 2$, allowing precise control over the system's chaotic behavior.

To provide a more comprehensive analysis of the chaotic dynamics, Poincaré maps are constructed for the 8-scroll attractor on the ϕ-x_1 and ϕ-x_3 phase planes, respectively, under the condition of $x_2 = 0$. These maps, depicted in Fig. 11.25a and b, serve as 2-D cross-sections of the system in a higher-dimensional phase space. The intricate and irregular structures observed from these maps highlight the rich and complex behavior of the chaotic system. These patterns not only emphasize on the sensitivity of the system to the initial conditions but also reveal the nonlinear dynamical interactions within the chaotic attractors.

Moreover, to quantify the chaotic nature of the 8-scroll attractor, the first four Lyapunov exponents are calculated obtaining $0.1135, 0, -0.6442$, and -0.9825. As is well known, the presence of a positive Lyapunov exponent confirms the system's sensitivity to initial conditions, a hallmark of chaos, whereas the zero exponent corresponds to the neutral stability along the trajectory. The negative exponents indicate contraction in the corresponding directions in the phase space. Furthermore, the Kaplan-Yorke dimension, computed as 2.1762, provides a fractal dimension for the attractor, reflecting the system's complexity and the fractal-like structure of its

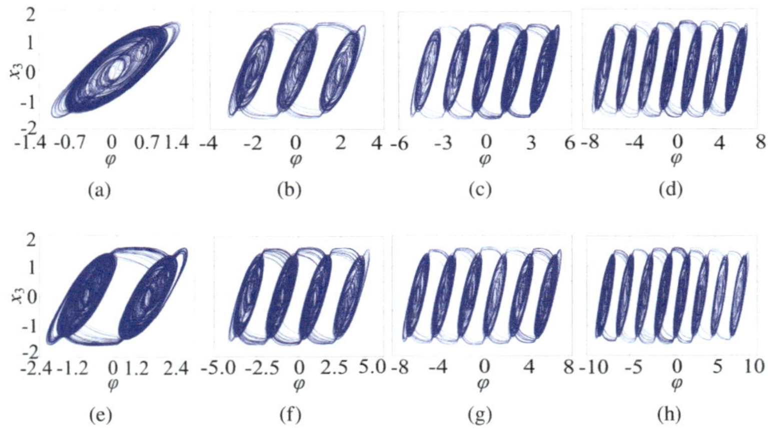

Fig. 11.24 n-scroll attractors with different values of N and M. **a** 1-scroll attractor, with $N = 0$. **b** 3-scroll attractor, with $N = 1$. **c** 5-scroll attractor, with $N = 2$. **d** 7-scroll attractor, with $N = 3$. **e** 2-scroll attractor, with $M = 0$. **f** 4-scroll attractor, with $M = 1$. **g** 6-scroll attractor, with $N = 2$. **h** 8-scroll attractor, with $N = 3$. Reprinted from [19], with permission by the IEEE

chaotic attractor. Together, these results offer a clear demonstration of the chaotic dynamics governing the 8-scroll attractor.

A memristor circuit is constructed, as depicted in Fig. 11.26, which is composed of a nonlinear function generator, an integration circuit, a multiplication circuit, and an absolute value function circuit.

Different circuit configurations are established using selection switches S_1, S_2, $\ldots$, S_n and control voltages $e_1, e_2, \ldots, e_i$. By adjusting these switches and control voltages, it is possible to generate multiscroll attractors with varying numbers of scrolls.

The circuit equations corresponding to the circuit shown in Fig. 11.26 are expressed as follows:

$$\begin{cases} i = \left(\dfrac{1}{R_a} + \dfrac{g \left| v_\varphi \right|}{R_b} \right) v, \\ C \dfrac{dv_\varphi}{dt} = \dfrac{v}{R_c} - \dfrac{h\left(v_\varphi\right)}{R_d}, \end{cases} \tag{11.38}$$

where $g = 0.1$ is the gain of the analog multiplier M, and the resistances are $R_a = R/a$, $R_b = gR/b$, $R_c = R/c$, and $R_d = R/d$.

Based on Eq. (11.36), the circuit for the magnetized HNN is designed, as shown in Fig. 11.27.

In this setting, the output capacitor voltages v_1, v_2, and v_3 correspond to the membrane potentials x_1, x_2, and x_3, respectively. The resistors R_1 through R_9 simulate nine fixed synaptic weight coefficients, while the memristor labeled "Mem" accounts for the influence of electromagnetic induction. From the configuration of the mag-

Fig. 11.25 Poincaré maps of the 8-scroll attractor. **a** φ-x_1 plane. **b** φ-x_3 plane. Reprinted from [19], with permission by the IEEE

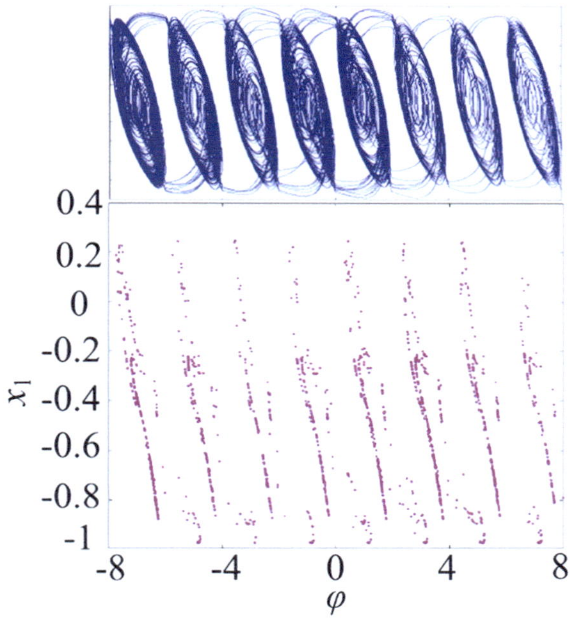

(a)

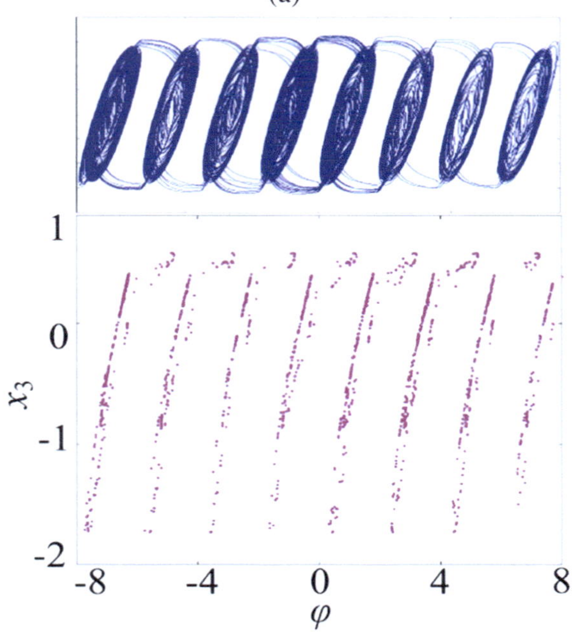

(b)

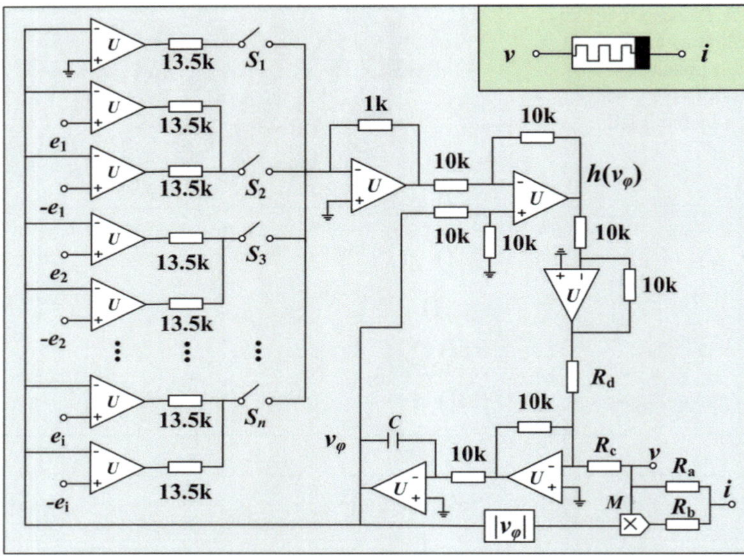

Fig. 11.26 Memristor circuit. Reprinted from [19], with permission by the IEEE

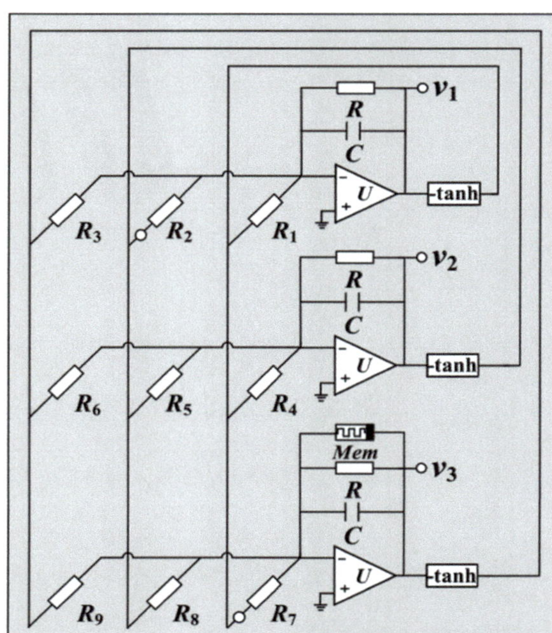

Fig. 11.27 Magnetized HNN circuit. Reprinted from [19], with permission by the IEEE

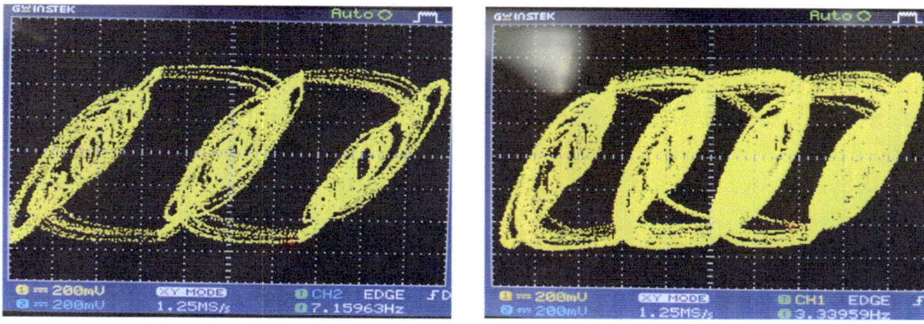

(a) (b)

Fig. 11.28 Experiment results. **a** 3-scroll attractor. **b** 4-scroll attractor. Reprinted from [19], with permission by the IEEE

netized HNN circuit shown in Fig. 11.27, the state equations of the whole circuit are as follows:

$$\begin{cases} Rc\dot{v}_1 = -v_1 + \dfrac{R}{R_1}\tanh(v_1) - \dfrac{R}{R_2}\tanh(v_2) + \dfrac{R}{R_3}\tanh(v_3), \\ Rc\dot{v}_2 = -v_2 + \dfrac{R}{R_4}\tanh(v_1) + \dfrac{R}{R_5}\tanh(v_2) + \dfrac{R}{R_6}\tanh(v_3), \\ Rc\dot{v}_3 = -v_3 - \dfrac{R}{R_7}\tanh(v_1) - \dfrac{R}{R_8}\tanh(v_2) + \dfrac{R}{R_9}\tanh(v_3) - Rv_3\left(\dfrac{1}{R_a} + \dfrac{gv\varphi}{R_b}\right), \\ Rc\dot{v}_\varphi = \dfrac{R}{R_c}v_3 - \dfrac{R}{R_d}h\left(v_\varphi\right). \end{cases}$$

The magnetized HNN circuit is implemented using the Multisim simulation platform. With $C = 1\,\text{nF}$ and $R = 10\,\text{k}\Omega$, and based on the fixed synaptic weight coefficients, the resistor values R_1 through R_9 are determined: $R_1 = 5\,\text{k}\Omega$, $R_2 = 12.5\,\text{k}\Omega$, $R_3 = 14.3\,\text{k}\Omega$, $R_4 = 2.85\,\text{k}\Omega$, $R_5 = 6.25\,\text{k}\Omega$, $R_6 = 6.66\,\text{k}\Omega$, $R_7 = 3.84\,\text{k}\Omega$, $R_8 = 8.33\,\text{k}\Omega$, and $R_9 = 10\,\text{k}\Omega$.

Additionally, the parameters R_a, R_b, R_c, and R_d are defined as

$$R_a = \frac{R}{a\mu}, \quad R_b = \frac{gR}{b\mu}, \quad R_c = \frac{R}{c}, \quad R_d = \frac{R}{d}.$$

The system (11.36) is physically implemented using discrete active and passive electronic components, as shown in Figs. 11.26 and 11.27. The components used include the M/AD633JN and U/TL082CP, with circuit parameters matching those used in the Multisim simulation. The 3-scroll and 4-scroll attractors generated by the system and observed on the oscilloscope are shown in Fig. 11.28.

11.12 Conclusion

This chapter explores various methodologies for generating multiscroll chaotic attractors in both autonomous and nonautonomous systems. By modifying some classical systems, such as the Lorenz system and the proto-Lorenz system, and utilizing techniques like PWL functions, oscillatory potential wells, and through thresholding, the study show cases the versatility and flexibility of chaos generation by various means including neural networks.

In particular, using arctangent function series, combined with nonautonomous forced systems, highlights some nontraditional approaches to designing complex dynamical systems and circuits for generating multiscroll chaotic attractors, especially $n \times m$-grid chaotic attractors. These methods have been verified through theoretical analysis, numerical simulations and circuit implementations, providing solid frameworks for advancing applications in secure communications, nonlinear control and signal processing.

References

1. Chen, G., Lü, J.: Dynamics of the Lorenz System Family: Analysis, Control and Synchronization. (in Chinese). Science Press, Beijing (2003)
2. Elwakil, A.S., Ozoguz, S., Kennedy, M.P.: Creation of a complex butterfly attractor using a novel lorenz-type system. IEEE Trans. Circuits Syst.-I **49**, 527–530 (2002)
3. Lü, J., Chen, G., Yu, Y.: Asymptotic analysis of a modified Lorenz system. Chinese Phys. Lett. **19**, 1260 (2002)
4. Elwakil, A.S.: Nonautonomous pulse-driven chaotic oscillator based on Chua's circuit. Microelectron. J. **33**, 479–486 (2002)
5. Miranda, R., Stone, E.: The proto-Lorenz system. Phys. Lett. A **178**, 105–113 (1993)
6. Lü, J., Chen, G.: A new chaotic attractor coined. Int. J. Bifur. Chaos **12**, 659–661 (2002)
7. Lü, J., Chen, G., Cheng, D.: A new chaotic system and beyond: the generalized Lorenz-like system. Int. J. Bifur. Chaos **14**, 1507–1537 (2004)
8. Lü, J., Chen, G., Zhang, S.: Dynamical analysis of a new chaotic attractor. Int. J. Bifur. Chaos **12**, 1001–1015 (2002)
9. Liu, W., Chen, G.: Can a three-dimensional smooth autonomous quadratic chaotic system generate a single four-scroll attractor? Int. J. Bifur. Chaos **14**, 1395–1403 (2004)
10. Aziz-Alaoui, M.A.: Differential equations with multispiral attractors. Int. J. Bifur. Chaos **9**, 1009–1039 (1999)
11. Aziz-Alaoui, M.A.: Multispiral Chaos. In: 2nd International Conference Control of Oscillations and Chaos, vol. 1, July 5–7, St. Petersberg, Russsia, pp. 88–91 (1999)
12. Brindley, J., Kapitaniak, T.: Existence and characterization of strange nonchaotic attractors in nonlinear systems. Chaos Solit. Fract. **1**(4), 323–337 (1991)
13. Grebogi, C., Ott, E., Pelikan, S., et al.: Strange attractors that are not chaotic. Phys. D **13**(1–2), 261–268 (1984)
14. Cheng, G., Li, D., Yao, Y., Gui, R.: Multi-scroll chaotic attractors with multi-wing via oscillatory potential wells. Chaos Solit. Fract. **174**, 113837 (2023)
15. Lü, J., Murali, K., Sinha, S., et al.: Generating multi-scroll chaotic attractors by thresholding. Phys. Lett. A **372**, 3234–3239 (2008)

16. Bao, H., Ding, R., Chen, B., Xu, Q., Bao, B.: Two-dimensional non-autonomous neuron model with parameter-controlled multi-scroll chaotic attractors. Chaos Solit. Fract. **169**, 113228 (2023)
17. Li, C., Chen, G., Liao, X., Yu, J.: HOPF bifurcation and chaos in tabu learning neuron models. Int. J. Bifur. Chaos **15**, 2633–2642 (2005)
18. Chen, Z., Wen, G., Zhou, H., Chen, J.: A new M× N-grid double-scroll chaotic attractors from Rucklidge chaotic system. Optik **136**, 27–35 (2017)
19. Lin, H., Wang, C., Sun, Y., Wang, T.: Generating *n*-scroll chaotic attractors from a memristor-based magnetized Hopfield neural network. IEEE Trans. Circuits Syst.-II **70**, 311–315 (2022)

Chapter 12
Implementation and Applications of Multiscroll Chaotic Attractors

Abstract This chapter explores the implementation and some practical applications of multiscroll chaotic attractors. It introduces digital circuit design and FPGA-based systems for generating attractors, along with their application in secure communication using chaotic phase shift keying (CPSK) and M-synchronization techniques. Moreover, the chapter discusses unidirectional synchronization of coupled chaotic systems, control strategies for multiscroll Chua's circuits, and the modeling of chaotic circuits using recurrent neural networks. Additionally, it proposes a true random bit generator (TRBG) based on chaotic dynamics, highlighting the role of chaotic systems in secure communication and cryptographic applications. These contributions emphasize the versatility and potential of multiscroll chaotic attractors in engineering and information security applications.

12.1 Generating Multiscroll Chaotic Attractors via Digital Circuit Implementation

Eguchi et al. [1] introduced a digital circuit designed to generate multiscroll chaotic attractors.

By defining the variables $x = v_{C1}$, $y = v_{C2}$, $z = i_L$, and constants $A = 1/R$, $B' = C_1$, $C' = C_2$, $D' = L$, the Chua's circuit Eq. (1.27) can be reformulated as follows:

$$\begin{cases} B'\dfrac{dx}{dt} = A'\left(y - x\right) - g\left(x\right), \\ C'\dfrac{dy}{dt} = A'\left(x - y\right) + z, \\ D'\dfrac{dz}{dt} = -y, \end{cases} \tag{12.1}$$

where $g(x) = m_1 x + 1/2(m_o - m_1)\left(|x + b_1| - |x - b_1|\right)$.

J. Lü et al., *Chaotic Systems with Multiscroll Attractors*,
https://doi.org/10.1007/978-3-032-09854-2_12

Then, consider the following difference transformation:

$$\begin{cases} \dfrac{dx}{dt} = \dfrac{X_{n+1} - X_n}{2^E}, \\ \dfrac{dy}{dt} = \dfrac{Y_{n+1} - Y_n}{2^E}, \\ \dfrac{dz}{dt} = \dfrac{Z_{n+1} - Z_n}{2^E}, \end{cases} \tag{12.2}$$

where E is an integer parameter. The corresponding difference equations of the general Chua's circuit are described by

$$\begin{cases} X_{n+1} = B\dfrac{A\frac{(Y_n - X_n)}{2^m} - G(X_n)}{2^{m-E}} + X_n, \\ Y_{n+1} = C\dfrac{Z_n - A\frac{(Y_n - X_n)}{2^m}}{2^{m-E}} + Y_n, \\ Z_{n+1} = -D\dfrac{Y_n}{2^{m-E}} + Z_n. \end{cases} \tag{12.3}$$

In this system, the parameters $A = [A'2m]$, $B = [1/B']$, $C = [1/C']$, and $D = [1/D']$ are integer values, where the brackets represent the Gauss integer function, which is the greatest integer less than or equal to the argument. The parameter m is an integer taking values from the interval $[0, 2m + 1]$, and $G(X_n)$ is a PWL function with k breakpoints. For simplicity, let $X_{n+1} = F_1(X_n, Y_n, Z_n)$, $Y_{n+1} = F_2(X_n, Y_n, Z_n)$, and $Z_{n+1} = F_3(X_n, Y_n, Z_n)$. Finally, the nonlinear function $G(X_n)$ is determined electronically using supervised learning under a neuro-fuzzy framework.

Figure 12.1 presents the general structure of the digital chaos circuit (12.3). This diagram includes a nonlinear function block, three operational blocks and three delay blocks.

Eguchi et al. [1] conducted numerical simulations to verify the above circuit design, where the membership functions are defined as follows:

MS1: $\mu_{A1}(x) = 2^{m+1} \ominus 10x.$

MS2: $\mu_{A2}(x) = \begin{cases} 10x, & \text{if } x < \frac{2^m}{5}, \\ 3 \times 2^m - 5x, & \text{if } x \geq \frac{2^m}{5}. \end{cases}$

MS3: $\mu_{A3}(x) = \begin{cases} 5x \ominus 2^m, & \text{if } x < 3 \times \frac{2^m}{5}, \\ (5 \times 2^m) \ominus 5x, & \text{if } x \geq 3 \times \frac{2^m}{5}. \end{cases}$

MS4: $\mu_{A4}(x) = \begin{cases} 5x \ominus (3 \times 2^m), & \text{if } x < 2^m, \\ 2^{m+2} \ominus 2x, & \text{if } x \geq 2^m. \end{cases}$

MS5: $\mu_{A5}(x) = 2x \ominus 2^{m+1}.$

Here, x denotes the input variable and $\ominus$ represents the bounded difference operator.

Figure 12.2 presents the numerically simulated 3-scroll and 4-scroll chaotic attractors generated by the digital circuit.

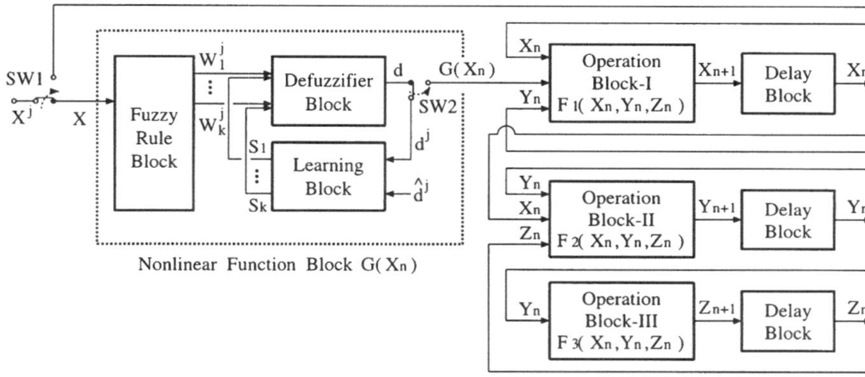

Fig. 12.1 General circuit architecture of the digital chaotic circuit. Reprinted from [1], with permission by the IEICE

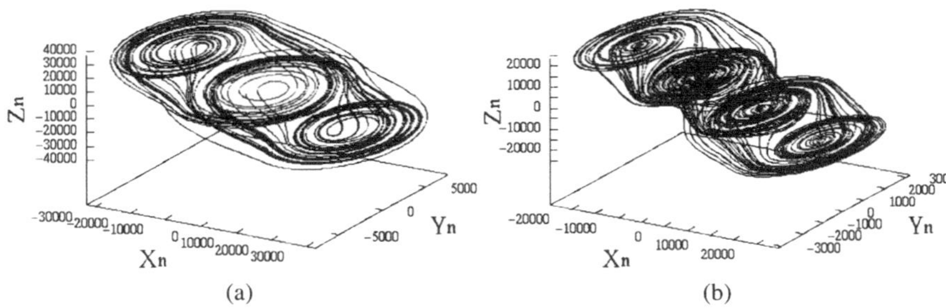

Fig. 12.2 Numerical simulation of the digital circuit. **a** 3-Scroll chaotic attractor. **b** 4-Scroll chaotic attractor. Reprinted from [1], with permission by the IEICE

Furthermore, Eguchi et al. [1] developed an FPGA-based chaotic circuit, which can generate n-scroll chaotic attractors, $n = 1, 2, 3, 4$, utilizing CAD tools and Verilog-HDL. Figure 12.3 illustrates the FPGA functional block distribution for the synthesized circuit. In this setting, the parameter values used are $k = 5$, $m = 14$, and the SRAM is excluded from the chip, used as an external memory.

12.2 Digitized Multiscroll Chaotic Attractors Model for Secure Communication

Tang et al. [2] introduced a secure digital communication system that uses digitized n-scroll chaotic attractors to protect against eavesdroppers during the transmission of signals and data. This methodology for designing secure communication systems can potentially be applied to various online communication applications, such as Internet telephone services, e-commerce transactions and digital data transfers.

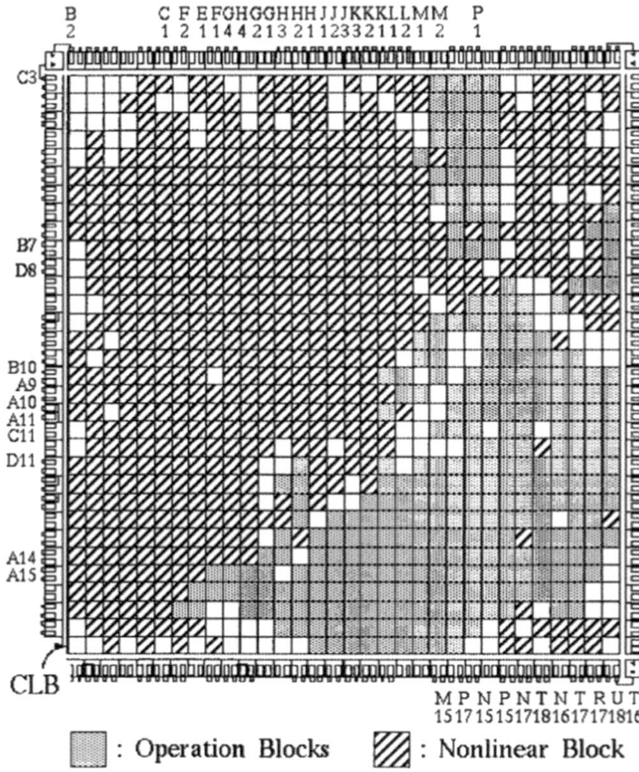

Fig. 12.3 The functional block allocation view of the FPGA implementing the digital circuit. Reprinted from [1], with permission by the IEICE

Apply the following bilinear transformation to Chua's circuit with $h(x_1) = ax_1 + bx_1|x_1|$,

$$\begin{pmatrix} x_1(k+1) \\ x_2(k+1) \\ x_3(k+1) \end{pmatrix} = \begin{pmatrix} \phi_{11} & \phi_{12} & \phi_{13} \\ \phi_{21} & \phi_{22} & \phi_{23} \\ \phi_{31} & \phi_{32} & \phi_{33} \end{pmatrix} \begin{pmatrix} x_1(k) \\ x_2(k) \\ x_3(k) \end{pmatrix} + \begin{pmatrix} \gamma_1 \\ \gamma_2 \\ \gamma_3 \end{pmatrix} h(x_1(k)), \tag{12.4}$$

where $(x_1, x_2, x_3)^{\mathrm{T}}$ is the state vector,

$$\begin{aligned} \phi_{11} &= \tfrac{4\alpha T^2+\zeta_1}{\zeta_1}, & \phi_{12} &= \tfrac{8\alpha T}{\zeta_1}, & \phi_{13} &= \tfrac{4\alpha T^2}{\zeta_1}, \\ \phi_{21} &= \tfrac{8\alpha T^3+4T\zeta_1}{\zeta_1\zeta_2}, & \phi_{22} &= \tfrac{16\alpha T^2+(8-\zeta_2)\zeta_1}{\zeta_1\zeta_2}, & \phi_{23} &= \tfrac{8\alpha T^3+4T\zeta_1}{\zeta_1\zeta_2}, \\ \phi_{31} &= \tfrac{4\alpha\beta T^4+2\beta T^2\zeta_1}{\zeta_1\zeta_2}, & \phi_{32} &= \tfrac{8\alpha\beta T^3+4(\zeta_2-2\beta T^2)\zeta_1}{\zeta_1\zeta_2}, & \phi_{33} &= \tfrac{8\alpha\beta T^4+4(\zeta_2-2\beta T^2)\zeta_1}{\zeta_1\zeta_2}, \end{aligned}$$

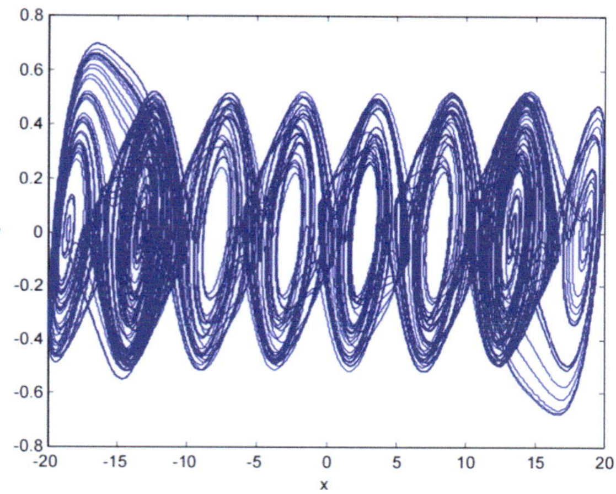

Fig. 12.4 Digital 8-scroll attractor from the modified Chua's circuit. Reprinted from [2], with permission by the IEEE

$\gamma_1 = -2\alpha T\zeta_2/\zeta_1$, $\gamma_2 = 4\alpha T^2/\zeta_1$, $\gamma_3 = -2\alpha\beta T^3/\zeta_1$, $\zeta_1 = 8 + 4T(1 + \alpha a) + 2T^2(\beta + \alpha + \alpha a) + \alpha\beta a T^3$, $\zeta_2 = 4 + 2T + \beta T^2$, $h(x_1(k)) = ax_1(k) + bx_1(k)$ $|x_1(k)|$, and sampling time $T > 0$, with constant parameters α, β, a, b.

Denote Eq. (12.4) as

$$x(k+1) = \Phi x(k) + \Gamma h(x_1(k)),$$

where Φ and Γ are constant matrices.

To generate n-scroll chaotic attractors, Tang et al. [2] extended the above nonlinear function to

$$h(u(k)) = au(k) + bu(k)|u(k)|,$$

where $u(k) = x(k) + 2ma/b$ or $u(k) = x(k) + (n-2)a/b$, obtaining

$$x(k) \in [-(2m-1)|a|/b, (2m-1)|a|/b],$$

for $m = 0, \pm 1, \ldots, \pm(n/2 - 1)$.

Figure 12.4 displays a digital 8-scroll chaotic attractor generated from Eq. (12.4) and $h(u(k))$.

Now, consider a two-coupled discrete-time autonomous chaotic systems, described by

$$\begin{aligned} S_1 &: w(k+1) = \Phi_1 w(k) + \Gamma_1 h_1(w(k)), \\ S_2 &: v(k+1) = \Phi_2 v(k) + \Gamma_2 h_2(v(k)), \end{aligned} \tag{12.5}$$

where $w(k), v(k) \in \mathbf{R}^3$.

If there exists a matrix $M \in \mathbf{R}^{3\times 3}$ satisfying

$$\lim_{k\to\infty} (w(k) - Mv(k)) = 0,$$

then the subsystems S_1 and S_2 are said to achieve M-synchronization.

Based on M-synchronization principle, Tang et al. [2] developed a secure digital communication scheme utilizing the CPSK method. The method is as follows.

Design a transmitter observer, denoted T_r, as

$$w(k+1) = M_i^{-1}\Phi M w(k) + M_i^{-1}\Gamma u(k), \tag{12.6}$$

where $w(k) \in \mathbf{R}^3$ is the state vector of T_r at time k, $\Phi \in \mathbf{R}^{3\times 3}$ is a stable matrix, $\Gamma \in \mathbf{R}^{3\times 1}$, and $M_i \in \mathbf{R}^{3\times 3}$ is a nonsingular synchronization matrix, and $u(k) \in R$ is a nonlinear feedback function.

Design a receiver R_c, as

$$v(k+1) = \left(\Phi + H_i \eta_i M_i^{-1}\right) v(k) + \Gamma u(k) - H_i s(k), \tag{12.7}$$

where $H_i \in \mathbf{R}^{3\times 1}$.

For each i, define the error function as

$$\varepsilon_i^2 = \sum_{k=\tau_m}^{\tau} \left((s_k) - \eta_i M_i^{-1} v(k)\right)^2, \tag{12.8}$$

where τ is the signal interval and τ_m is the minimum number of steps to realize synchronization.

Let

$$\eta_1 = \eta_2 = [1 \quad 0 \quad 0], \quad M_1 = \begin{pmatrix} 2 & 0 & 0 \\ 0 & 4 & 0 \\ 0 & 0 & 2 \end{pmatrix}, \quad \text{and} \quad M_2 = \begin{pmatrix} -2 & 0 & 0 \\ 0 & 3 & 0 \\ 0 & 0 & -2 \end{pmatrix}.$$

At the receiver end, $H_i \in \mathbf{R}^{n\times 1}$ is designed such that the desired eigenvalues are $(0, 0.001, -0.001)$ and $(0, 0.002, -0.002)$ for H_1 and H_2, respectively. Figure 12.5a shows the error between $\eta_i \times M_i^{-1} v$ and s. Figure 12.5b shows the relationship of the probability bit error rate (BER) with the signal-to-noise ratio (SNR), where additive white Gaussian noise (AWGN) noise is contaminated.

12.3 Unidirectionally Coupled Synchronization of Multiscroll Chaotic Attractors

Yu et al. [3] developed a unidirectional coupling synchronization scheme for interacting n-scroll chaotic attractors generated by a 4-D modified Chua's circuit, with additional breakpoints in a modified Chua's circuit, as shown below.

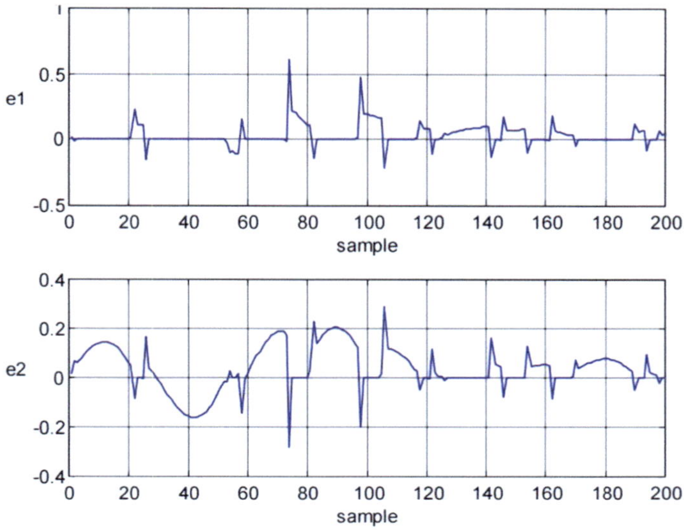

(a)

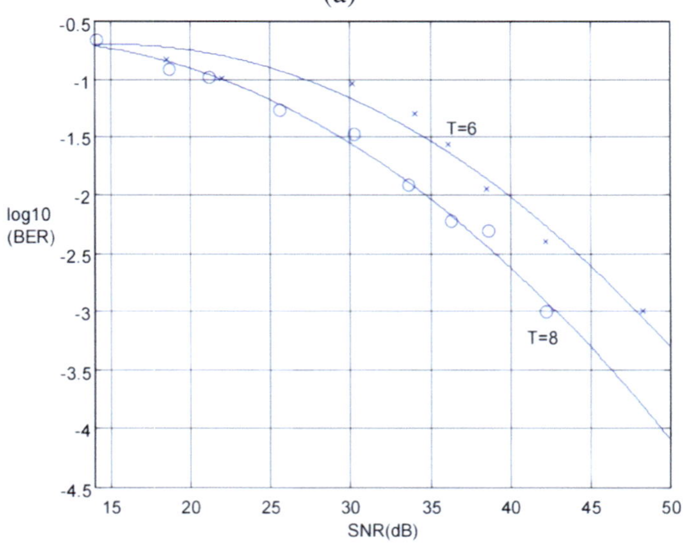

(b)

Fig. 12.5 Numerical simulations of the M-CPSK scheme. **a** Errors between $\eta_i \times M_i^{-1}v$ and s. **b** BER under AWGN transmission channel. Reprinted from [2], with permission by the IEEE

The drive system, with a 6-scroll attractor, is designed as

$$\begin{cases} \dot{x}_1 = \alpha\left(y_1 - x_1 - g\left(x_1\right)\right), \\ \dot{y}_1 = x_1 - y_1 + z_1, \\ \dot{z}_1 = -\beta\left(y_1 - w_1\right), \\ \dot{w}_1 = -\gamma_2\left(z_1 + \gamma_1 w_1\right), \end{cases} \tag{12.9}$$

where $g\left(x_1\right) = m_5 x_1 + \frac{1}{2}\sum_{i=1}^{5}\left(m_{i-1} - m_i\right)\left(\left|x_1 + b_i\right| - \left|x_1 - b_i\right|\right)$.

The response system is designed as

$$\begin{cases} \dot{x}_2 = \alpha\left(y_2 - x_2 - g\left(x_2\right)\right) + \delta_x\left(x_1 - x_2\right), \\ \dot{y}_2 = x_2 - y_2 + z_2 + \delta_y\left(y_1 - y_2\right), \\ \dot{z}_2 = -\beta\left(y_2 - w_2\right), \\ \dot{w}_2 = -\gamma_2\left(z_2 + \gamma_1 w_2\right), \end{cases} \tag{12.10}$$

where

$$g\left(x_2\right) = m_5 x_2 + \frac{1}{2}\sum_{i=1}^{5}\left(m_{i-1} - m_i\right)\left(\left|x_2 + b_i\right| - \left|x_2 - b_i\right|\right).$$

In the above two systems, the parameter values are: $\alpha = 9.934$, $\beta = 14.334$, $\gamma_1 = 27.333$, $\gamma_2 = 0.0497$, $m_0 = m_2 = m_4 = m_a = -1.246$, $m_1 = m_3 = m_5 = m_b = -0.6724$, $b_1 = 10$, $b_2 = 29.2466$, $b_3 = 49.2466$, $b_4 = 68.4931$, and $b_5 = 88.4931$.

The conditions for chaos synchronization is as follows.

Lemma 12-1 [3]. Suppose that there exist coupling coefficients $\delta_x > 0$ and $\delta_y \geqslant 0$ such that the inequality

$$\alpha m_a + \delta_x + (\alpha + \alpha m_a + \delta_x)\delta_y > 0$$

holds for the unidirectionally coupled systems (12.9) and (12.10). Then, the response system (12.10) will globally asymptotically synchronize with the drive system (12.9).

On the one hand, when the coupling coefficients are set as $\delta_x = 23$ and $\delta_y = 0$, then one obtains

$$\alpha m_a + \delta x + (\alpha + \alpha m_a + \delta_x)\delta_y = 10.6222 > 0.$$

From Lemma 12-1, this condition indicates that systems (12.9) and systems (12.10) can achieve synchronization.

On the other hand, when both δ_x and δ_y are set to zero, i.e., $\delta_x = \delta_y = 0$, synchronization between systems (12.9) and Eq. (12.10) cannot be possible, because there is no coupling between the two systems therefore they have no interactions.

Yu et al. [3] designed a circuit, as shown in Fig. 12.6, to experimentally verify unidirectional synchronization of the two coupled systems. The circuit contains two

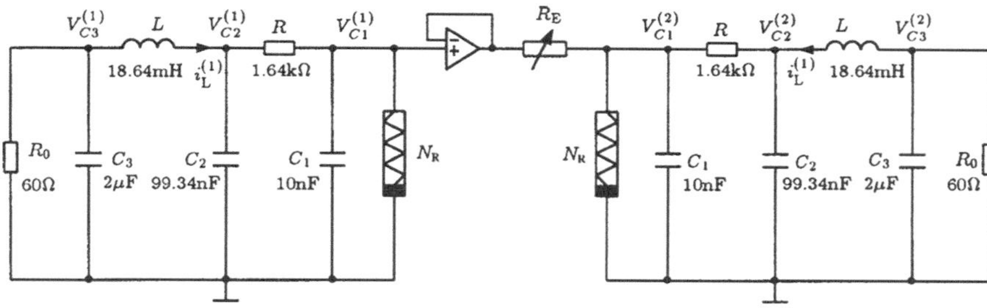

Fig. 12.6 Circuitry for unidirectionally coupled synchronization between two 6-scroll chaotic attractors

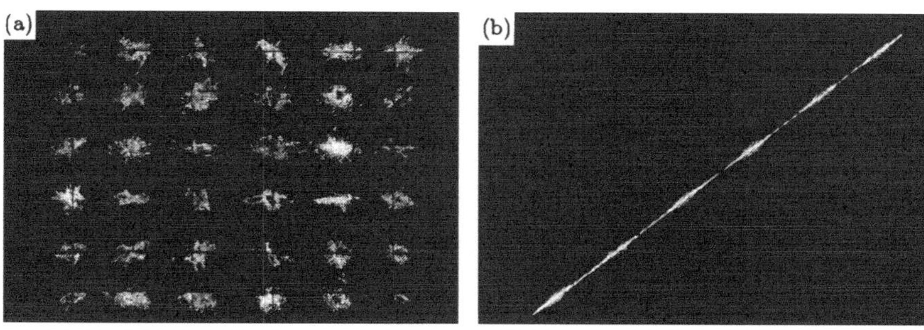

(a) (b)

Fig. 12.7 Experimental observation of unidirectionally coupled synchronization between two 6-scroll chaotic attractors. **a** Asynchrony when there is no coupling. **b** Synchronization when $R_E = 0.7\,k\Omega$

unidirectionally coupled identical 6-scroll chaotic systems [3]. When the coupling resistor $R_E = 0$, the coupling coefficients are

$$\delta_x = \left(\frac{C_2 R}{C_1}\right) R_E = 0 \quad \text{and} \quad \delta_y = 0.$$

Under these conditions, the two 6-scroll chaotic attractors cannot synchronize to each other, as illustrated by Fig. 12.7a.

However, when the coupling resistor is set as $R_E = 0.7\,\text{k}\Omega$, the coupling coefficients become

$$\delta_x = \left(\frac{C_2 R}{C_1}\right) R_E = 23 \quad \text{and} \quad \delta_y = 0.$$

With these parameters, the two 6-scroll chaotic attractors successfully synchronize, as shown in Fig. 12.7b.

Yu et al. [3] also explored a unidirectional synchronization scheme for two-coupled n-scroll hyperchaotic attractors generated from the following 6-scroll hyperchaotic circuit:

$$\begin{cases} \dot{x}_1 = \alpha\left(g\left(y_1 - x_1\right) - z_1\right), \\ \dot{y}_1 = \beta\left(-g\left(y_1 - x_1\right) - w_1\right), \\ \dot{z}_1 = x_1 + z_1, \\ \dot{w}_1 = \gamma y_1, \end{cases} \tag{12.11}$$

where

$$g\left(y_1 - x_1\right) = m_5\left(y_1 - x_1\right) + \frac{1}{2}\sum_{i=1}^{5}\left(m_{i-1} - m_i\right)\left(\left|\left(y_1 - x_1\right) + c_i\right| - \left|\left(y_1 - x_1\right) - c_i\right|\right).$$

The corresponding response system is given by

$$\begin{cases} \dot{x}_2 = \alpha\left(g\left(y_2 - x_2\right) - z_2\right) + \delta_x\left(x_1 - x_2\right), \\ \dot{y}_2 = \beta\left(-g\left(y_2 - x_2\right) - w_2\right) + \delta_y\left(y_1 - y_2\right), \\ \dot{z}_2 = x_2 + z_2 + \delta_z\left(z_1 - z_2\right), \\ \dot{w}_2 = \gamma y_2 + \delta_w\left(w_1 - w_2\right), \end{cases} \tag{12.12}$$

where $\delta_x, \delta_y, \delta_z, \delta_w$ are the coupling coefficients, and

$$g\left(y_2 - x_2\right) = m_5\left(y_2 - x_2\right) + 1/2\sum_{i=1}^{5}\left(m_{i-1} - m_i\right)\left(\left|\left(y_2 - x_2\right) + c_i\right| - \left|\left(y_2 - x_2\right) - c_i\right|\right).$$

Denote the state errors by

$$e_x = x_1 - x_2, \quad e_y = y_1 - y_2, \quad e_z = z_1 - z_2, \quad e_w = w_1 - w_2.$$

Then,

$$g(y_1 - x_1) - g(y_2 - x_2) = m(e_y - e_x),$$

where $-0.9 = m_a \leq m \leq m_b = 2.9$. Thus, the synchronization error system is obtained as follows:

$$\dot{\mathbf{e}} = (A - K + M)\,\mathbf{e}, \tag{12.13}$$

where $\mathbf{e} = \left(e_x, e_y, e_z, e_w\right)^{\mathrm{T}}$, and

$$A = \begin{pmatrix} 0 & 0 & -\alpha & 0 \\ 0 & 0 & 0 & -\beta \\ 1 & 0 & 1 & 0 \\ 0 & \gamma & 0 & 0 \end{pmatrix}, \quad K = \begin{pmatrix} \delta_x & 0 & 0 & 0 \\ 0 & \delta_y & 0 & 0 \\ 0 & 0 & \delta_z & 0 \\ 0 & 0 & 0 & \delta_w \end{pmatrix}, \quad M = \begin{pmatrix} -\alpha m & \alpha m & 0 & 0 \\ \beta m & -\beta m & 0 & 0 \\ 0 & 0 & 0 & 0 \\ 0 & 0 & 0 & 0 \end{pmatrix}.$$

Set the parameters be $\alpha = 2$, $\beta = 20$, $\gamma = 1.5$, $m_0 = m_2 = 0.9$, $m_4 = -0.8$, $m_1 = m_3 = m_5 = 2.9$, $c_1 = 1$, $c_2 = 3.1$, $c_3 = 5.4$, $c_4 = 8.193$, and $c_5 = 12.0019$.

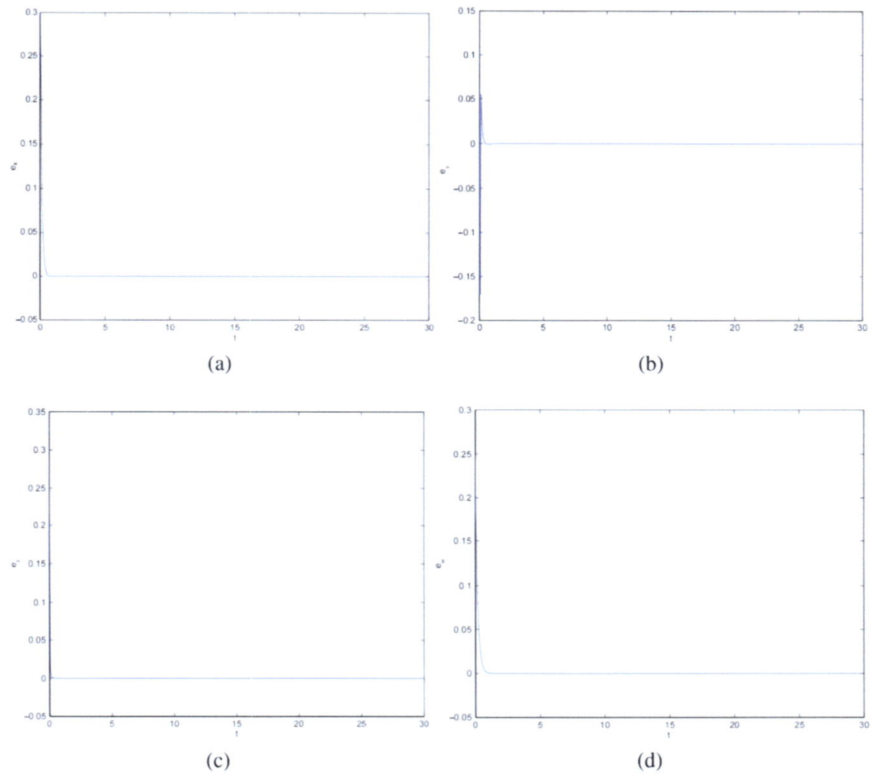

Fig. 12.8 Numerical simulations of unidirectionally coupled synchronization between two-coupled hyperchaotic systems with 6-scroll attractors. **a** Error e_x. **b** Error e_y. **c** Error e_z. **d** Error e_w. Reprinted from [4], with permission by World Scientific Pub. Co.

The condition for the unidirectional coupling synchronization is as follows.

Lemma 12-2 [3]. If there exists a positive definite symmetric constant matrix P such that

$$(A - K + M)^T P + P(A - K + M) \le \mu I < 0$$

uniformly for all x_1, x_2, y_1, y_2 in the phase space, where I is the identity matrix and $\mu < 0$, then the error system (12.13) is globally exponentially stable at zero.

For the values $\delta_x = 5$, $\delta_y = 60$, $\delta_z = 60$, and $\delta_w = 6$, one can select $P = \text{diag}(0.15, 0.15, 0.15, 0.15)$ and $\mu = -0.2$ to satisfy the condition of Lemma 12-2.

Figure 12.8 presents the numerical simulation results.

12.4 Controlling Multiscroll Chua's Circuit

Zou et al. [5] proposed a simple proportional and derivative controller to regulate the n-scroll Chua's circuit, driving it towards a stable equilibrium point or a stable periodic orbit.

Based on the eigenvalue characteristics of the system Jacobian matrix at the equilibria, the equilibria can be categorized into two types.

1. Type I: It occurs when n is even and k is odd, with $k \leq |c|$, or when n is odd and k is even, with $k \leq |c|$. The corresponding eigenvalues consist of one negative real root and one pair of complex conjugate roots with positive real parts.
2. Type II: It occurs when n is even and k is even, with $k \leq |c|$, or when n is odd and k is odd, with $k \leq |c|$. The corresponding eigenvalues consist of one positive real root and one pair of complex conjugate roots with negative real parts.

A controller $u = k_1 \dot{x}$ is designed, where k_1 is an adjustable parameter. The resulting controlled system is as follows:

$$\begin{cases} \dot{x} = \alpha\left(y - f(x)\right), \\ \dot{y} = x - y + z + u, \\ \dot{z} = -\beta y, \end{cases} \tag{12.14}$$

where

$$f(x) = \begin{cases} \dfrac{b\pi}{2a}(x - 2ac), & \text{if} \quad x \geq 2ac, \\ -b \sin\left(\dfrac{\pi x}{2a} + d\right), & \text{if} \quad -2ac < x < 2ac, \\ \dfrac{b\pi}{2a}(x + 2ac), & \text{if} \quad x \leq -2ac. \end{cases}$$

Theoretical analysis shows that the equilibria of Type I can be stabilized using the controller u described above. However, the same controller cannot stabilize the equilibria of Type II.

For the equilibria of Type I, the characteristic equation of the controlled system (12.14) is

$$\lambda^3 + (2.4373 - 10.814k_1)\lambda^2 + 4.6233\lambda + 20.1226 = 0. \tag{12.15}$$

According to the Routh–Hurwitz criterion, the equilibria of Type I can be stabilized if $k_1 < -0.1771$. By activating the controller as soon as the orbit enters the subspace corresponding to the ith scroll, the orbit can be stabilized at the equilibrium within that subspace.

On the other hand, if the complex conjugate roots of Eq. (12.15) meet the following conditions,

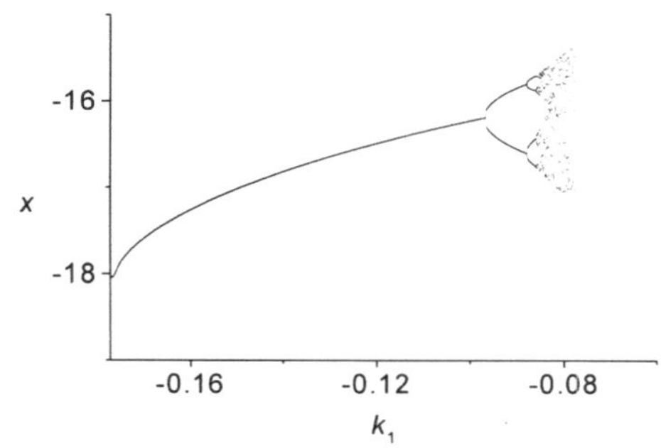

Fig. 12.9 Period-doubling bifurcation from the controlled Chua's circuit with a 8-scroll attractor, where $-20 < x < -15$. Reprinted from [5], with permission from World Scientific Pub. Co.

$$\mathrm{Re}\,(\lambda)\Big|_{k_1=k_0} = 0, \quad \mathrm{Im}\,(\lambda)\Big|_{k_1=k_0} \neq 0, \quad \frac{d\lambda}{dk_1}\Big|_{k_1=k_0} \neq 0,$$

where k_0 is a critical value, then the controlled system (12.14) undergoes a Hopf bifurcation. For $k_0 = -0.1771$, these conditions are satisfied, resulting in the occurrence of a Hopf bifurcation, as depicted in Fig. 12.9.

12.5 Recurrent Neural State-Space Model for Chua's Circuit

Suykens and Vandewalle [6] proposed a simple discrete-time autonomous neural state-space model for Chua's circuit, which is capable of producing a double-scroll chaotic attractor.

Consider the equation $x_{k+1} = f(x_k)$, where $x_k \in \mathbf{R}^n$ is the state vector and f is a continuous function. By approximating f with a multilayer feedforward neural network having a single hidden layer, one obtains

$$\hat{x}_{k+1} = W \tanh\left(V\hat{x}_k + \beta\right), \quad \hat{x}_0 = x_0, \tag{12.16}$$

where the interconnection matrices are defined as $W \in \mathbf{R}^{n \times n_h}$ and $V \in \mathbf{R}^{n_h \times n}$, with the bias vector $\beta \in \mathbf{R}^{n_h}$, where n_h represents the number of hidden neurons. The system can be represented as

$$\hat{x}_{k+1} = \Phi(\hat{x}_k, \alpha),$$

where α is a component of the unknown parameter vector $\theta = (W(:), V(:), \beta)$, where "(:)" indicates the conversion of a matrix into a vector by scanning it column by column.

This approach enables the representation of the system dynamics by encapsulating the interaction between input and output through the neural network. The function Φ maps the current state $\hat{x}_k$ and the parameters α to the next state $\hat{x}_{k+1}$.

The cost function is minimized off-line, as follows:

$$\min_{\theta} J(\theta) = \frac{1}{2N} \sum_{k=1}^{N} \left(x_k - \hat{x}_k(\theta)\right)^{\mathrm{T}} \left(x_k - \hat{x}_k(\theta)\right).$$

The steepest descent algorithm is used for dynamic backpropagation:

$$\hat{\theta}_{t+1} = \hat{\theta}_t - \eta \frac{\partial J}{\partial \theta},$$

where η is the learning rate and $\hat{\theta}_t$ is the tth iteration of the parameter vector.

For Chua's circuit, the following parameters are used: $C_1 = 1/9, C_2 = 1, L = 1/7, R = 10/7, m_0 = -0.8, m_1 = -0.5$, and $b_1 = 1$. The simulation is carried out with a time step of $h = 0.05$ and an initial state $x_0 = (0.9365, -0.0610, -0.1889)^{\mathrm{T}}$.

The neural network training set consists of the first 1000 data points generated from the system. In the neural network state-space model, three hidden neurons are used and the bias vector β is set to zero.

Then, an *ad hoc* optimization procedure is carried out, as follows.

(1) Get a random initial parameter vector θ_0, and set $a := 1$.
(2) Do, when $a < a_{final}$,

$$-\ \theta^* = \arg\min_{\theta} J(\theta) = \frac{1}{2a\Delta} \sum_{k=1}^{a\Delta} \left(x_k - \hat{x}_k\right)^{\mathrm{T}} \left(x_k - \hat{x}_k\right),$$

where $-\theta_0 = \theta^*$, $\Delta = 50$, $a_{final} = 20$, θ_0 is the starting point for the optimization problem, and θ^* is the local optimal solution.

The neural network state-space model is established follows:

$$W = \begin{pmatrix} 3.191701795026490e+00 & -3.961031505875602e+00 & -2.544300729387972e+00 \\ 6.302937463967251e-01 & 2.746315947131907e+00 & 8.024248038305574e-01 \\ -1.411085817901605e+00 & 8.436161546347900e+00 & 3.174868294146957e+00 \end{pmatrix},$$
$$V = \begin{pmatrix} -2.446514424620466e-01 & 1.557093499163188e+00 & -6.192223155626144e-01 \\ 4.935534636051888e-01 & -9.111907179654775e-01 & 6.785493890595847e-01 \\ -1.711046461663226e+00 & 3.794433297552783e+00 & -2.105183558255941e+00 \end{pmatrix}.$$

Figure 12.10 shows a basic recurrent neural network emulator designed to produce a double-scroll attractor through the neural network state-space model. The error sharply increases after the vertical line, which is expected due to the inherent unpredictability of chaotic systems.

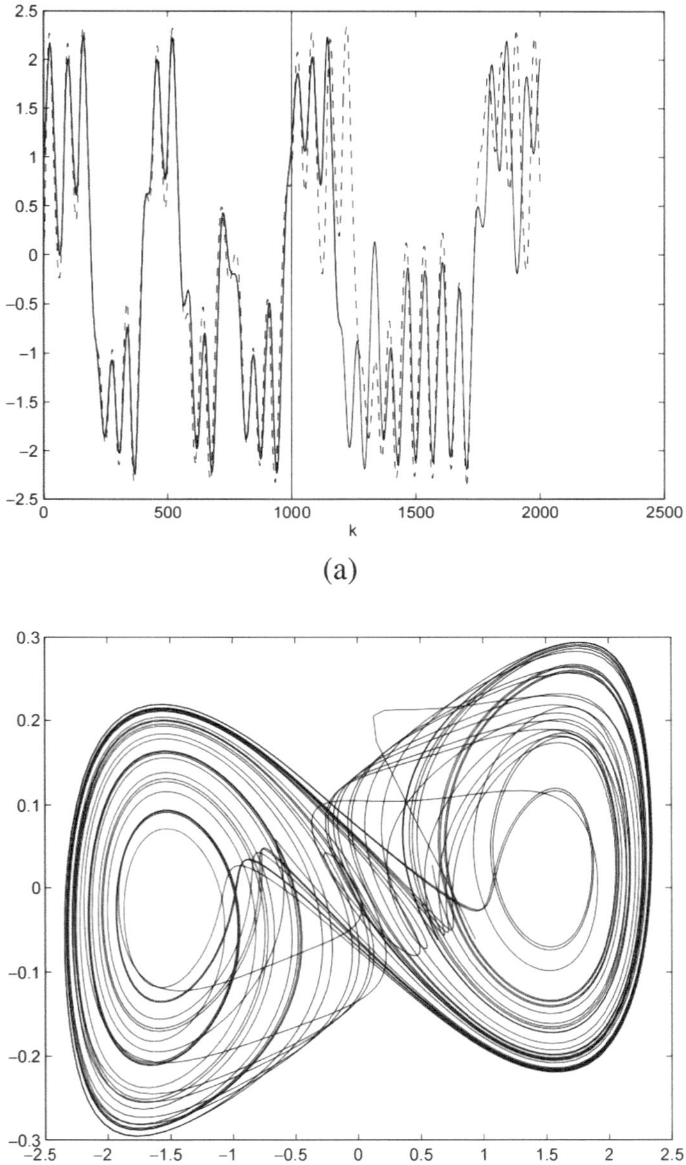

(b)

Fig. 12.10 A simple recurrent neural network emulator for Chua's circuit. **a** The x variable and its emulating series. **b** The reconstructed double-scroll attractor (x_k, y_k). Reprinted from [6], with permission by the IEEE

12.6 True Random Bit Generation from Double-Scroll Chaotic Attractors

Yalcin et al. [7–9] introduced a true random bit generator (TRBG) based on a double-scroll chaotic attractor. This approach takes advantage of the inherent randomness of chaotic systems to produce genuinely random bits.

Consider a system [9], which can generate multiscroll attractors, in the following form:

$$S_1 : \dot{\mathrm{x}} = A\mathrm{x} + Bf\left(C\mathrm{x}\right), \tag{12.17}$$

with

$$A = \begin{bmatrix} 0 & 1 & 0 \\ 0 & 0 & 1 \\ -a & -a & -a \end{bmatrix}, \quad B = \begin{bmatrix} 0 \\ 0 \\ a \end{bmatrix}, \quad C = \begin{bmatrix} 0\,0\,1 \end{bmatrix},$$

where

$$f\left(\zeta\right) = \begin{cases} 1, & \text{if } \zeta \geq 0, \\ -1, & \text{if } \zeta < 0, \end{cases}$$

and $\mathrm{x} = [x; y; z] \in R^3$ and $\zeta \in R$.

The generalized nonlinearity for Eq. (12.17) is given by

$$f\left(x\right) = \sum_{i=1}^{M} g_{(-2i+1)/2}\left(x\right) + \sum_{i=1}^{N} g_{(-2i-1)/2}\left(x\right)$$

with

$$g_\theta\left(\zeta\right) = \begin{cases} 1, & \text{if } \zeta \geq 0, \theta > 0, \\ 0, & \text{if } \zeta < 0, \theta > 0, \\ 0, & \text{if } \zeta \geq 0, \theta < 0, \\ -1, & \text{if } \zeta < 0, \theta < 0. \end{cases}$$

With parameters $a = 0.8$, $M_x = 1$, and $N_x = 0$, this system generates a double-scroll chaotic attractor.

A direct method to derive a bit sequence from a real-valued chaotic signal is to use a threshold function [10],

$$\sigma_c\left(x\right) = \begin{cases} 0, & \text{if } x < c, \\ 1, & \text{if } x \geq c. \end{cases}$$

Since the chaotic system generates S_1 in continuous time, define thresholds c_1 and c_2. For specific values of c_1 and c_2, the state space of the double-scroll attractor is divided into three regions: $V_0 = \{(x, y, z) \mid x \leq c_2\}$, $V_T = \{(x, y, z) \mid c_1 < x < c_2\}$, and $V_1 = \{(x, y, z) \mid x \geq c_1\}$.

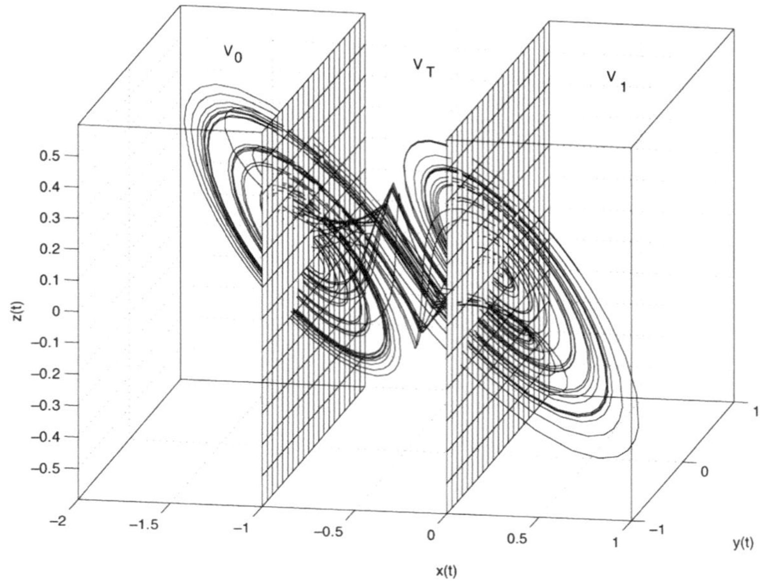

Fig. 12.11 Discretizing the state space, when $c_1 = 0$ and $c_2 = -1$. Reprinted from [9], with permission by the IEEE

Figure 12.11 illustrates the discretized state space [8, 9], for $c_1 = 0$ and $c_2 = -1$. In this context, the block S_2 is defined as

$$S_2 : \begin{cases} \sigma^1 (x(t)) = \begin{cases} 0, & \text{if } x(t) < c_1, \\ 1, & \text{if } x(t) \geq c_1, \end{cases} \\ \sigma^0 (x(t)) = \begin{cases} 0, & \text{if } x(t) < c_2, \\ 1, & \text{if } x(t) \geq c_2. \end{cases} \end{cases} \tag{12.18}$$

A bit is generated when the signal passes the subspace V_0 or the subspace V_1 region through the subspace V_T. The bit generator block S_3 is the initial block in the software section, and it is defined by

$$S_3 : \sigma_i \left(\sigma^0, \sigma^1\right) = \begin{cases} 0, & \text{if } \sigma^0 = 0, \ \sigma^1 : 0 \uparrow 1, \\ 1, & \text{if } \sigma^1 = 0, \ \sigma^0 : 0 \uparrow 1, \end{cases} \tag{12.19}$$

where $\sigma_0 : 0 \uparrow 1$ represents the rising edge, which is the transition from logic value 0 to logic value 1 of σ_0, with $i \in \{0, 1, 2, \ldots\}$.

Essentially, the bit generation process is used to detect transitions of the signal $x(t)$, whether it shifts from one scroll to another or stays within the same scroll.

The de-skewing block S_4 is described by

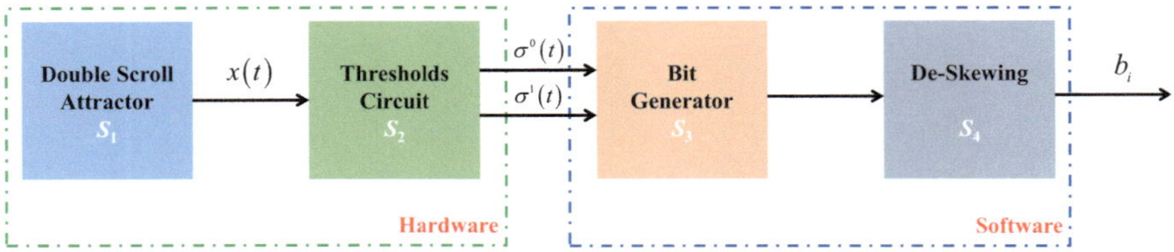

Fig. 12.12 Diagram of the TRBG

$$S_4 : b_i\,(\sigma_i,\ \sigma_{i-1}) = \begin{cases} 0, & \text{if } \sigma_{\mathrm{i}} = 0 \wedge \sigma_{\mathrm{i}-1} = 1, \\ 1, & \text{if } \sigma_{\mathrm{i}} = 0 \wedge \sigma_{\mathrm{i}-1} = 0. \end{cases} \tag{12.20}$$

To obtain an unbiased bit sequence, the bit generator is combined with a simple subroutine from the de-skewing block S_4.

Figure 12.12 illustrates the TRBG, which comprises both hardware and software components. The state space of the chaotic system with a double-scroll attractor is divided into three subspaces by block S_2 of the hardware component. Two of these subspaces are associated with the positions of the scrolls in the state space. The bit generator (block S_3) is the first block of the software component. The last block S_4 in the software component is a de-skewing process that eliminates bias in the output of the bit generator.

The TRBG has a single output, denoted as $B = \{b(1), \ldots, b(i-1), b(i), b(i+1), \ldots\}$, where $b(i) \in \{0, 1\}$. Since noise has a maximum entropy, the threshold value c_2 is chosen to maximize the measured entropy of the TRBG. The relationship between the measured and theoretical entropy of the proposed TRBG, based on a partition defined by c_2, is given by

$$h^{c2} = \lim_{n \to \infty} \frac{H_n^{c2}}{n},$$

where

$$H_n^{c2} = -\sum_{B^n} P\left(B^n\right) \ln P\left(B^n\right).$$

Let $P(B^n)$ represent the probability of occurrence of the binary subsequence (or words) $\{\ldots, \{b(i), b(i+1), \ldots, b(i+n-1)\}, \{b(i+n), \ldots, b(i+2n-1)\}, \ldots\}$ of length n. The entropy H_n^{c2} achieves its maximum value when the n-word sequences in the output B of the TRBG are uniformly distributed.

Figure 12.13 shows the values of H_n^{c2}/n for $n = 3, 4, \ldots, 10$, demonstrating how the TRBG approaches the maximum entropy of ln 2. In this experiment, the threshold c_1 is set to 0 V, and c_2 is set to -1.44 V.

Yalcin et al. [8, 9] also verified the randomness of the TRBG through experimental tests using two widely recognized cryptographic test sets: FIPS-140-1 and Diehard.

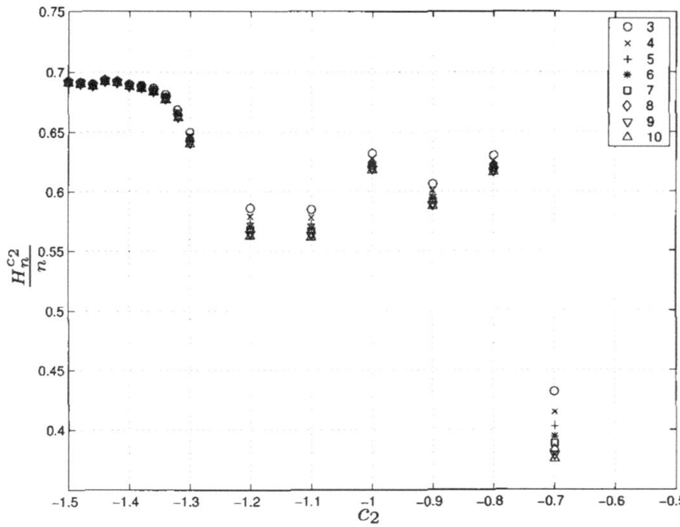

Fig. 12.13 H_n^{c2}/n as a function of the threshold c_2, $n = 3, \ldots, 10$. Reprinted from [9], with permission by the IEEE

12.7 Conclusion

This chapter offers an overview of the implementation and various applications of multiscroll chaotic attractors. It discusses the design of digital circuits, including FPGA-based implementations, which effectively generate multiscroll attractors with improved scalability and functionality. The use of these attractors in secure communication systems, such as chaotic phase shift keying (CPSK) and M-synchronization, highlights their potential applications for protecting data transmission security.

Moreover, the study examines advanced synchronization techniques for coupled chaotic systems, control strategies for various multiscroll Chua's circuits, and neural network modeling of chaotic systems. The development of a true random bit generator (TRBG) utilizing chaotic attractors further emphasizes the relevance of chaos theory to cryptographic systems. These findings underscore the importance of multiscroll chaotic attractors in enhancing secure communications, nonlinear control and cryptographic technologies.

References

1. Eguchi, K., Inoue, T., Tsuneda, A.: Synthesis and analysis of a digital chaos circuit generating multiple-scroll strange attractors. IEICE Trans. Fund. **E82-A**, 965–972 (1999)
2. Tang, W. K. S., Man, K. F., Chen, G.: Digitized n-scroll attractors model for secure communications. In: IEEE International Symposium on Circuits and Systems, vol. 3, May 6–9, Sydney, Australia, pp. 787–790 (2001)

3. Yu, S., Ma, Z., Qiu, S., et al.: Generation and synchronization of n-scroll chaotic and hyperchaotic attractors in fourth-order systems. Chinese Phys. B **13**, 317 (2004)
4. Lü, J., Chen, G.: Generating multiscroll chaotic attractors: theories, methods and applications. Int. J. Bifur. Chaos **16**, 775–858 (2006)
5. Zou, Y., Luo, X., Jiang, P., et al.: Controlling the chaotic n-scroll Chua's circuit. Int. J. Bifur. Chaos **13**, 2709–2714 (2003)
6. Suykens, J.A.K., Vandewalle, J.: Learning a simple recurrent neural state space model to behave like Chua's double scroll. IEEE Trans. Circuits Syst.-I **42**, 499–502 (1995)
7. Yalcin, M.E., Ozoguz, S., Suykens, J.A.K.,Vandewalle, J.: Scroll maps from n-scroll attractors. In: 10th International Workshop on Nonlinear Dynamics of Electronic Systems, June 21–23, Izmir, Turkey, pp. 45–48 (2002)
8. Yalcin, M.E., Suykens, J.A.K., Vandewalle, J.: A double scroll based true random bit generator. In: IEEE International Symposium on Circuits and Systems, vol. 4, May 23–26, Vancouver, Canada, pp. 581–584 (2004)
9. Yalcin, M.E., Suykens, J.A.K., Vandewalle, J.: True random bit generation from a double-scroll Attractor. IEEE Trans. Circuits Syst.-I **51**, 1395–1404 (2004)
10. Kohda, T., Tsuneda, A.: Statistics of chaotic binary sequences. IEEE Trans. Inform. Theory **43**, 104–112 (1997)

Index

J. Lü et al., *Chaotic Systems with Multiscroll Attractors*,
https://doi.org/10.1007/978-3-032-09854-2